AVIATION VISUAL PERCEPTION

Aviation Visual Perception

Research, Misperception and Mishaps

RANDY GIBB
United States Air Force Academy, USA

ROB GRAY
Arizona State University, USA

&

LAUREN SCHARFF
United States Air Force Academy, USA

ROUTLEDGE
Routledge
Taylor & Francis Group
LONDON AND NEW YORK

First published 2010 by Ashgate Publishing

Published 2016 by Routledge
2 Park Square, Milton Park, Abingdon, Oxon OX14 4RN
711 Third Avenue, New York, NY 10017, USA

Routledge is an imprint of the Taylor & Francis Group, an informa business

British Library Cataloguing in Publication Data
Gibb, Randy.
Aviation visual perception : research, misperception and mishaps. -- (Ashgate studies in human factors for flight operations)
1. Airplanes--Piloting--Human factors. 2. Visual perception. 3. Airplanes--Piloting--Safety measures.
I. Title II. Series III. Gray, Rob. IV. Scharff, Lauren.
629.1'3252-dc22

ISBN 9780754674979 (hbk)

Library of Congress Cataloging-in-Publication Data
Gibb, Randy.
Aviation visual perception : research, misperception and mishaps / by Randy Gibb, Rob Gray and Lauren Scharff.
p. cm. -- (Ashgate studies in human factors for flight operations)
Includes index.
ISBN 978-0-7546-7497-9 (hardback)
1. Flight--Physiological aspects. 2. Visual perception. 3. Aeronautics--Human factors. I. Gray, Rob. II. Scharff, Lauren. III. Title.
RC1075.G53 2010
616.9'80213--dc22

2009038272

Contents

List of Figures

List of Tables

Foreword

There is an old saying that if humans were meant to fly, they would have been born with wings. If I had any choice in the matter, I'd be more realistic and trade the wings for a bit more reliable sensory system; one that was fine-tuned for flying and played fewer tricks on me on dark and stormy nights. When it comes to the evolution of the human visual processing system, 100 years of manned flight doesn't begin to scratch the hourglass. Eyes that have evolved for bipedal hunting and gathering have some catching up to do to handle the three dimensional variations of manned flight. But we are who we are—and that is why this book is so vitally important.

Comprehending the complexities and limitations of human vision in aviation is essential to operators of all types of aircraft. In the past, the importance of this subject has been understated, neglected or overlooked altogether by most aviators. Perhaps we can blame this oversight on the fact that there has not been a good single point of reference on the subject of visual perception in aviation. Thanks to Dr. Gibb, Dr. Gray, and Dr. Scharff, that problem no longer exists. *Aviation Visual Perception* combines deep pools of evidence-based and science-based research without losing the key messages for those who just want to learn to fly better. That is no small feat.

Currently, aviators learn to adapt their ground-based vision to the aviation environment through trial and error, using techniques offered by their instructors or shared pilot to pilot. Vital topics such as *composite crosscheck, see and avoid scanning* and *visual illusions* are informally passed along generation to generation, evolving nearly as slowly as we are. It is not that the scientists aren't doing their jobs; they certainly are. Each year there are dozens of advances made in key areas regarding the human machine interface. However, up until now you had to comb through dozens of scientific journals to find these studies. Even then, the relevant material was not always user-friendly or easily understood, at least not for knuckle-dragging pilots like myself. But beyond pulling the essential materials together in a single reference, there is another reason this book is a true breakthrough.

Pilots, as well as other aviation professionals, are a skeptical lot. We are only as committed as we are convinced. It is not enough to be offered new tools and techniques for improvement; we want to know *why* something works. Even then, we still have to prove it to ourselves. Toward that end, this book crosses a panoramic landscape of material essential for pilots as well as others associated with the design and engineering of aircraft to provide the baseline comprehension of how our visual systems work—and sometimes *don't* work—owing to our human limitations.

There are far too many highlights in this book to list here, but the crystallized and powerful improvement messages inside these covers compel me to mention a few. The authors do an exceptional job of linking the physical with the psychological in Chapters 2 to 5 in the discussion of the deceptive nature of visual cues owing to the *expectations* of the pilot. In addition, Chapter 6 provides the finest compilation and discussion of visual illusions I have ever read, and pilots who internalize this information will be immediately safer. You will also find relevant and actionable information on visual flying, night flying and a list of those fatal factors that have resulted in too many funerals owing to avoidable aviation mishaps.

Most readers of this book will understand that much of what we know about safety is a direct result of the loss of thousands of our brothers and sisters who failed to return from what should have been routine flights. The draconian nature of aviation is such that a *single* visual illusion at the wrong moment can mean the difference between life and death. Most seasoned pilots, including this one, will admit to moments of terror when we temporarily lost situational awareness owing to spatial disorientation when our sensory systems could not keep up. This book will make those moments less frequent, and in so doing will save lives.

Only the right mix of operational and scientific background could have produced this book and it is exceedingly rare to find that captured by a trio of authors. Their contribution provides the aviation community with not only a stunningly comprehensive reference text on the physiological and psychological aspects of vision in the aviation environment, but also offers the entire industry a vision of the future where enhancements in the human-machine interface will improve both safety and operational effectiveness for decades to come.

The book of Proverbs (29:18) reminds us, "*where there is no vision, the people perish*," and with *Aviation Visual Perception,* the authors provide a clear articulated vision of how important it is to understand the nuances of our most dominant and important sense for everyone who flies.

Tony Kern, Ed.D.
Retired Air Force pilot and author of *Redefining Airmanship*, *Flight Discipline* and *Darker Shades of Blue: The Rogue Pilot*

Chapter 1
Vision in Aviation

Veridical perception of visual cues is necessary for spatial orientation and controlling our movements as we navigate within our environment. Driving and athletics are two arenas with which everyone can associate in terms of visually guided behavior and successful execution of desired goals. In typical daily life our interface with the environment consists of our feet on or near the ground, movements in the left-right and/or fore-aft direction as well as a one-gravitational force (1-G) acting vertically on our bodies with the horizon straight ahead. The interpretation of visual cues from the environment and the perception of vestibular inputs as we maneuver ourselves are founded on these typical constants, leading to confidence about where our feet are, where the horizon is, and which is "up."

Aviation, however, allows the human operator to accomplish visually guided actions not experienced anywhere else. With the additional spatial dimension of altitude and the possibility of extreme vertical movement, combined with potential extreme velocities and accelerations in the left-right and fore-aft directions, flying poses challenges to humans that are not faced in other domains. When flying, obtaining and maintaining spatial orientation is predominantly accomplished by the visual system. Thus, while flying 30 m (100 ft) above the ground at high speeds or controlling the aircraft for a night visual approach to landing, pilots must accurately perceive and interpret environmental cues with their eyes. Herein lies the problem; the human body is not physiologically prepared to cope with these extreme and sometimes violent movements and forces that occur in aviation. These visual perception challenges must be recognized and appreciated by pilots and aviation researchers.

Despite physiological limitations for sensing and perceiving their aviation environment, pilots can often make the required visual judgments with a high degree of accuracy and precision. At the same time, however, visual illusions and misjudgments have been cited as the probable cause of numerous aviation accidents, and in spite of technological and instructional efforts to remedy some of the problems associated with visual perception in aviation, mishaps of this type continue to occur. Clearly, understanding the role of visual perception in aviation is key to improving pilot performance and reducing aviation mishaps. Furthermore, with the implementation of enhanced and synthetic visual systems, the next generation of aviation is banking heavily on knowledge of visual perception.

Over the years numerous articles, pamphlets, and books have been written on the topics of spatial disorientation and visual illusions in aviation (e.g., Benson, 1988; Cocquyt, 1953; Kern, 2002; Kraft and Elworth, 1969; Lessard, 2000; Newman, 2005; Ostinga et al., 1999; Pitts, 1967; Previc, 2004; Schiff, 1990 and 1994;

Wulfeck, Weisz, and Raben, 1958; as well as the Federal Aviation Administration flying safety pamphlets, military flying manuals). In 2007 the Australian Transport Safety Bureau published *An overview of spatial disorientation as a factor in aviation accidents and incidents*. Also, magazines have dedicated entire issues to the subject. For instance, *IEEE Engineering in Medicine and Biology*, in their March/April 2000 edition, addressed aeronautical illusions, and the Naval Aviation magazine, *Approach*, in May/June 2004, did the same.

This book intends to update and synthesize the previous work to provide the reader with a single resource for comprehensive and detailed explanations of visual disorientation as well as the physiological and perceptual background of the visual system associated with aviation-related perceptual illusions. Vestibular physiology and disorientation is also presented as it is highly integrated with our body's spatial orientation system. Examples of aircraft accidents are included to illustrate failed visual perception and spatial orientation and to demonstrate that pilots have been and are still today far too confident in their limited visual perceptual capabilities when flying. It is not the intent to disparage any of the pilots involved but rather to ensure that others learn from their experiences. The objective of this book is to help educate pilots and others regarding the seduction of visual misperception, with the intention to provide not only a resource for pilots but also a starting point for further research into aviation visual perception.

The Challenge

Visual perception within aviation is not well understood (Calvert, 1950; Havron, 1962; Warren and Owen, 1982; Mulder, et al. 2000), and there is still much to be learned about visual perception in general. While our subjective experience leads us to believe that our brain has access to a perfect, high-definition image of the outside world, Smallman and St. John (2005) point out that this "naïve realism" is simply not consistent with what is known about the human visual system. In actuality, visual perception is "sparse and sewn together" and assumptions must be made to simplify complex scenes that "distort interpretation and result in imperfect, just-in-time, just-good-enough approximations of reality" (p. 8). Perception can be thought of as a series of educated guesses regarding the outside world rather than the detection of what is there with 100 percent accuracy and certainty. The brain sometimes guesses wrong, resulting in visual illusions. Small and St. John summarized these points nicely:

> The illusion of objectivity is that the ubiquity of these errors goes unobserved, thereby fostering and maintaining naïve realism. The brain is a master at concealing its tricks, and only occasionally does one get to glimpse the real Wizard of Oz behind the curtain. (p. 9)

This lack of veridicality makes the challenge of understanding visual perception an incredibly complex endeavor. Because perception does not involve a perfect 1:1 representation of the external environment, we cannot understand how vision works simply by investigating the basic physical characteristics of the world. Knowing that a runway is 3,000 m long × 50 m wide tells us little about how a pilot will perceive this object, because the visual system does not even use these units. Thus, while the physiology of the eye and the study of optics are straightforward and well understood, higher-level visual processes such as understanding how a three-dimensional (3D) world is recreated from a two-dimensional (2D) retinal image are still not completely understood (Smallman and St. John, 2005).

A further challenge to understanding the role of vision in aviation arises from the fact that the pilot is a part of a complex human-machine-environment system (illustrated in Figure 1.1), in which there are numerous, complex interactions between the factors of the environment, the pilot, and the aircraft. Environmental factors such as weather, time of day, geographic location, and G-forces can directly alter a pilot's perception of the world (e.g., by reducing visibility), and/or change the way in which visual cues are used to control the aircraft (e.g., by altering the amount airspeed changes for a given stick movement). The pilot adds both capabilities and potential liabilities into the system. The pilot's skill, proficiency, training, and experience are all factors that increase overall system success. However, if a pilot is not prepared for a particular flight (e.g., is lacking in motivation, personal life stressors or fatigued), the environment can

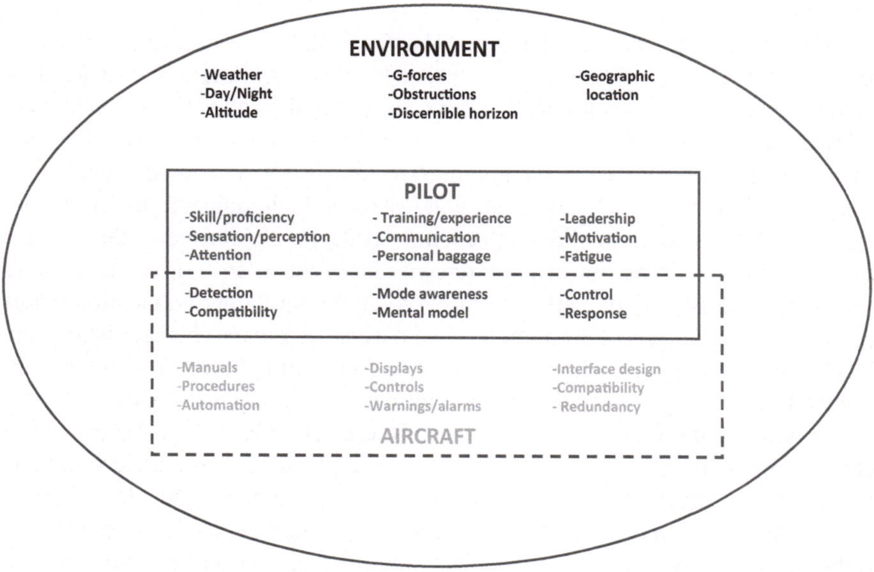

Figure 1.1 Environment, pilot, and aircraft interaction

quickly expose a pilot's vulnerabilities. The aircraft in the system has certain characteristics and limitations given its particular design, including properties such as aircraft manuals, procedures, the level of automation, displays, controls, warnings and alarm systems, redundancy in emergency systems, and the overall interface design.

Chapter 6 presents a detailed summary of visual perception illusions and spatial disorientations. Aviation accident examples involving visual misperception are given in Chapter 7. To help introduce misperception in aviation, below is a fairly recent mishap that highlights several issues within aviation visual perception: the misperception of height and distance, the meteorological limitation of night (lack of visual cues), and the physiological limitation of a color deficiency. Increased awareness of both the visual phenomenon of misperception of height and distance as well as night flying hazards are major themes of this book.

Aviation Mishap Involving Visual Misperceptions

July 26, 2002, prior to sunrise, a Boeing 727, operated by a commercial freight carrier, struck trees during a short final approach and crashed 472 m (1,550 ft) short of runway 09 at Tallahassee Regional Airport (National Transportation Safety Board [NTSB] report, 2002). Three crew members were seriously injured and the airplane was destroyed. Although there were many interesting aspects of this particular mishap, detail will only be presented on aspects of the visual factors contributing to the accident.

The flight had departed Memphis, TN, for Tallahassee and operated on an instrument flight plan. The forecast weather for the arrival destination was night visual meteorological conditions. The crew had debated whether to land on runway 27 with an Instrument Landing System (ILS, precision glide-path approach) or the more conveniently aligned visual approach to runway 09, and had decided to use runway 09. Runway 09 did have Precision Approach Path Indicator (called PAPIs) lights available to assist in glide-path control. As the pilot maneuvered the airplane into alignment with the runway, the descent rate of the aircraft increased beyond the 3-degree desired glide-path. According to the Flight Safety Foundation report (2005), the profile view of the approach had the concave shape characteristic of the black-hole illusion as illustrated in Figure 1.2. The PAPI lights are also depicted in Figure 1.2, showing how they visually inform pilots of their glide-path.

Although a more detailed discussion of the black-hole illusion occurs in the chapter on visual illusions and misperception (Chapter 6), to better understand this mishap it is briefly explained. When a pilot approaches a runway that lacks terrain features and other ambient visual cues during a dark night, the only visual referent is the lighted runway shape. This approach-and-landing environment makes it very difficult for a pilot to estimate height above and distance to the runway. Due to the lack of terrain features the pilot perceives the plane to be higher and farther from the runway than it actually is and, consequently, initiates an unwarranted descent

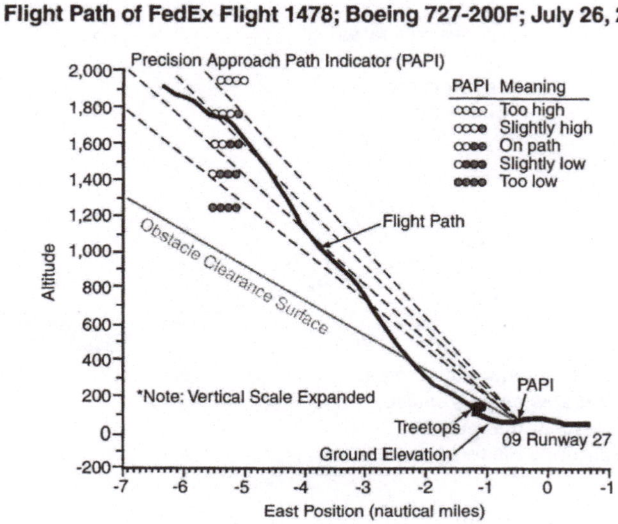

Figure 1.2 Mishap glide-path

Source: From NTSB report, 2002.

below the normal glide-path. In this scenario a pilot will realize far too late that the plane is on an extremely shallow approach angle to the runway and dangerously low; controlled flight into terrain often results. Other phrases used are "landing short" or "under-shooting" the runway. The profile view of this type of approach glide-path has a concave shape due to the excessive descent rate that then shallows out as the pilot approaches the landing runway (shown in Figure 1.3).

In the case of the accident in Florida, upon examining the profile of the aircraft's descent, one could argue the black-hole illusion caused the pilot to misperceive the glide-slope starting at 10.2 km (5.5 nautical miles) from the runway until impact. The final approach to runway 09 required the aircraft to fly over a national forest area which had no lights or terrain features (FSF report, 2005). Prior to impact the airplane was flying at 270 km/hr (146 knots) airspeed with a descent rate of 161 m/min (528 ft/min), but 20 seconds earlier it had a descent rate of 380 m/min (1,248 ft/min), nearly twice what it should have been.

The PAPI lights for the approach (shown in Figure 1.2) signaled "below glide-path" from a point 8.3 km (4.5 nm) from the runway to "well below glide-path" at the 5.6 km (3 nm) point. Procedurally, any indication of a "too-steep" glide-path should be immediately followed with a positive correction. All crew members stated they were shocked upon hitting the ground (NTSB report, 2002). Despite the PAPI indications, none of the pilots perceived their glide-path to be below normal and had not imagined *the accident that was about to occur*. Figure 1.4 provides a better description of how the lights inform the pilots of their position relative to the desired 3-degree glide-path.

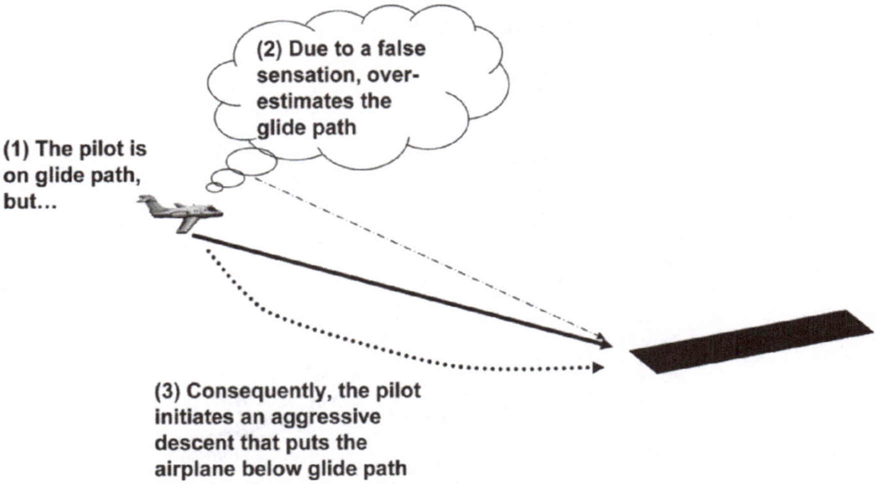

Figure 1.3 Black hole illusions

Source: From Gibb, 2007.

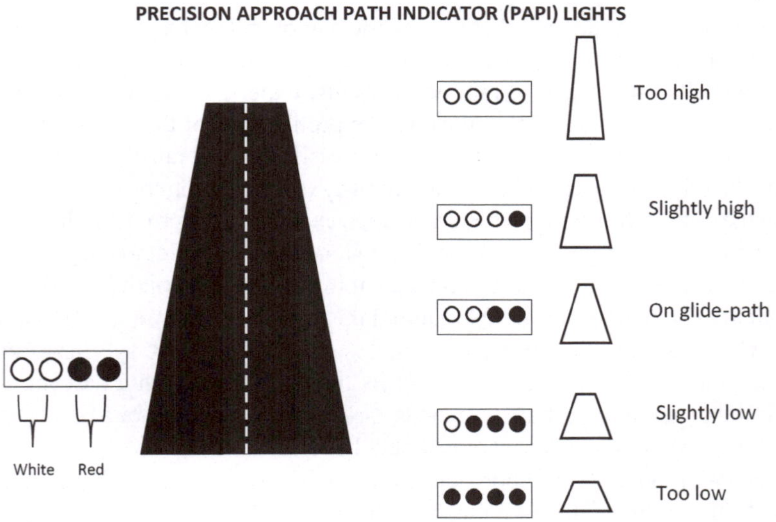

Figure 1.4 Precision approach path indicator lights

The colored lights were developed to assist pilots regarding their glide-path to the runway. Granted, each pilot may have his own perceptual interpretation of the lights and manner by which he controls the aircraft relative to the lights; regardless however, all white indicate an approach that is too steep and all red signals a flight path that is dangerously low regardless of technique or aircraft. Consequently, immediate and appropriate control inputs are needed by the pilots in these extreme situations.

To better understand the accident sequence, the description of events will begin at approximately 5:13 AM local time, about 24 minutes prior to the aircraft landing short. At that time the flight engineer, after coordinating with the airfield for their parking plans upon landing, briefed both the captain and first officer, as required by the commercial freight carrier's procedures, that Tallahassee was a "moderate" risk for controlled flight into terrain. What follows are the voice cockpit recordings and summaries of their conversations up to the point of terrain impact taken from the 2002 NTSB report.

At 5:16 AM local time, the first officer (right seater or copilot), who was going to fly the approach and landing, thoroughly briefed the other two crew members on the approach into runway 27 and the captain concurred with the briefing details.

At 5:19 AM, the first officer, although having just briefed an approach and landing to runway 27, suggested to the captain runway 09 instead of 27, being that they were more conveniently aligned to land on runway 09. Runway 27 required a longer flight path to the far side of the airfield to get aligned for landing compared with already being somewhat aligned to land on runway 09. Their flight path was coming from the northwest and heading towards the east, thus landing runway 27 would require flying east for some time to then turn back to the west and then land heading 270 degrees. Because it was prior to sunrise with no traffic conflicts, landing to the east was more convenient given their position.

At 5:24:03 AM, the controller advised the crew to expect a visual approach into Tallahassee and to report when they had the airport in sight.

At 5:24:23 AM, the captain queried, "Runway nine … PAPI on the left side … I don't know, you wanna try for nine?" The first officer responded, "We're pointed in the right direction, I don't know, like you said. Kinda long … taxiback." The runway debate continued with the captain saying, "… the only advantage you have, landing to the west you have the … glide-slope … which you don't have to the east." (The decision to not land on runway 27 with the glide-slope, needless to say, was a key link in the chain of events leading to the mishap.)

At 5:26:41 AM, the captain asked, "You familiar with the airport here at Tallahassee?" to which to first officer replied, "No, I'm not."

At 5:28 AM, the crew finally decided upon runway 09 for their landing.

At 5:30 AM, the first officer reported to the crew that he had the airport's beacon in sight. The light he saw, however, was that of a power plant. According to the NTSB report, pilots flying in from the northwest direction often misperceive the power plant's light to be that of the airport. The captain corrected the first officer. The crew then configured the aircraft for landing, accomplished the before-landing check, and the remaining transcript contains the pilots discussing their approach.

At approximately 5:35 AM, the landing gear was extended and the "before-landing check" was initiated.

At 5:36:20 AM, the first officer apologized for his final approach and said, "Sorry, 'bout that … I was lining up on the paper mill or something." At about the same time the aircraft's ground proximity warning system sounded an alert, announcing passing through 305 m (1,000 ft) above the ground. With that warning and then in response to the first officer's apology, the captain responded, "That's all right, no problem."

At 5:36:37 AM, the aircraft was 4.6 km (2.5 nautical miles) from the runway and correcting to final. The investigative analysis determined at this time the aircraft's spatial position relative the PAPIs was three red lights and one white light. This is slightly low in regard to glide-path, or in other words, given their distance from the runway, their vertical position (altitude) was just below of the desired 3-degree glide-path.

At 5:36:40 AM, the PAPI would have showed all four lights as red. This indication is too low or well-below the desired 3-degree glide-path.

AT 5:36:43 AM, the ground proximity warning system sounded an altitude alert; the captain replied, "Stable." The safety board determined at this time the aircraft was 152 m (500 ft) above the ground and 3.3 km (1.8 nautical miles) from the runway. Their vertical descent speed was 380 m/minute (1,248 feet per minute). The aircraft's glide-path at this point in the visual approach was acceptable (just slight low), approximately 2.6 degrees to the runway. The problem was that their descent rate was twice what it ought to have been and they had already gone so far below the desired 3-degree glide-path so that, although their current 2.6 degree glide-path was no longer excessive, an immediate level-off or climb was needed to correct and avoid a hazardous situation. Due to the low altitude of their approach, their problem was exacerbated by the elevated terrain prior to the runway.

At 5:36:49 AM, the first officer said, "I'm going to have to stay just a little bit higher, or I'm going to lose the end of the runway." Power was slightly increased and the rate of descent decreased from 427 m/minute to 293 m/minute (1,400 ft/min to 960 ft/min); however, they remained below glide-path.

At 5:36:56 AM, the flight engineer, finishing the before-landing check, asked about landing clearance; the captain replied, "Clear to land runway … nine."

The captain stated at 5:37:09 AM, "It's startin' to disappear a little bit in there, isn't it?" It is noteworthy that during the visual approach and while the discussion centered on losing sight of the runway, no discussion was made regarding the need to accomplish a go-around.

At 5:37:13 AM, the flight engineer announced the completion of the before-landing checklist. This was the last recorded communication on the voice cockpit recorder.

At 5:37:14 AM, the ground proximity warning system sounded a 30 m (100 ft) above the ground altitude alert. According to the flight-data recorder, the aircraft was 1.3 km (0.7 nautical miles) from the runway. This equated to approximately 1.3-degree glide-slope to the runway, which was dangerously shallow, not accounting for the higher than runway terrain between the aircraft and the runway. Thus, the 1.3-degree glide-path combined with the high terrain made it certain that terrain impact prior to the runway would occur. It is unclear why the pilots did not initiate a go-around at this point. Perhaps they could not see the terrain due to the dark night as well as confusion in terms of a mismatch of where they thought they were on the visual approach contradicting with the audible warning sound of their low altitude.

At 5:37:19 AM, the ground proximity warning altitude alert announced descending through 15 m (50 ft) above the ground. Then at 5:37:20 AM, the cockpit voice recorder had the sound of a "crunch" (most likely due to hitting trees) as the aircraft passed through 12 m (40 ft) above the ground.

At 5:37:26 AM, the data recording ended due to the crash; approximately 13 seconds of silence marked the final stages of the approach to landing. Roughly 74 minutes prior to sunrise and resting some 472 m (1,556 ft) from runway 09, the aircraft came to a burning stop. It is noteworthy that in the final seconds of the flight no conversation took place and no other comments were recorded since the "cleared to land … runway nine" by the captain. This is another indication of confusion by the flight crew because when things get busy or pilots get task-saturated, they often stop talking. In this case the silence may actually have said a lot.

The captain was 55 years old with between 13,000 and 14,000 flying hours. Records indicated that he had not received the company's required "black-hole" and "controlled flight into terrain" training that took place in 1995 and 1999. The captain also had not gotten much sleep in the time prior to the flight. The captain monitoring the approach never challenged the pilot flying regarding his descent rate and/or his spatial position relative the PAPIs. Also of note was that the first officer readily admitted he was not familiar with the airport and then followed up with the visual confusion of perceiving the power plant's lights to be that of the airports. Further, according to the NTSB report, the captain did not recall seeing any form of a "low" PAPI warning. Passing through 244 m (800 ft) above the ground he recalled perceiving a "white red" PAPI … which is equivalent to being "on" glide-slope (about 3 degrees). About five weeks prior to this flight, the captain had his flying medical exam and the only limitation was that he was required to wear corrective contact lenses when he flew.

The first officer was 44 years old and had between 7,500 and 8,500 total flying hours. He admitted to having trouble adjusting to the night-flying cycle that his flight schedule required, but he had received company's "black-hole" training in 1999. In the post-crash interview, the first officer did not recall seeing any "low" indications on the PAPIs.

The flight engineer was 33 years old and totaled approximately 2,600 flying hours. Like the captain, he had never received the "black hole" training. Similar to the captain and first officer, the flight engineer in his NTSB post-accident interview stated when he first saw the runway, the PAPI lights were a white, a pink, and two red lights; and the runway was in clear sight, "plain as day." The flight engineer's interpretation of the "crunch" of the trees was that they had encountered some turbulence.

As alluded to previously, the final 13 seconds of the cockpit voice recorder was silent, implying that the pilot flying, the first officer, was not the only one that perceived the visually controlled night approach as perfectly safe. Given some of the discussion centered on momentarily losing sight of the runway it is hard to imagine that they were so confident to fly a night visual approach without being cognitively primed to accomplish a go-around. The silence by the more senior pilot, the captain, approved the manner in which the first officer was flying the approach. The NTSB report stated (p. 55):

> The Safety Board concludes that the approach to runway 9 at TLS [Tallahassee] (which was flown over unlighted terrain and in night visual conditions) resulted in black hole conditions, which likely contributed to the flight crew's failure to properly perform the approach. However, the Safety Board also concludes that PAPI lights, such as those installed at runway 9 at TLH, are a recognized countermeasure for use in black hole conditions and should have been, but were not effectively used to maintain an appropriate glidepath by the first officer (who was the flying pilot) or by the captain and flight engineer (who, under the principles of basic crew coordination, were in a position to receive this information and initiate a corrective response).

The confusing aspect of this portion of the mishap description/assessment is how the pilot monitoring the approach, the captain, medically cleared to fly by all standards, also did not perceive the need to amend their approach angle to the runway. In terms of cockpit dynamics, the captain is ultimately "in charge" and responsible for the aircraft and should not hesitate to correct the first officer's flying. In fact, research has shown that when the captain is flying, the lesser ranking, non-flying pilot at times fails to correct the more senior pilot but usually not the other way around (NTSB report, 1994). The recommendation in this 1994 report suggested having the captain monitor challenging approaches—as was the case in the present accident. The captain, however, was fully aware that the first officer was unfamiliar with the Tallahassee runway environment. Despite these facts, the captain never challenged the first officer's flying of the approach to landing, implying that from the captain's perspective the approach was being flown safely in terms of visually guided aircraft control inputs as well as the interpretation of the PAPIs.

The NTSB report (2002) further presented counter-arguments regarding the handling of the aircraft by the first officer. In attempting to de-conflict interpretations of pilot perceptions and aircraft inputs, the following was presented:

> It is possible that the first officer interpreted the uniform PAPI light indications as "white" because that was consistent with available visual indications (for example, the black hole illusion and the slight runway upgrade) that would lead him to perceive that the airplane was higher on the approach than it was. Such interpretations would be consistent with the first officer's conduct of earlier portions of the approach, with occasionally excessive rates of descent and lower-than-normal engine power settings. However, just after the airplane descended

> through 500 feet agl [above ground level], the first officer stated, "I'm gonna have to stay just a little bit higher, (or) I'm gonna lose the end of the runway." About this time, the FDR [Flight Data Recorder] data indicated that the airplane's descent rate began to decrease from about 1,400 to 900, then to 500 fpm [foot per minute]. It was not clear exactly why the first officer moderated the descent rate at this time; however, it is possible that he was trying to reconcile a conflict between the 500-foot GPWS [Ground Proximity Warning System] callout and a mistaken illusion of the airplane's elevation above the field. (pp. 61–62)

As had been mentioned previously, had the first officer flying, or the captain monitoring the approach, recognized four red PAPI lights, an aggressive go-around would/should have been initiated. Also, as presented in later chapters, at this point in a black-hole illusion, the runway begins to appear flat. The concave approach shape brings the pilot in rather shallow, consistent with the decreased descent rate. In retrospect, it was odd that the captain (the pilot not flying with any color vision deficiency) failed to notice the four red PAPI lights or that neither pilot noticed the flattening of the runway. Either visual cue should have been salient enough to prompt an immediate go-around; however, the pilots failed to perceive that their current position relative to the runway and terrain was dangerous, warranting a go-around. This inaction may be the result of cognitive overload in attempting to understand their false perception and the unfolding reality of their situation.

Mishaps in general rarely are the result of a single cause. Aviation accidents especially come about due to a complex temporal sequence of organizational, supervisory, and individual contributing factors. The accident discussed here in Chapter 1, as well as those presented later in Chapter 7, is no exception. Multiple factors contributed to the eventual landing short mishap. In this particular situation, fatigue and poor crew resource management (lack of monitoring flying performance) also contributed to the accident, per the NTSB report. More relevant to the topic of this book, the investigators unveiled a visual perception factor in addition to the black-hole illusion and deemed it also as contributing to the accident. The formal report found that the first officer had a color vision deficiency which prevented him from properly interpreting the color changes of the PAPI lights from white to red. This finding alone spurred a new national and international interest in color vision assessment in both the civil aviation and military aviation communities.

The NTSB report stated that the first officer's vision throughout his 16-year naval aviation career was reported as 20/20 in terms of Snellen acuity. Also, his color vision was assessed annually and that he passed the Farnsworth Lantern test, which is the US Navy's standard color vision screening test. The first officer passed the test 13 times, 10 with a perfect 9/9 score, twice with no documentation, and once with "passed-by history" (p. 30). According to the NTSB report, the first officer reported he had no identified color-vision deficiencies while flying for the Navy. In 1995, however, the first officer had failed a color-vision assessment, one using Pseudo-Isochromatic Plates (referred to as PIP). These plates consist of colored dots and embedded within the dots are more colored dots forming a number. The individual taking the assessment must identify the colored number embedded within the colored dots. If a color vision deficiency exists, the individual fails to sense/perceive (detect/recognize) the number. (See Chapter 3 for more detailed information on color vision testing.)

The first officer had indeed failed the PIP test, but despite the identified "mild red-green defect," he was issued a flying certificate and a Statement of Demonstrated Ability based on his past operational experience/capability (NTSB report, 2002). After the 2002 mishap, a color vision examination of the first officer was conducted by the US Air Force Aerospace Medicine specialists. Although he again passed the Farnsworth Lantern test, he proceeded to fail seven other red/green color vision assessments. It was concluded that the first officer had a "severe congenital deuteranomaly" that could result in difficulties differentiating reds, greens, and whites (NTSB report, p. 61). Deuteranomaly is a common form of color deficiency, affecting 4.6 percent of males and stands for "green weak", meaning that a person requires more green stimulation than normal to acquire a color match (Tredici and Ivan, 2008). The Young-Helmhotlz theory of color vision describes color as a perceptual experience defined by the individual based on a combination of the three different cone types in the retina. (The topic of color vision is addressed in more detail in Chapters 2 and 3.) The NTSB report cited a letter from the US Air Force Aerospace Medicine specialists (p. 61):

> We believe that he [first officer] would definitely have had problems discriminating PAPIs … because the red light would appear not to be red at all, but more indistinguishable from white than red … it would be extremely unlikely that he would be capable of seeing even the color pink on the PAPI … more likely a combination of whites and yellows and perhaps, not even that difference.

The NTSB report continued in reference to the USAF Aerospace Medicine specialists in describing that the first officer had successfully perceived his aviation world based on other visual strategies that, up until this mishap, had worked. For instance, brightness, location, color shades all may have helped differentiate color perception. The NTSB report continued examination of the role of the color vision and its role in this mishap (p. 61).

However, during the approach to runway 9 at TLH [Tallahassee], the first officer had to rely more heavily on his color vision because the PAPI lights were the only reliable source of glide-path information in the black-hole approach environment leading to runway 9. The first officer's interpretation of the PAPI lights would have been even more challenging because all four lights were red during most of the final approach. As a result, there would have been no differing levels of brightness for the first officer to perceive across the lights (as might have been apparent if both white and red PAPI lights were visible), nor would there have been a change in brightness to observe (as there might have been when a PAPI light transitioned from white to red during the descent). Either of these would likely have assisted the first officer's color interpretation.

Although one could argue the major or minor role that color vision deficiency played as a contributor in this specific mishap, there is no argument that this accident highlighted the problems of color vision assessment. This accident spurred research regarding color vision and color vision tests for pilots as well as attempts to standardize assessments between the Federal Aviation Administration, the US Army, the US Navy, the US Air Force, and international flying organizations. The NTSB issued a 2004 Safety Recommendation to follow up this mishap and further emphasized the role of color vision assessment problems in aviation. In that report the NTSB cited an Australian study that examined color vision in pilots with a similar deficiency to the first officer in this mishap. This study found that approximately 29 percent such pilots mis-identified a red light signal with a white light signal. The NTSB Safety Recommendation concluded with a call for research into color vision assessment to include color differentiation tests in time-critical situations and mild hypoxic conditions.

The accident investigation (NTSB report, 2002) concluded that the probable cause of the accident was "failure to establish and maintain proper glide-path during the night visual approach to landing" (p. 68). Thus, key aspects of this mishap were night conditions, visual misperception, a black-hole approach environment, and color vision—all applicable to this book.

Visual Perception Allows for Heroism

The human visual system is a phenomenal combination of physiological and psychological processes that allow us to interact with our environment. Because of our abilities to sense and perceive, detect and discriminate, and recognize and react, we more often than not are able to successfully negotiate dynamic and challenging activities. Acts of human expertise in aviation are difficult to quantify and thus are rarely reported and researched. Every pilot has had situations that *almost* resulted in tragic or destructive consequences; however, due to quick reactions a disaster or an incident was avoided. These "almost accidents" are rarely documented; no headline news reports were made of them. As far as the organizations (e.g., military, commercial airline company) are concerned, they operate safely based upon the lack of any known problems (e.g., mishaps)—but are they truly safe or just lucky? Reason (2008, p. 265) said it best: "safety is a term defined more by its absence than its presence." Quantifying safety is difficult. Of course just because accidents are not occurring does not necessarily equate to safe operations; articulated best by Dekker (2005, p. 26), "past success does not guarantee future safety."

We use examples of aviation spatial-disorientation mishaps to demonstrate that what killed pilots and destroyed airplanes decades ago is still occurring today. Respect and awareness of human limitations in terms of spatial orientation has not significantly improved over the years. Consequently, our intent is to inform the pilot or the aviation researcher of the need to be significantly more aware of the risk in trusting visual perceptual systems. Our depiction of research data and accidents spanning decades is to show clearly that the aviation community has not made sufficient progress, and consequently lives and resources are still lost due to the contributing factor of visual misperception. Occasionally, however, an incident occurs in which it is obvious that all the aviation training, education, skill, and proficiency came to fruition, allowing human capability to outshine any human limitations. An incident occurs that the aviation community can take pride in regarding their proactive safety efforts. An incident occurs that despite all odds are against success safety clearly was present.

While nearing the completion of writing this book a heroic event occurred in aviation that few will forget. On 15 January 2009, the crash-landing and successful ditching of US Airways Flight 1549 in the Hudson River amazed the world. The pilots and crew of Flight 1549 did a superb job in handling the emergency and controlling the evacuation of the aircraft as water filled the cabin (Figure 1.5). Even though the final NTSB report has not been released, enough information has been reported through NTSB preliminary reports that allow for a visual perception discussion.

Flight 1549 departed LaGuardia International Airport, New York, bound for North Carolina. The airplane was an Airbus A320 carrying 150 passengers and 5 crew members (NTSB Preliminary report, 2009). It was estimated that as the aircraft climbed through 900 m (3,200 ft) both engines ingested Canadian Geese,

Figure 1.5 Hudson River ditching

Source: Permission granted, photographer Greg Lam Pak Ng, Flickr.com.

rendering them no longer able to produce the required thrust for flight. Bird strikes are common in aviation and migratory birds are one of the most dangerous hazards that airports try to mitigate. Large flocks of birds are often observed on radar screens by traffic controllers; however, in terms of visual perception, it is very difficult to perceive a bird and react prior to it striking the aircraft. Even if the aircrew of Flight 1549 had detected the birds in time, an aircraft the size of the Airbus would have had difficulty maneuvering to avoid the birds. Given the number of birds, regardless of Flight 1549's maneuvering, there was a good chance they would have had a bird strike somewhere on the aircraft. Thus, in terms of visual perception, the detection of a hazard (birds) while flying is often left to chance, and if birds are sensed and perceived they are very difficult to evade. One pilot technique is to initiate some type of pull-up maneuver as birds normally dive below and away from an aircraft.

Captain Sullenberger, the pilot in command of Flight 1549, is a 1973 graduate of the Air Force Academy. In *Checkpoints*, a magazine published by the US Air Force Academy Association of Graduates, an article summarized an interview that he gave for a television news program. When asked about the actual bird strike, Captain Sullenberger responded, "… about 90 seconds after takeoff I noticed there were birds filling the entire windscreen—from top to bottom, left to right—large birds, too close to avoid. It felt like the airplane was being pelted by heavy rain or hail" (2009, p. 18). Thus, in terms of detection, sensing, and perceiving the birds, the pilot flying Flight 1549 had no visual indication of a threat until the birds were within very close range because of the small size of the birds relative to the backdrop of the sky and the closure rate between the Airbus and the birds. Visually detecting such a small retinal image is a physiological limitation at those speeds.

The second aspect of visual perception involved in Flight 1549 was the accomplishment of the ditching maneuver, an emergency water landing. Every pilot in emergency training has read about ditching an aircraft with the realization that the odds of that happening being extremely remote. Aircraft emergency procedures discuss at length how to safely ditch, but pilots realize that the engineers designing the aircraft really don't know what is going to happen to the aircraft upon water impact.

Configuration and airspeed are the two primary aspects for a ditching maneuver. Normally for a given weight, the airspeed of an aircraft prior to "landing" on water is 10 knots below normal landing speed. For an aircraft right after initial takeoff, landing speed would be very high. So, although the target speed is 10 knots slower than a normal approach, such a landing would still be quite fast. At high speeds, landing on water is still similar to concrete, hence the need for a "soft" water landing.

On that day the winds were calm and the water was fairly smooth (as shown in Figure 1.5), the pilot of Flight 1549 was able to accomplish a perfect ditching maneuver and gently settle the aircraft onto the water during the flare. The pilot had to smoothly put the aircraft into the Hudson River in order to avoid excessive structural damage that could have torn the fuselage of the aircraft. A real danger in a ditching maneuver is catching a wing-tip and throwing the aircraft into a tumble in the water. Any of these events would have most certainly resulted in significant loss of life. But not a single life was lost. Captain Sullenberger explained his landing:

> I needed to touch down with the wings exactly level. I needed to touch down with the nose slightly up. I needed to touch down at a descent-rate that was survivable. And I needed to touch down just above our minimum flying speed but not below it. And I needed to make all these things happen simultaneously. (*Checkpoints*, 2009, p. 19)

In many cases, a smooth water surface for "landing" causes perceptual errors due to a lack of features for height and distance perception. Specifically, an open body of water provides few referents for global and local visual cues regarding the aircraft's glide-path toward impact. Fortunately, the Hudson River and its shoreline provided plenty of ambient cues regarding height above the water's surface to the pilot of Flight 1549. Additionally, the calm winds and smooth water surface benefited the ditching maneuver (water impact). Had the ditching maneuver occurred on smooth water in a large open area, like an ocean or large lake, height and distance perception may not have been as accurate as they were on the Hudson River.

Based on Captain Sullenberger's explanation of the ditching, his thoughts during the last few seconds of flight were on a smooth descent for the slowest possible water impact while maintaining a wings-level attitude. To accomplish a safe water landing the pilot's ambient vision system was unconsciously computing horizon perspective (to control wings level and bank angle) and sink rate (to provide input to the pilot's control surface movements). He may have been using focal vision (conscious attention) to perceive airspeed readings and radar altimeter information or to determine where on the water ahead the airplane would splash down. The ambient visual system, however, provided the unconscious but critical environmental inputs for global perception as the aircraft glided into the water, leading to a success story for aviation and for the human visual perception system's capabilities.

To understand the raw visual perception mechanisms at work on this January day, it is worth emphasizing that the pilots accomplished this maneuver with no instrument approach procedures to help guide them. There was no runway outline or runway shape to assist in visual glide-path guidance. There were no approach lighting systems or PAPIs; all that was available to the pilots were the environmental cues of the water surface (which were not much) and the shoreline.

Previously it was mentioned that documenting expert performance is difficult and not often discussed in research. One author has tackled the topic and his work outlines many parallels to the Hudson River ditching incident. In Reason's (2008) book, *The Human Contribution*, the author articulated factors for heroic human action in high risk activities; that is, aspects of the situation and of the individuals involved and how they successfully overcame adversity. He examined how many total disasters have been avoided due to phenomenal actions by operators within complex systems. All of the attributes of such incidents described by Reason apply to Flight 1549, and those were training, discipline, and leadership; sheer unadulterated professionalism; luck and skill; and inspired improvisation. Reason addressed how humans have to learn and train to cope with both expected and unexpected scenarios. For instance, pilots routinely practice and rehearse takeoff emergency actions. Every pilot prior to releasing brakes for takeoff should have an emergency return scenario in mind if an engine fails, an unlikely but possible scenario.

The pilots of Flight 1549 had minimal time to assess the situation and take appropriate action. The water landing choice, however, came about due to limited options and a quick risk assessment of any possible alternatives other than the Hudson River. On another day with the same crew or a different crew, the Hudson River option might not have been as successful. If it had happened at night, the pilots would have had significantly increased difficulty in perceiving their height above the water and the ditching maneuver could have been tragic. So yes, luck played a role; however, the pilot, given the situation and the environment, had the appropriate environmental cues for veridical perception to accomplish the ditching. Often "luck" only comes to those that are prepared. And without a doubt, the entire crew, and especially the pilot at the controls, had Reason's "sheer unadulterated professionalism."

Our Approach

We strongly believe that spatial disorientation's contribution to aviation mishaps can be significantly decreased through awareness and through education utilizing research-based operational/realistic training scenarios. Spatial disorientation is attributed to nearly one quarter of all mishaps in military aviation; thus it is greatly beneficial to chip away at this source of accident factors. The US Naval Aviation Safety Center (2009) reported that between the years 1990 and 2008 the number one human factor involved with nearly 80 Class A mishaps (those accidents involving a fatality, destroyed aircraft, or >$1M in damage) was spatial disorientation. While people often associate spatial disorientation with vestibular processing misperceptions, there are actually substantial visual perception influences in spatial disorientation. The role of visual spatial disorientation happens more often than expected and is greatly underreported as a contributor to mishaps. Veronneau and Evans (2004) summarized the main objective of this text while addressing their own thoughts on spatial disorientation:

> Successful SD (spatial disorientation) countermeasures will impact all types of aviation operations, including those in the civil community, as there will always

> be an essentially universal susceptibility of human pilots to SD illusions. To achieve measureable success in improving the overall aviation safety climate, such as the FAA's goal of an 80 percent reduction in the commercial accident rate, then many of the so-called minor contributors to aviation risk must be addressed. Spatial disorientation is one such factor that can be clearly identified and specifically addressed. (p. 220.)

Indeed the problems of aviation visual perception are not going to go away. In *Wingman*, the US Air Force's safety magazine, an article by Sabric (2009) shared a very recent F-16 near-mishap story. The instructor pilot concluded that, "the student had convinced himself that the access road was Runway 03L, and instead of trusting his instruments, he trusted his visual perception" (p. 29). Interesting choice of words by the F-16 pilot, but fitting and accurate.

Throughout the chapters of this book it will become apparent to the reader that all pilots are susceptible to spatial disorientation and that mishap reports regarding the incidence of spatial disorientation are underreported. Even less reported and less respected in accident investigation reports is the role of visual misperception as a contributor to spatial disorientation and the mishap sequence. This book intends to highlight visual spatial disorientation to academics, researchers, accident investigators and to the pilot community. As Veronneau and Evans (2004) advocated, the potential of spatial disorientation awareness/education for reducing mishap rates is significant; therefore we also promote this area ripe for aviation research and safety enhancement. By bringing forth both research and historical mishap examples, visual misperception's presence is undoubtedly demonstrated. It is not simply "pilot error" but a human perceptual limitation that needs to be better accounted for in aviation education, training, and technological advances as well as in accident investigation.

It is our belief that understanding the role of vision in aviation requires an interdisciplinary approach. Our main goal in writing this book is to provide a comprehensive, single-source document encompassing all the aspects of aviation visual perception. Thus this book includes the foundations of visual and vestibular sensation and perception, and how visual perceptual abilities are assessed in pilots (Chapters 2 and 3), the pilot's perspective of visual flying (Chapter 4), a summary of human factors research on the visual guidance of flying (Chapter 5), examples of specific visual and vestibular illusions and misperceptions (Chapter 6), mishap analyses from military, commercial, and general aviation (Chapter 7), and, finally, how knowledge from these other disciplines is being used to create the next generation of aviation visual perception (Chapter 8).

References

Benson, A. J. (1988). Spatial disorientation—general aspects. In J. Ernsting and P. King (Eds.), *Aviation Medicine* (pp. 277–96). London: Butterworths.

Calvert, E. S. (1950). Visual aids for landing in bad visibility with particular reference to the transition from instrument to visual flight. *Transactions of the Illumination Engineering Society of London, 6*, 183–219.

Checkpoints Editorial Staff (2009, March). Service Before Self: Sullenberger, '73, saves the lives of 155 people in a watery crash landing. *Checkpoints*, 18–19.

Cocquyt, P. (1953). Sensory illusions. *Shell Aviation News, 178*, 19–24.

Dekker, S. W. A. (2005). *Ten Questions About Human Error*. London: Lawrence Erlbaum Associates.

Gibb, R. W. (2007). Visual spatial disorientation: re-visiting the black hole illusion. *Aviation, Space, and Environmental Medicine, 8*, 801–808.

Havron, M. D. (1962). Information available from natural cues during final approach and landing. Federal Aviation Agency, National Aviation Facilities Experimental Center, HSR-RR-62/3-MK-X.

Kern, T. (2002). *Controlling Pilot Error: Approach and Landing*. New York: McGraw-Hill.

Kraft, C. L. and Elworth, C. L. (1969, March–April). Night visual approaches. *Boeing Airliner*, 2–4.

Lessard, C. S. (2000). Spatial disorientation: dealing with aeronautical illusions. *IEEE Engineering Medicine and Biology, 19*, 25–27.

Mulder, M., Pleijsant, J., van der Vaart, H., and van Wieringen, P. (2000). The effects of pictorial detail on the timing of the landing flare: Results of a visual simulation experiment. *The International Journal of Aviation Psychology, 10*, 291–315.

National Transportation Safety Board. (1994). Safety Study: A review of flightcrew-involved, major accidents of U.S. air carriers, 1978 through 1990, Department of Commerce, National Technical Information Service, NTSB/SS-94/01.

National Transportation Safety Board. (2002). Collision with trees on final approach: Federal Express Flight 1478, Tallahassee, FL.

National Transportation Safety Board. (2008, June 10). Safety Recommendation.

National Transportation Safety Board. (2009). Preliminary Report, NTSB ID DCA09MA026.

Newman, D. G. (2005, Sep-Oct). Eyeball error. *Flight Safety Australia*, 27–32.

Newman, D. G. (2007). An overview of spatial disorientation as a factor in aviation accidents and incidents. Australian Government, Australian Transport Safety Bureau, Aviation research and analysis report—B2007/0063.

Ostinga, J., Wolff, M., Newman, D., and White, S. (Nov-Dec, 1999). What killed JFK Jr? Lessons for VFR pilots. *Flight Safety Australia*, 22–29.

Pitts, D. G. (1967). Visual Illusions and Aircraft Accidents. USAF School of Aerospace Medicine Division, Brooks AFB, TX, Report No. SAM-TR-67–28.

Previc, F. H. (2004). Visual illusions in flight. In F. H. Previc and W. R. Ercoline (Eds.), *Spatial Disorientation in Aviation, Progress in Astronautics and Aeronautics, Vol. 203* (pp. 283–322). Reston, VA: AIAA.

Reason, J. (2008). *The Human Contribution*. Burlington, VT: Ashgate.

Sabric, G. (2009). It's not over… Wingman: United States Air Force Journal of Aviation, Ground, Space, and Weapons Safety, 2(Fall), 28–29.

Schiff, B. (1990). Visual illusions can spoil your whole day. *Flight Safety Foundation Accident Prevention*, *47*, 1–4.

Schiff, B. (1994, January-March). Black hole approach. *Boeing Airliner*, 16–20.

Smallman, H. S. and St. John, M. (2005). Naïve realism: Misplaced faith in realistic displays. *Ergonomics in Design*, Summer, 6–13.

Tredici, T. J. and Ivan, D. J. (2008). Ophthalmology in aerospace medicine. In J. R. Davis, R. Johnson, J. Stepanek, and J.A. Fogarty (Eds.), *Fundamentals of Aerospace Medicine*, (pp. 349–379). Philadelphia: Lippincott, Williams, and Wilkins.

United States Naval Aviation Safety Center (2009). Raw mishap data provided from the Navy's safety center.

Veronneau and Evan (2004). Spatial disorientation mishap classification, data, and investigation. In F. H. Previc and W. R. Ercoline (Eds.), *Spatial Disorientation in Aviation*, Vol. 203, Progress in Astronautics and Aeronautics (pp. 197–241). Reston, VA: American Institute of Aeronautics and Astronautics, Inc.

Warren, R. and Owen, D. H. (1982). Functional optical invariants: A new methodology for aviation research. *Aviation, Space, and Environmental Medicine*, *53*, 977–983.

Wulfeck, J. W., Weisz, A., and Raben, M. W. (1958, November). Vision in Military Aviation. Wright Air Development Center, United States Air Force, Wright-Patterson Air Force Base, OH, TR 58–399.

Chapter 2
Sensation and Perception Foundations

Our sensory and perceptual systems do an amazing job of informing us about the world in an almost real-time manner, so that we can safely and efficiently maneuver through the environment. These processes happen so quickly and easily that, for many years, researchers believed it would soon be possible to build a machine that mimicked human perceptual capabilities. Several decades later that expectation has still not come to fruition; however, we do have a much more detailed understanding of the sensory and perceptual processes.

One major take-away point based on what we have learned during the past several decades is not simply that the sensory and perceptual systems are complex, but that they rely on and take advantage of the rich perceptual environments in which they typically operate. Most errors in perception occur in impoverished environments (such as often occur when flying at night or in fog), or with carefully constructed stimuli (i.e., illusions) that have limited cues available (e.g., they are only two-dimensional or they do not contain other "real world" cues such as texture gradients, motion, or shading).

A second fundamental concept that should be kept in mind is that our perception of the external environment is *constructed* by our sensory and perceptual systems, and while it corresponds well enough to external reality for us to survive in most circumstances, it does not perfectly match reality. General unawareness of this mismatch was referred to in Chapter 1 as naïve realism. The mismatch is the result of several processes (Scharff, 2008): our brains receive limited information from our senses (e.g., we cannot directly perceive electromagnetic energy beyond the visible spectrum), as the system processes the sensory information, some of it is enhanced beyond reality (e.g., edge enhancement), some of the information is altered (e.g., due to context effects), and some of it is further filtered (often due to lack of attention). Further, in order to give us near real-time perception of the environment, our brains rely on heuristics (mental shortcuts) which were developed based on consistent interactions with the environment (e.g., light comes from above, objects tend to be opaque). As with any heuristics, they usually serve us well, but sometimes they can lead to inaccurate perceptual conclusions (which, in aviation, can be deadly). Finally, we must also acknowledge that we do not become consciously aware of much of the information that activates our senses, although sometimes, even information we are not consciously aware of can impact our behavioral responses to the environment.

This chapter overviews the basics of visual (main emphasis) and vestibular sensory and perceptual processes because they have special relevance to aviation. To begin, we will set a framework by outlining some of the overarching theories of visual perception and by discussing how what we consciously perceive is largely dependent on attention. This framework is crucial because it reminds us that, although some components of our sensory and perceptual systems are fairly well understood, they do not work in isolation, which in turn means that the real-life impact of changing environmental stimuli may not be easy to predict. We will then review some of the basic physiological and perceptual components of vision and the vestibular systems. Knowledge of how a pilot judges depth, motion, color, size and shape is necessary to fully appreciate the act of flying. Finally, a basic understanding of visual perception is needed to understand where the next generation of aviation is heading, for example synthetic and artificial visual systems.

Bottom-up versus Top-down Processes and Theories of Vision

In our quest to understand visual perception, several theories have been proposed that attempt to explain the general strategies used by the visual system. None of these theories are complete, but they all have made contributions to our understanding of how the visual system might work (see Rock, 1983 for a more thorough discussion).

Two terms are important to clarify as we introduce these theories: bottom-up perception influences and top-down perception influences. Bottom-up perception influences refer to stimulus information that activates the sensory neurons and travels through the system "up" to the brain, where it ultimately leads to the conscious awareness (perception) of the stimulus. Top-down perception influences refer to the alterations in perception caused by memories, expectations, and so on that already exist in the brain. Dreams and hallucinations essentially are completely top-down perceptual processes where there is no real external stimulus corresponding to the perception. In normal, awake perception, however, the two processes interact, with the top-down processes increasingly dominant as the stimulus becomes more impoverished. Even without impoverished stimuli, top-down processes interact with the bottom-up flow of information by directing attention, which in turn exerts a filtering process on the bottom-up, sensory-driven flow of information.

The different perception theories differ in their focus on bottom-up and top-down processes. Purely stimulus-driven theories are bottom-up and focus exclusively on perception as determined by the external stimulus' characteristics. Generally such theories are regarded as too simplistic because of the lack of a possible one-to-one correspondence between an external stimulus and the image of that stimulus that falls on the light-detecting neurons (photoreceptors) on the retina at the back of the eye. For example, consider a simple stimulus such as a rectangular door. Only if the door is viewed from a perfectly straight-on perspective will its image at the back of the eye have a perfectly rectangular shape. As the door is opened, the image becomes trapezoidal, with the closer vertical side of the door having a longer length in the image than the further vertical side of the door. Somehow, however, our perceptual systems consistently interpret the shape of the door as a constant rectangle (shape constancy). In other words, what we perceive is more similar to our mind's perception of the object that what we sense in the raw stimulus from the retinal image. In Chapter 6 we discuss an analogy of this door example in aviation: the runway illusion. This type of non-veridicality is true for all objects because the real world is three-dimensional, and the retinal image is two-dimensional.

J.J. Gibson (1979), however, argued that in the real world, due to the large number of stimulus cues that redundantly signal stimulus characteristics, a purely stimulus-driven explanation of perception is possible (a *direct* approach to perception). His Ecological Perception Theory especially focused on optic flow information and rich texture gradients that tend to occur as a person interacts with the real environment because he believed that these cues led to environmental invariants, that is, information that is consistent regardless of the path of observer motion. Optic flow is the systematic change in the retinal image as a person scans or moves through the environment. The geometry of the environment would not be accurately captured in a single, static image due to the three-dimensional-to-two-dimensional translation discussed above. However, across multiple images as occurs in optic flow, the real world geometry and shapes of objects become disambiguated. Texture gradient, the density of a pattern in the environment, is another systematic environmental cue; patterns (e.g., the texture of a grassy field) will become more densely arranged as distance increases or as the surface declines away from the viewer. Both these cues can provide accurate information about objects in the environment without the need to rely on top-down processes.

Flach and Warren (1995) described ecological perception related to aviation as "a continuously traversed dynamic loop or cycle of perception and action" (p. 66). An example of an invariant would be the aimpoint of a landing. The aimpoint at the beginning of the runway, the threshold, should not change once a 3-degree glide-path has been established. Movement of the aimpoint within the optic flow of a pilot's field of view provides information about which vertical and lateral direction the aircraft has deviated from the desired target.

Even with rich and redundant cues, however, other research shows that one cannot fully explain human perception solely through bottom-up processing. Even Gibson himself, the father of Ecological Perception, tempers the direct approach of perception with acknowledgment that perceptual learning must occur to know how to properly interpret the depth perception cues in an ambient array (Gibson, 2002). Beyond the influence of such learning, however, other information such as our expectations influence our perception of even highly salient stimuli. Further, in the real world, stimuli are often degraded (e.g., at night or in the fog or due to partial occlusion). Our experiences with the world and objects in it lead to memories and expectations that allow us to complete such stimuli and make sense of the degraded signal reaching our sensory system. Generally we only become aware of these top-down processes when they lead us to make an error based on an inaccurate assumption. Otherwise, our perceptions occur so smoothly and seemingly accurately that we don't given them a second thought (as colorfully articulated in Smallman and St. John, 2005).

Perception theories that are not completely bottom-up can incorporate two additional types of processes that can both nominally be referred to as "top-down." The first of these refers to signal processing mechanisms that become "hardwired" in the system due to consistent experiences with the environment (e.g., perceptual learning). Most humans will show similar processes of this type due to the fact that we all experience the same general consistencies in our environments (e.g., light from above, object occlusion). Rock (1983) referred to theories that focused on such processes as being "Spontaneous Interaction Theories". Many of the Gestalt principles of organization (e.g., grouping by similarity, good continuation, figure-ground processes) fall in this category. See Figure 2.1 for an example of the interpretation of shading. This image shows several shaded circles, and at first glance, most people immediately interpret the top-right and bottom-left ones to be "craters" and the bottom-right one to be a "mound." The top-left image looks like a mound with a small crater in the middle. These perceptions are based on the assumption of lighting from above. It is possible to "flip the perception", especially if the viewer physically turns the picture upside down. Often, pictures of real craters will appear to be "mounds" due to an assumption of lighting from above, which

Figure 2.1 Crater shading illusions

Source: With permission from Irani, Slonowsky, and Shajahan, 2006.

often is not true when the craters are viewed from space. Thus, in the case of moon craters, perception does not match the stimulus (so it's not completely bottom-up perception), but the inaccurate perception is fairly consistent and doesn't change based on expectations or other knowledge of the illusion; the craters still look like mounds even once we know they are really craters. Such learned assumptions are important to understand because they will consistently lead to misinterpretations. With new consistent experiences, new assumptions can become "hardwired" in the system (see below for examples of the interaction of optic flow and height and of the interaction of size and distance of runways). By making people (e.g., pilots) aware of such hardwired perceptual influences, they can learn to take measures to counteract their inaccurate assumptions.

The second type of top-down processes is those that occur when we use our conscious memories and expectations to help process the stimuli. Although it's really a continuum, these top-down processes are distinguished from the more "hardwired" processes in that we can rapidly alter our perception of the scene by changing our conscious expectations or by drawing on a different memory to help us understand what we are viewing. For example, a pilot may initially misperceive his glide-path to a runway based on a runway's size and shape relative to the runway with which a pilot is most familiar. However, upon interpretation of environmental cues as well as knowledge of the runway's dimensions, the pilot may then re-guage his perception and "see" the glide-path for what it is. There are times however, as presented in Chapters 6 and 7, when pilots fail to re-guage their perception because of impoverished visual cues.

A final "theory" that should be acknowledged is the Information Processing Theory. This approach is really not a true theory of perception, but a summary of what is known about how the stimulus is transformed as it is processed through the different levels of the system. For vision, there is a very good understanding of the optics of the eye, so the information processing theory can very accurately predict the characteristics of the image that will fall on the retina and stimulate the photoreceptors. As we gain more and more knowledge about the neural processes in the retina and the different areas of the brain involved in visual processing, we can more and more accurately predict the final perception that might be experienced. The information processing theory takes into account both bottom-up and top-down processing influences. The processes that are currently least well understood are those mediating the top-down influence of the more conscious memories and expectations, although recent advances are rapidly expanding our understanding of the physiological mechanisms involved in those processes. The major parts and pathways of the visual system are described below in the section covering the Physiology of Vision. Figures 2.2 below and 2.3 in the following section show the major known parts/pathways of the visual system and the eyeball, respectively, and indicate general flow of information (both top-down and bottom-up). In Figure 2.2, the weight of the arrows showing input to the LGN indicates the amount of input from each of the sources. The dashed lines leaving the primary visual cortex indicate a non-direct pathway with other areas not shown in the diagram. The Information Processing Model is also often portrayed using box flow

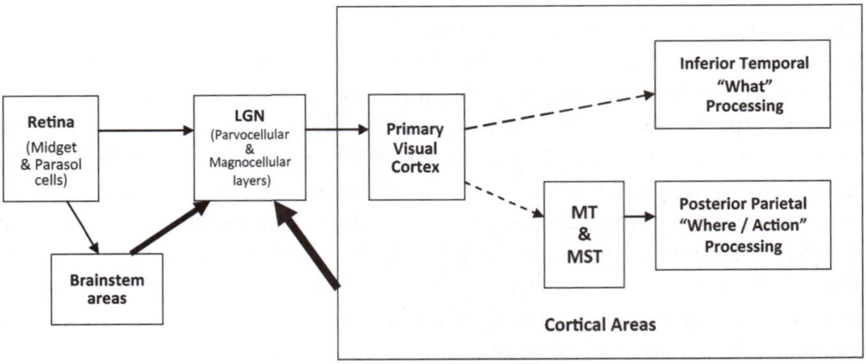

Figure 2.2 Major components and pathways of the visual system

charts (e.g., Wickens, Lee, Liu, and Becker, 2004), which are useful for highlighting both the top-down and bottom-up processing stages (e.g., sensory register, attention, memory, response selection), without trying to localize those functions to specific physiological pathways or processing areas.

Attention

Attention is a top-down influence that enhances the neural responses to those objects or areas being attended (Murray and Wojciulik, 2003), and inhibits the response to stimulus information not attended so that it often does not achieve conscious awareness. Thus, attention can both help and hinder the perception of items in the visual field. There are several distinct attentional effects that have been recognized, and each can potentially lead to errors in perception and decision making.

Inattentional blindness (sometimes also referred to as selective attention), is a direct result of the narrowing of awareness to those objects that are attended. Unattended objects in our environment can be highly salient and well above detection thresholds, but a person can often be completely unaware of them. Thus, subsequent behavioral decisions will not take into account the unattended information. This ability to focus attention is beneficial so that distracting information can be ignored. For example, during a challenging instrument approach flown in the weather, ambient cues of aircraft lights reflecting off fog may be sensed by the pilot; however, the pilot's complete attention is on the attitude indicator and glide-slope needles. Thus, this attentional selectiveness is a benefit by helping the pilot not to fall prey to looking outside and becoming disoriented. Selective attention becomes a liability when cognitive selectivity excludes vital sensory input from being perceived. For instance, failure to "see" the gear is *not* down and locked as indicated by the position of the gear handle and the lack of green gear lights can occur due to attention focused elsewhere in the cockpit or outside the aircraft. This example has the same modality, vision, competing for attentional resources but often it can also occur across modalities. For instance, in the gear-up landing example, the lack of gear being down and locked (visual indication) is also accompanied by an auditory warning (alarm) and this sensory input is also often heard but not listened to.

Inattentional blindness often occurs due to attentional "capture". A current aviation challenge with respect to attentional capture is the implementation of head-up displays (HUDs). These displays were intended to reduce the loss of attention to objects outside the cockpit window that occurs when a pilot has to physically look down to read cockpit gauges. HUDs place essential gauge information in a pilot's direct line of sight out the window by using transparent projected light letters and numbers. In order for these HUD displays to be easily read, they must have fairly high contrast with the background environment outside the window. Often a pilot's attention is drawn by the salient detail of the HUD's displayed information (e.g., attitude indicator bars, airspeed and altitude quantities, etc.) and the external, ecological cues are not perceived. This topic will be addressed again later in Chapter 8.

Conscious attention is generally directed toward objects that are fixated upon (e.g., where the eyes are focused so that the object of interest falls on the foveal area of the retina, where the most detailed vision is possible). This is referred to as overt attention, because other observers are able to determine a person's object of attention by noting at what that person is directly looking. In contrast, covert attention occurs when people focus their attention to an area where they aren't directly looking. Because the visual image will be processed with less resolution outside the foveal area, covert attention often additionally recruits information from other senses. For example, if a pilot is attending to what the copilot is saying, even though the pilot is perhaps looking directly at the attitude indicator and not the copilot, he might not notice a dangerous change in their attitude. Attention given to internal thoughts can also disrupt the processing of otherwise highly salient visual stimuli. For example, many people report driving for miles and not remembering any of the drive because they were "thinking about something." In most cases this is not a problem, but if a pilot misses a progressive change in a gauge reading due to this type of inattentional blindness, what could have been an easy adjustment early on may become an emergency situation.

The inattentional blindness issues discussed above are categorized as "Channelized Attention" by the DoD Human Factors Analysis and Classification System (2005) for use during accident investigations. Channelized Attention is a factor when the individual is focusing all conscious attention on a limited number of environmental cues to the exclusion of others of a subjectively equal or higher detectability, or of a more immediate priority. Still more terms used in aviation that describe the concept of either selective or channelized attention are "attentional narrowing" or "cognitive tunneling."

A second type of attentional influence is change blindness. This attentional influence is primarily due to the limited amount of information that a person can hold in working memory. Attention to a stimulus allows us to become consciously aware of that stimulus. Conscious awareness places information into working memory, where it is held and can be compared to other information previously encoded into long-term memory in order to allow identification and understanding of the current stimulus. Information that is not brought into working memory initially can cause a neural response, but that neural response fades and become inaccessible within a short period of time. Iconic memory (the visual sensory store) fades in less than one second, and echoic memory (the auditory sensory store) fades in 3–4 seconds. This fading of unattended stimuli affects perception in that observers become likely to miss obvious changes in stimuli across different, and *non-continuous*, scans of a scene. If an observer is continuously looking at a scene, changes often "pop out" due to our perceptual system's sensitivity to change (bottom-up processing). However, if a break occurs in the viewing of the scene, the items that were not directly attended in the first viewing can undergo major changes (e.g., appearance/disappearance, color changes, etc.) and not be noticed. For example, if a pilot on the tarmac looks out the cockpit window, looks down to the gauge displays, and then looks back out the window, she might completely miss another plane that has taxied into view. Because the route was clear in the first glance out the window, and she did not notice the change in the scene, she might continue on a collision course without noticing the obvious stimulus.

Inattentional blindness interacts with change blindness, and together they can have major consequences on what is perceived and acted upon by observers. These effects are compounded in situations of multi-tasking, because a person is shifting their attention between multiple tasks. Thus, change blindness is more likely to occur (due to the shifts in focus), and working memory capacities are more likely to be exceeded. Unfortunately, multi-tasking is common for pilots. During an instrument cross-check, for example, visual scan is centered on the attitude indicator, which as an object display provides both pitch and bank information and is centrally located. Other displays, such as airspeed, altimeter, vertical velocity indicator, radar altimeter, and the primary instrument display (ILS, GPS, TACAN, etc.) depicting lateral and vertical displacement all compete with the each other for visual attention. Training and experience can reduce the negative effects of divided attention; for example, the instrument cross-check pattern will be driven by the pilot's experience in terms of where to look when and for how long (dwell time). Experience will develop a deterministic, coherent strategy in terms of the information displayed (Mosier, Sethi, McCauley, Khoo, and Oransanu, 2007).

Figure 2.2 on page 26, illustrated a simplified diagram that shows major components of the visual system and indicates some fundamental bottom-up and top-down flows of information. The following section will step through these parts and outline some of their basic functions, especially those that impact aviation. More detailed information on the visual pathways can be found in texts that completely focus on visual perception (e.g., Bruce, Green, and Georgeson, 1996; Spillmann and Werner, 1990; Wandell, 1995).

The Physiology of Vision

The Eye

Bottom-up visual perception begins with the process of converting electromagnetic energy into electrical impulses (referred to as the process of transduction) that are then transmitted to the brain for interpretation. Transduction of the electromagnetic energy begins with the photoreceptors in the retina (neural tissue) at the back of the eyeball. However, before the light image reaches the photoreceptors, the structures of the eye interact with the light rays. The major function of the eyeball is to focus the light rays onto the retina so that a clear image will be transduced. Several structures in the eyeball (roughly 25 mm in diameter) work together to focus the image. Figure 2.3 shows those structures most involved in the focusing process, and they are briefly described below.

The *cornea* is the transparent front of the eye that bends the incoming light (*refraction*) to begin the focusing process. The cornea does most of the focusing; its optical power is 43 diopters of the eye's total 60 diopters (Wandell, 1995). Any deviation of the cornea's surface from a spherical shape leads to image distortions (astigmatism). PRK and LASIK surgeries (see Chapter 3) reshape the cornea to alter its refraction; generally these surgeries are used to correct myopia (nearsightedness), but they can also correct some mild astigmatism.

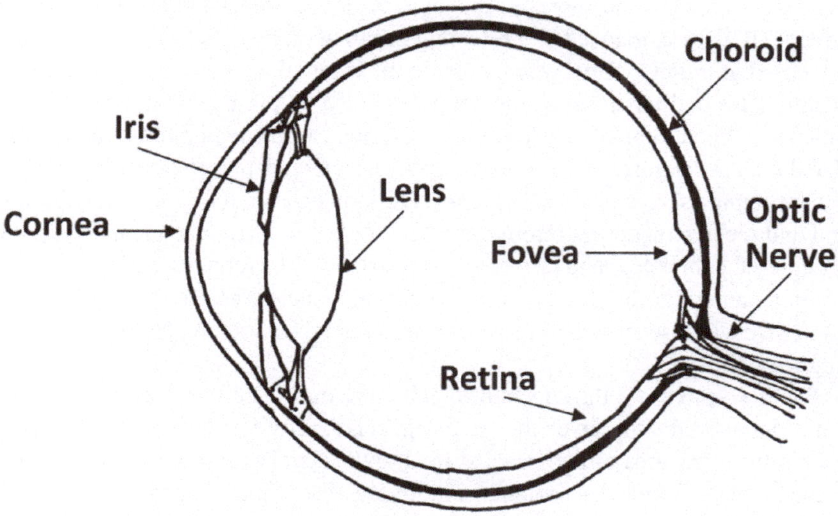

Figure 2.3 A schematic cross-section of the human eye

The *pupil* is the hole in the *iris*; its size is regulated by the ciliary muscles in the iris and varies from 2 to 8 mm in diameter to regulate the total amount of light entering the eye. The smaller the pupil, the sharper the image on the retina, because fewer lens aberrations are able to distort the image. Biological lenses are imperfect, so some distortions are unavoidable. However, in dim lighting, the system opens the pupil, sacrificing better acuity for the ability to detect objects the visual scene. As a person ages, the pupil range decreases, reducing the amount of light entering the back of the eye in dim lighting conditions. By age 50 the maximum dark-adapted pupil diameter is approximately 5.5 mm, and by age 75 it is approximately 3 mm (Benjamin, 2006).

The *lens* further refracts incoming light to achieve accommodation, the ability to change refraction in order to focus on objects of different depths. For farther objects the lens becomes thinner (less refraction) and for near objects the lens becomes more spherical (more refraction). With increasing age the lens progressively becomes less transparent and less flexible, so that it cannot change its shape as effectively for near objects (presbyopia). By the age of 40 years, many people require reading glasses, and by the age of 60, approximately 70 percent less light reaches the back of the eye compared with a child (due to changes in both the pupil and the transparency of the lens; Turner and Mainster, 2008). These and other effects of aging on a pilot's ability to safely fly an aircraft are discussed in detail in Chapter 3.

Accommodation is a relevant topic in aviation displays due to the attempt to minimize a pilots' change of accommodation when transitioning from head-down displays to viewing the more distant objects outside the window. The head-up display (HUD) was founded on presenting information to a pilot that is collimated, or focused at infinity. Thus, when viewing the external aviation environment vital aircraft status and navigation information can also be presented. However, visual clutter, attentional capture, and inattention blindness are all cognitive attention issues related with the engineer's attempts to relieve accommodation for the pilot.

The *retina* is composed of several layers of neurons (specialized cells that send and receive messages through the use of electrical and chemical signals). It is roughly the thickness and size of a postage stamp. The retina is analogous to the film in a camera in that it contains material that reacts when hit by light. In this chapter we will discuss two of these neuron layers: the photoreceptor layer and the ganglion cell layer.

Photoreceptors are the specialized cells that transduce the light photons into neural signals. After this layer, the visual signal is strictly a pattern of neural activity flowing from one area of the visual system to the next (with processing occurring at each area). Somewhat counter intuitively, the photoreceptors are located against the back of the eyeball rather than being in the first layer of the retina to be reached by the light image. This organization is due to high metabolic needs of the receptors, so they can easily access nutrients through the blood network in the *choroid* layer that lines the back of the eye behind the retina. In humans, the choroid layer is black so it absorbs stray light and reduces glare in the eye. There are two types of receptors, *rods* and *cones*. In the *fovea* (the area of the retina along the line of sight), the other layers of neurons are pushed to the side so that the light can travel directly to the receptors, which reduces light scatter and increases the clarity of the image. The *ganglion cells* are the final processing layer of neurons in the eye, located closest to the front of the eye. Their axons (the message-sending part of neurons) leave the eye in an organized bundle (the *optic nerve*) to transmit information to the brain.

Rods and Cones and the Duplex System

The *rods* are named for their long, thin and cylindrical shape. There are roughly 120–130 million rods in the human retina and they are most heavily concentrated in the periphery of the retina. There are no rods at all in the central foveal area (where the image of what is directly being focused on falls on the retina). The *cones* are named for their short, conical shape. There are about 6–8 million cones in the human retina and they are primarily concentrated in the *fovea*, which is roughly 1.5 mm in diameter.

Rods function in dim lighting (scotopic conditions). In fact, they can respond to a single photon of light under ideal, dark-adapted conditions. Cones require many photons in order to be activated, so they are used in bright lighting (photopic) conditions. These properties and the location of rods and cones in the retina explain why dim light sources are best seen if not directly looking at them; peak sensitivity occurs 15–20 degrees to the side of the fovea, where the concentration of rods is the highest (Tredici and Ivan, 2008). The transition between rods and cones occurs in lighting conditions such as those experienced at dusk or dawn (mesopic conditions). In the photopic light levels, the rods become completely saturated and can no longer modify their outputs to signal different light levels being reflected from objects in the environment.

Both rods and cones contain chemicals that react to the light photons (photopigment molecules). In rods it is called rhodopsin (a derivate of vitamin A). There are three different photopigment molecules in cones. The different photopigment molecules respond differently to different wavelengths of light. Because they only have one photopigment molecule type, all rods have the same wavelength sensitivity (with the maximum response occurring for roughly 500 nm). The three cone photopigment molecules lead to maximal responses for wavelengths of 440 nm, 540 nm, and 565 nm, for the short, medium, and long-wavelength sensitive cones, respectively. For each receptor type, the strength of the response decreases roughly monotonically as the wavelength is moved away from the maximum value. As further explained below in the color vision section, rods do not support color vision (due to their single photopigment molecule), whereas cones do support color vision (due to more than one type of cone photopigment molecule available to respond at the same time).

In addition to the above differences between rods and cones, they also differ in their ability to support visual *acuity*, the ability to accurately localize spatial detail in the visual image. Rods are connected to the ganglion cells (through other connections with the intervening layers of the retina), in a "many-to-one" fashion, that is, many rods (up to 1000) ultimately send messages to a single ganglion cell. While this converging pattern of connectivity makes rod-connected ganglion cells respond effectively at low levels of light (many rod signals pool together to activate the ganglion cell), it also results in low acuity. When a rod-connected ganglion cell sends a message to the brain, the brain is not able to determine precisely which rod(s) actually responded to the light stimulus. In contrast, cones have low-to-no convergence on their ganglion cells. This pattern of connectivity makes the cones ineffective at low light levels (since no pooling of signals occurs). However, when a cone-connected ganglion cell sends a message to the brain, the brain knows precisely where the light stimulus fell on the retina and how that pattern of light changed across the different cones that responded. The foveal area's good acuity is further supported by the organization of the visual cortex; half of the area in the primary visual cortex is devoted to processing the central ten degrees (2 percent) of the entire visual field (Wandell, 1995).

The above differences in rods and cones essentially provide us with dual visual systems: *ambient* and *focal* vision. The rods are responsible for ambient vision, which works in low light levels, is color-blind, and which primarily involves peripheral vision. Two of the primary jobs of this system are to detect the presence of objects (e.g., by detecting motion) and to provide us with information about our spatial orientation in the environment. The term "ambient" describes the unconscious nature of this system (Previc, 2004). In other words, we process this peripheral information without awareness of actually doing it and without thinking about it; however, we make visually guided action inputs based on it. For example, as one walks down a crowded sidewalk and avoids collision with other pedestrians, the ambient visual system guides the movement of our feet, legs, and torso in an unconscious manner. An aviation example is formation flying. To an experienced pilot, the aircraft becomes an extension of his body, and maintaining proper three-foot, finger-tip position becomes a matter of inputs guided by unconscious visual positioning.

Processing ambient visual system cues for motion and spatial position helps orient people within their environment (Parmet and Gillingham, 2002). Previc (2004) described ambient vision as providing stable perception of Earth-fixed coordinates and that it "provides us with a veridical three-dimensional spatial frame, including the distance and slant of the world, our tilt relative to it, and our motion within it" (p. 99). When these ambient visual cues to orientation are not available, *visual spatial disorientation* can occur. Spatial disorientation has been cited as the cause of numerous aviation accidents, Chapter 7, and is discussed in detail in Chapter 6.

The cones are responsible for *focal vision*, which occurs in bright light levels, is sensitive to color, and primarily involves the area of the visual field directly in front of the eye. Focal vision is primarily responsible for object identification and relies on the high-acuity of the cone-signal pathway. Individually measured acuity can vary from person to person for several reasons, the most common being the variation in the quality of the optics. If a blurred image falls on the retina, acuity will be poor. However, Curcio, Sloan, and Packer (1987) found that the retinal cone densities in human foveas vary by up to a factor of three; peripheral cone density did not vary across individuals. This suggests that even with optical corrections, some individuals will not be able reach maximum human acuity. It also has implications for the design and functionality of virtual retinal displays (VRDs); a current challenge is that the lasers used in such displays bypass the optics of the eye and present images that are too high frequency (too detailed) to be captured by the receptor arrays and the resulting image becomes aliased. Thus, some individuals will have greater problems with aliasing using VRDs.

The duplicity of vision (rod/ambient and cone/focal) also explains why acuity degrades as the overall *luminance level* of the scene decreases: when light levels are very low the rods will account for the majority of image processing because cones cannot function at low light levels. It is important to note that not all visual functions are impaired at low luminance levels, however (Owens, 2003). In particular, the ability to detect, discriminate and identify objects (which is done by the focal vision system because it requires high acuity) is impaired, while the ability to process motion and crude spatial orientation information (which is done by the ambient vision system) is relatively unaffected. Owens further proposed that one of the reasons why there are so many night driving accidents is the fact that our ability to steer the vehicle effectively (supported by the ambient visual system) makes drivers overconfident about their ability to detect pedestrians, road signs and other objects (supported by the focal visual and greatly impaired at night). Therefore, most drivers do not alter their driving behavior at night (e.g., reduce speed).

In support of the above proposal, Owens (2003) reported a study in which subjects' vision was manipulated to assess the relative importance focal and ambient vision. Focal vision was restricted by myopic-inducing lenses and by reduced illumination levels. Ambient vision was restricted by viewing the display scenes through drinking straws, thus removing peripheral cues. The dependent measures were acuity (focal vision task) and steering task (ambient oriented task). Owens' results confirmed the functional differences between the two modes of vision. Tunnel vision hindered steering abilities but had no effect on acuity, while blur and low-light conditions reduced acuity but failed to effect steering control. As will be discussed in later chapters, these principles can also be applied to aviation.

Dark Adaptation of the Photoreceptors

Night flying is becoming more and more prevalent, especially in military aviation because many newer technologies (e.g., night vision goggles) are thought to turn "night into day." However, this belief is one of the problems facing aviation safety, because it results in pilot overconfidence. Many pilots experience a peaceful and calm feeling when flying at night. The air is smoother, the radios become a bit less congested, and a pilot at night may be lulled into a false sense of security. The lights of cities below and the stars above make flying at night a truly beautiful and serene environment. But don't be fooled … the human orientation system is not fully equipped to fly without risk during daylight in rich viewing conditions, and it is woefully outmatched to fly at night. A pilot's visual perceptual system is physiologically limited in perceiving a nighttime environment and, when combined with a pilot's overconfidence, can turn "night into death."

In the earlier discussion of rods and cones, it was mentioned that rods are more sensitive to light. However, after being saturated with light, they recover more slowly than cones. When light is immediately removed from the environment (lights turned off in a room), both rods and cones begin regenerating their photoreceptive chemicals to allow for absorption of more light. Cones regenerate much more quickly and, initially, we only have our cones available to "see" in the dark. Cones, however, are not sensitive to dim light sources, thus in a dark room you cannot "see" much at all for the first 5–7 minutes of dark adaptation. After about 7–10 minutes, rods have regenerated enough of their photopigment molecule, rhodopsin, so that they are functioning better than the cones. Rods reach their peak in sensitivity after about 30 minutes. Consequently, after 10 minutes you can begin to see better, and then after 20–30 minutes your "night vision" becomes considerably improved due to the sensitivity and regeneration of the rods.

Any exposure to bright light will again saturate the rods and necessitate that they again go through the time-consuming dark-adaptation process in order to support good night vision. To counteract this cycle but at the same time allow lighting to support focal vision using cones, Kern (2002) recommended using a dim red light because it won't degrade dark adaptation. While rods respond to most wavelengths in the visible spectrum (from ~400nm–750 nm), they are not at all sensitive (i.e., they don't respond) to light composed of very long wavelengths (above 650 nm). However, the long-wavelength cones can respond to those wavelengths and mediate detailed vision while not affecting dark adaptation. Unfortunately, with increased color cockpit displays and alerting/warning systems, a red light may reduce discrimination of color perception in the cockpit (see the *color constancy* discussion below for further explanation of this color vision effect). Thus, Kern as well as Tredici and Ivan (2008) suggest using a low-intensity white light for use in a dark cockpit.

Tredici and Ivan (2008) make several suggestions to improve night vision and the dark adaptation process. First, since the rod-dominant periphery does not sense light of long wavelengths, if aviators wore red-tinted glasses, their eyes would be dark-adapting even in a brightly lit room prior to a night sortie. To improve dark adaptation the authors suggest a diet which consists high levels of Vitamin A (because the primary chemical in the rod photopigment molecule, Rhodopsin, is a derivative of vitamin A). Sources of food high in Vitamin A are lettuce, carrots, cabbage, peaches, tomatoes, green peas, and bananas, as well as milk, eggs, butter, cheese, and liver. Supplemental oxygen can also positively influence dark adaptation, in order to combat the hypoxia that occurs when flying at increasing altitudes (due to the high metabolic needs of the receptors).

Tredici and Ivan (2008) presented the following altitudes along with their corresponding reductions in visual capability (see Table 2.1). Basically, a reduction in oxygen can be equivalent to excessive fatigue or smoking, all resulting in reduced visual capability. Smoking is especially bad for a pilot prior to night flying because it increases carbon monoxide levels. According to Tredici and Ivan, a 5 percent saturation level of carbon monoxide is similar to flying at 3,000 m (9,840 ft) without oxygen, and smoking 3 cigarettes prior to a night flight increases carbon monoxide saturation levels to 4 percent, reducing night vision performance by 18 percent. This value is significant when considering the difficulty in visual perception at night even when fully prepared and physiologically healthy.

Table 2.1 Altitudes along with their corresponding reductions in visual capability

Altitude	Reduction in dark visual capabilities
1,100 m (~3,300 ft)	5%
2,800 m (~9,200 ft)	18%
4,000 m (~13,120 ft)	35%
5,000 m (~16,400 ft)	50%

Any reduction in aviation performance should be avoided if possible. Thus, the first author (and other many pilots) routinely breathed 100 percent oxygen while awaiting takeoff clearance to improve night vision prior to flying training sorties in the T-38. Also, while flying multiple-hour strategic airlift missions across the ocean, breathing 100 percent oxygen greatly improved vision and overall disposition prior to landing. A 7+ hour flight at a cabin altitude of 2,100 to 2,700 m (7,000 to 9,000 ft) creates a slight hypoxic state. Thus, breathing 100 percent oxygen can improve dark adaptation prior to a night landing. This point is further highlighted in Chapter 7 during the presentation of aviation mishaps due to visual misperception, which often occur at night.

Post-receptor Processing

The final stage of processing within the retina occurs in the ganglion cells. Prior to these cells, the cells in previous layers of the retina have interacted so that a simple point of light no longer is the optimal stimulus in order to activate the ganglion cell response. More specifically, ganglion cells will respond best to simple contrast patterns (light on dark or dark on light). In fact, a major function of the neural network created in the retina is to enhance our ability to detect edges (e.g., the edge of a runway next to grass). There are several types of ganglion cells, with two types in particular feeding the flow of information that is ultimately processed in the visual cortex: the midget cells and the parasol cells. Midget cells are smaller (and thus are better able to process detail and support fine acuity), maintain color distinctions, require high contrast, and only respond to stationary or very slow-moving stimuli. Parasol cells show the opposite characteristics; they are larger (and thus process less detail), don't support color vision, are able to respond to low contrast stimuli, and respond best to moving or changing stimuli. Contrast sensitivity (the ability of the system to detect objects of low contrast) is thus mediated primarily by the pathway that does not support good detailed vision, which explains why small objects, letters, and so on, are difficult to distinguish when they are faint and of low contrast. Chapter 3 discusses measures of contrast sensitivity and how they provide a more complete measure of visual perception than traditional acuity measures that only use high contrast stimuli.

The electrical signals leaving the retina are carried to the brain via the optic nerve (comprised of the axons of the ganglion cells). At the location where the optic nerve leaves the retina there are no photoreceptors; thus any incoming light falling on this area will not be detected by the visual system. For this reason, this location is often called the *blind spot*. The blind spot sits 15 degrees nasal to the fovea and covers an area 7 degrees in height and 15 degrees in width. Visual angle is discussed later in this chapter, but for reference, a thumbnail held at arm's length is approximately 1.5 visual degrees in size.

Signals carried by the optic nerve primarily are carried to the lateral geniculate nucleus (LGN) in the thalamus of the mid-brain, and then onto highly specialized areas in the visual cortex. About 10 percent of the projections leaving the retina go to brainstem areas involved in eye movements, alertness levels, and biological clocks. In the LGN, the midget ganglion cells and parasol ganglion cells send signals to parvocellular and magnocellular cells, respectively. These cells show the same characteristics as the ganglion cells from which they receive signals, that is, cells in the LGN do not perform additional spatial processing on the signal. The major function of this structure seems to be as a center for the influence of attention and other top-down and regulatory processes. Note that in Figure 2.2 the thicker LGN input arrow is coming from the cortical areas rather than the retina. There is also significant input from brain-stem areas that control levels of arousal and alertness. In fact, it is estimated that only 10 percent of the LGN input is from the retina, and that the remaining 90 percent of the inputs are split evenly between the visual cortex, the thalamic reticular nucleus, and the brainstem, (Kastner, Schneider, and Wunderlich, 2006). Also reinforcing the top-down influences on perception, Soto, Hodsoll, Rotshtein, and Humphreys (2008) highlight the importance of working memory in modulating attentional resources and influencing the likelihood of awareness of environmental stimuli.

The interplay between bottom-up processing and top-down influences, especially with respect to working memory processing limitations can be appreciated when comparing the processing demands and abilities of experienced versus novice pilots. More experienced pilots will have more long term memory schemas (mental models of expectations about situations) to work from, which will reduce the cognitive load of working memory, and in turn leave more attentional resources for bottom-up environmental cue perception. A novice pilot, in contrast, has less complete schemas in long-term memory, and consequently has higher cognitive demands to interpret the situation. These cognitive demands leave fewer attentional resources for bottom-up cue perception and integration.

Following the LGN, visual signals travel to the primary visual cortex located at the back of the brain. In this area, several layers of cells further process the signal. These cells are selective for basic visual features such as size, color, orientation, direction of motion and depth. As the visual signals move to higher levels of the brain, the cells become more complex in their stimulus requirements. Two pathways become apparent as the signals leave the primary visual cortex: a path that flows toward the sides of the brain (inferior temporal area), and a path that flows toward the upper-back portions (posterior parietal areas) of the brain. The temporal areas are specialized for the conscious recognition of objects in the visual scene (often referred to as the *what* pathway, dominated by focal vision processes and inputs from the parvocellular pathway), and the parietal areas are specialized for coordinating actions with the visual environment (often referred to as the *where* or *action-oriented* pathway, dominated by ambient vision processes and inputs from the magnocellular pathway).

Specialized Visual Processing (Color, Depth, Motion)

Color Processing and Color Vision Deficiencies

The above discussion of rods and cones introduced the concept of photopigment molecules and receptivity to certain wavelengths of light. Generally speaking, light wavelength is the physical property that is associated with the perception of color. However, photons are not colored; the perception of color is created by our visual systems. Thus, individuals or animals of other species might perceive different colors depending on the physiological make-up of their visual systems.

Color helps us differentiate between objects, specifically objects of interest from a background. Proctor and Proctor (1997) described color as a means to highlight, emphasize, and code information. Even in the early years of aviation, color discrimination was needed to fly safely. Pilots had to interpret light signals from an airport tower or ground crew. Light signals marked the outline of the runway, water landing areas, and also landing clearance (green safe to land, red do not land or unsafe). Flags and smoke of different colors were also used in early aviation communication. Thus, color assessments for pilots came about due to the requirement for light-signal color differentiation.

Modern aviation has increased the reliance on color vision. Figure 1.1 from Chapter 1 presented the human factors model and listed were concepts such as compatibility, alarms/warnings, and displays. Color is regularly used as a means for getting information to a pilot regarding aircraft status, and to display map/

navigation information to depict terrain and weather systems. More importantly, enhanced technologies are leading to the development of synthetic vision systems and displays that use color as a prominent method of presenting ecological information as well as aircraft status information to the pilot.

There are three main theories of color vision, each of which explains different aspects of color vision that occur at different levels in the visual system. The first of these theories of color vision is the *Trichromatic theory* of color vision (also referred to as the Young-Helmholtz theory after its founders). This theory basically states that to perceive color, the brain compares the outputs of the different cone types. In most humans, there are three cones types (short, medium and long-wavelength as described above), and each one reacts slightly differently to the various wavelengths of lights, resulting in three different photon absorption functions (see Figure 2.4). Notice that they each respond to an overlapping range of wavelengths, and that the middle and long wavelength curves are highly similar (hence, the large impact on color vision if either curve is shifted toward the other due to anomalous trichromacy). These curves have been normalized so that their peaks are the same for comparison. In reality there are many fewer

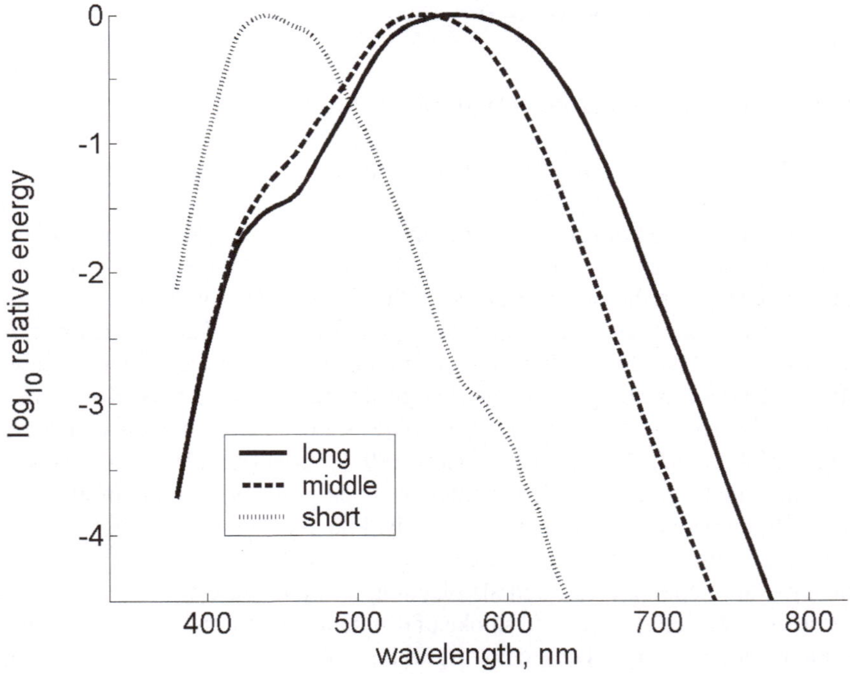

Figure 2.4 Photoreceptor absorption curves for the three cone types

Source: Graph generated from cone fundamentals data reported in Smith and Pokorny, 1975.

short wavelength cones, and they overall absorb fewer photons. By comparing the output of the three cones types, wavelength information can be disambiguated from intensity information (which also influences cone responses), and allow most people to distinguish millions of colors. Because of trichromatic vision, millions of colors can be created from the appropriate combination of three colored lights (e.g., as is done with a CRT monitor or television). If a person is missing one (or more) of the cone types, or if the wavelength absorption spectrum of a cone type is shifted due to genetic variation (anomalous trichromat), then a person would be able to distinguish fewer colors (or no colors if there is only one or no functioning cone types). Color deficiencies/blindness are described in greater detail below.

Hering's *Opponent Process theory* is a second accepted theory of color vision. The opponent process theory explains the occurrence of color afterimages. Color afterimages are experienced by looking at an object of a certain color for 20–50 seconds, and then looking at a neutrally-colored surface (white or gray). For instance, after looking at a green circle with a smaller white circle in the middle, the afterimage becomes a red circle with a smaller black circle in the middle. Physiological studies discovered that the processing of red and green, blue and yellow, and black and white are linked so that individual color-opponent neurons respond in an excitatory manner to one color in a pair (e.g., red) and in an inhibitory manner to the other color of the pair (e.g., green). The opponent color processes occur not at the receptor level but beginning in later neural processing stages in the retina.

The third color vision theory, the *Retinex Theory*, explains why surface reflectance colors appear to be constant despite the fact that the actual wavelengths that reach the eye can vary considerably, depending upon the light source. This consistency of color perception regardless of whether we are outside in sunlight (longer wavelength dominant light source) or inside under florescent light (shorter wavelength dominant light source) is called *color constancy*. While the Trichromatic theory and the Opponent Process theory were both proposed in the late 1800s, the Retinex theory was not proposed until 1977 by Edwin Land (camera inventor and Polaroid founder). Color constancy only occurs if the light source is some form of white light (contains all wavelengths, although the proportions of the wavelengths can differ), and if there are multiple objects in the visual field. For example, an aspirin tablet isolated on a piece of black velvet under florescent light will look faintly blue, and under incandescent light it will look faintly yellow. However, as soon as any other colored object is placed next to the aspirin tablet, it will immediately look white. This shift in color perception is an indication that the visual system components mediating color constancy are able to make comparative reflectance calculations, rather than using simple, absolute reflectance information. Physiological studies have shown that neurons in the secondary primary cortex show color constancy properties (Conway, 2003).

More relevant to aviation, color constancy also breaks down if the light source is of a restricted spectrum (i.e., not white light). Thus, when using red light sources (such as recommended above to avoid dark adaptation losses), color constancy no longer occurs, and colors cannot accurately be distinguished or identified. This negative consequence is the reason that dim white light is now often used instead of red light. Some dark adaptation will be lost when a dim white light is turned on, but accurate color vision to interpret gauge readings has greater usefulness than full dark adaptation.

Color Deficiencies

To highlight the influence of having normal, trichromatic color in aviation, recall the 2002 mishap described in Chapter 1. The freight carrier made a night visual approach, and the Precision Approach Path Indicator (PAPI) lights were improperly perceived by the pilot (NTSB report, 2002). The report stated that a color vision deficiency in the first officer's vision contributed to his low approach-path. Their theory was based on the pilot's inability to differentiate various shades of reds, pinks, and whites, and thus he failed to perceive the aircraft's spatial position as being below glide-path via the PAPI lights. Tredici and Ivan (2008) also addressed this same mishap and the subsequent re-evaluation of color vision assessment measures due to the lack of standardization. They specifically cited the problem of color-deficient pilots making more mistakes, taking more time for interpretation, and requiring closer color target matching than color normal pilots.

There are actually many variations of color deficiency/color blindness. Normal human color vision (trichromatic color vision) relies on the three cones types with absorption functions as shown above in Figure 2.4. Color vision anomalies can be due to one or more missing cone types, shifted cone absorption spectrums, disease or damage to one or more of the cone types, or damage to color processing areas in the brain.

Individuals who only have two functioning cone types are called *dichromats*. They can be missing the short-wavelength cone type (tritanope), the medium-wavelength cone type (deuteranope), or the long-wavelength cone type (protanope). Both deuteranopes and protanopes perceive the rainbow spectrum only as shades of blue (for the shorter wavelengths) and shades of yellow (for the longer wavelengths). These two types of color blindness are commonly referred to as "red-green" color blindness, which is misleading. For example, their receptors actually respond to wavelengths that trichromats perceive as green; their systems, however, cannot support as many color distinctions, and the final output in this case is the perception of a shade of yellow. The "red-green" forms of color blindness are more common in males than females because the genes for the medium and long wavelength cone types are on the X-chromosome. In order to be colorblind, females (who have two X chromosomes) need two copies of the color blindness gene (it is a recessive gene), while males need only one (they only have one X-chromosome). Tritanopia is very rare and is caused by a mutation on chromosome 7. Tritanopes perceive the rainbow spectrum of colors as all being in shades of green (for the shorter wavelengths) and shades of red (longer wavelengths). In other words, they can "see" something "yellow" (i.e., what a trichromat would perceive as yellow), but it would look reddish to them.

Anomolous trichromacy occurs when an individual technically has three different functioning cone types, but the spectrum sensitivity of one of them is shifted compared to "normal". The amount of shift depends on the number of amino acid changes in the gene coding for that cone type. Anomolous trichromacy can vary from being barely detectable to inaccurate perceptions similar to dichromats. Sometimes this condition is referred to as being "weak" in one of the cone types. Deuteranomaly (shifted medium wavelength cone) is what the first officer of the 2002 Florida mishap was identified as having.

Monochromatism is a condition when only one type of receptor is functioning at a given time. Because the brain only receives signals from one receptor type, the light variables of wavelength and intensity are unable to be disambiguated. Ultimately, such a person only perceives the world in shades of gray. Cone monochromats have one functioning cone type (used in brighter light) plus rods (used in dim light). Rod monochromats have no functioning cone types and are only able to visually function in dim lighting.

The above conditions are generally due to genetic causes, but color vision deficiencies can also be acquired. The short-wavelength cone type is more fragile and can be selectively damaged (e.g., the short wavelength cones are more likely to be damaged in retinitis pigmentosa than are the medium and long wavelength cone types; Swanson, Birch, and Anderson, 1993). Complete loss of color vision, or the perception of "washed out" color can be due to brain damage (achromatopsia).

Prevalence rates for the different types of color deficiency are summarized in Table 2.2. Prevalence rates are indicated separately for males and females because some types of color deficiency show large gender differences. As the numbers

Table 2.2 Prevalence rates of different types of color deficiency for males and females

Type of color deficiency	Prevalence in males	Prevalence in females
Protanopia	1%	.01%
Deuteranopia	1%	.01%
Tritanopia	.002%	.001%
Protanomoly	1%	.03%
Deutanomoly	5%	.4%
Tritanomoly	.01%	.01%
Rod monochromat	1in 50,000	1 in 50,000
Short wavelength cone monochromat	1 in 100,000	1 in 10 billion
Medium or long wavelength cone monochromat	1 in 100 million	1 in 100 million

Source: Prevalence rates from Huang, Wu, and Chen, (2008) and Sharpe, Stockman, Jagle, and Nathans (1999).

indicate, a substantial portion of the male population shows at least some amount of color deficiency. Chapter 3 will further discuss color vision testing with respect to aviation.

Depth Processing

The accurate perception of depth is crucial for people to safely and efficiently navigate through their environments. Depth perception, however, poses a challenge: while the real world is three-dimensional, the retinal images are two-dimensional. Thus, depth must be interpreted from the images.

In general there are two types of cues the visual system uses to interpret the environment and to help perceive depth and distance. *Binocular depth cues* refer to visual cues that require the combined input from the two eyes, while *monocular cues* can be processed using information from only one eye. According to Tredici and Ivan (2008), monocular cues are learned and can even be improved through training; however, it is often these monocular cues that are most susceptible to misperception. The neural system supporting the use of binocular cues, in contrast, seems to be innately predisposed and become hardwired through early visual experience; however, binocular cues are much more restricted in their use in terms of distance.

Binocular depth cues There are three binocular depth cues/processes: convergence, accommodation, and stereopsis. *Convergence* is the eye muscle movement required for the eyes to turn in or out to focus on an object. Although the estimates vary somewhat, there is agreement that this cue is not effective for distances greater than nine meters (Cutting and Vishton, 1995; McKee and Smallman, 1998; Proctor and Proctor, 1997). *Accommodation* is the amount of change in the thickness of the lens needed to refract light entering the eye. The lens becomes thicker to focus near objects and flattens out to focus far objects. Although it might be used in restricted, artificial environments, research shows that its contribution to perceived depth is minimal (Proffitt and Caudek, 2003).

Stereopsis is the process of using retinal disparity to determine depth. Retinal disparity is the difference between the retinal images in the two eyes, due to the fact that the eyes are laterally offset and, thus, view the world from slightly different perspectives. The object of focus in the visual scene will fall on corresponding points in the two retina (on the foveas of each eye), but objects at depths different from the focused object will not fall on corresponding points. The greater the depth between the focused object and other objects, the greater the corresponding disparity. The geometry of the perspectives is very consistent, and leads to a robust and accurate perception of depth. However, given that the distance between the two eyes is on average only 6 cm (2.5 inches), (Wandell,1995), the difference in perspectives is not great enough to provide useful depth information at very large distances, which limits the use of stereopsis for the large viewing distances involved in flying. Fortunately several studies have indicated that these cues are not necessary because monocular cues are more dominant and completely sufficient throughout all normal phases of aviation (Benson, 1988; Calvert, 1950; Harris, 1977; Hasbrook, 1971; Mertens, 1981).

Monocular depth cues Monocular depth cues fall into two categories: static cues and motion-related cues. The static cues refer to visual scene cues also used by artists to imply depth in static, two-dimensional pictures; these cues are often referred to as pictorial or artistic cues. Pictorial cues are very good at implying the depth order of objects in the visual scene, but they do not give good information about the precise amount of depth between the objects. There are many pictorial cues available. Several are listed below along with aviation-related examples.

Pictorial cues

1. *Occlusion (interposition)* occurs when more distant objects are hidden from view by closer objects. Another aircraft between a pilot and their runway will occlude or hide a portion of the landing runway, indicating that the runway is more distant than the other aircraft.
2. *Relative height* is based on the idea that more distant objects appear higher in field of visual scene.
3. *Relative size* relates to the idea that, as distance increases, an object of fixed size will take up less space in the visual scene and on the retinal image. Thus, if there are two objects of known equal physical size, the one that appears smaller in the scene will be perceived to be farther away. This cue becomes difficult to accurately use if the real physical sizes of the objects in the scene are not known or if an error is made in object identification. For example, sometimes birds are mis-identified as planes and the resulting perceived depth is much greater than reality.
4. *Atmospheric/aerial perspective* results from distant objects having less detail and appearing hazy (due to the accumulation of water molecules, dust, and so on, in the air between the observer and the distant object). Also, the perception of detail is lost because as depth increases the details become too small to be resolved by the visual system.
5. *Linear perspective* occurs when parallel lines appear to converge as they get farther away. This cue was mastered by the Renaissance artists and led to the amazingly realistic paintings of that time period. Of relevance to pilots, a runway appears to converge at the far end when viewed from a distance.
6. *Texture gradient* cues provide depth information because more distant textures become more densely packed.
7. *Shadows* provide depth information because the portions of an object nearer the light source block the light from portions of the object more distant from the light. Light sources are usually assumed to be from above, although this assumption can lead to errors of perception as illustrated in the crater illusion above.

8. *Foreshortening* is a cue related to texture gradient but focuses on the change in the medial aspect rather than the lateral aspect of an object (i.e., a change in object shape as if the object had been "squashed"). Foreshortening more clearly indicates angle of viewing rather than distance per se. Angle of viewing can change because the surface itself slants away from or toward the viewer, or because the viewer changes his angle of viewing. For example, Figure 2.5 shows a runway from four perspectives: directly overhead, on a steep glide-path, on the desired 3-degree glide-path, and on too low a glide-path. Unfortunately, because both surface slant and changes in viewer perspective can lead to foreshortening, errors in perception can occur.

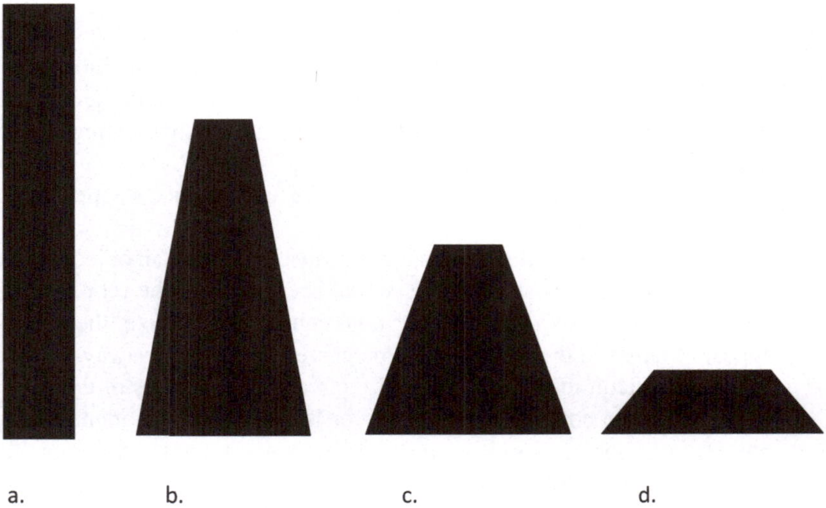

Figure 2.5 Foreshortening examples of a runway

Motion-related depth cues In general, a compelling 3D perception of objects in a visual scene is possible if multiple pictorial cues are available. However, the addition of monocular, motion-related cues can greatly increase the accuracy of perceived depth in a scene. The first motion-related cue is *accretion/deletion*. This cue is related to the static pictorial cue of occlusion. As a person moves through the environment, those objects that are nearer will seem to pass in front of farther objects and block their view until the viewer's perspective changes enough that the far object is again unblocked. Similarly if a closer object is moving through a visual scene, it will systematically block and then unblock the view of more distant objects.

The second motion-related cue is *motion parallax*. Similar to stereopsis, motion parallax is based on systematic changes in the retinal image based on the geometry of the visual scene. While for stereopsis the two eyes are simultaneously viewing the visual scene from two slightly different perspectives, for motion parallax the different perspectives of the scene occur across time as the observer or objects in the scene move. Objects that are nearer the observer will shift farther across the retina in a given amount of time than will more distant objects. Motion parallax will lead to very good information about the amount of depth between objects or between objects and the observer. Motion parallax is effective at large distances and can be used by pilots; however, once in the air, a pilot will have few if any other objects in the air space around her, so the cue is not a primary source of depth information while flying.

Motion Processing

Both an observer and other objects in the environment can move, and it's crucial for accurate navigation to be able to process motion cues. Generally speaking, objects moving in the environment will cause a discrete, local shift in the image on the retina. Neurons in the primary visual cortex are sensitive to local movement in specific directions. In contrast, if an observer moves her eyes or her body, then the entire visual image will shift on the retina. Eye movements alone will cause a simple left-right, up-down, and so on, type of shift in the image on the eye. If a person walks forward or moves in reverse, the image will expand or contract, respectively, from the point of focus. If the person tilts his head, then the image will rotate. Neurons in MT (medial temporal) and MST (medial superior temporal) areas process large field unidirectional shifts and expansion/contraction/rotation shifts, respectively. Expansion, contraction and rotation of the entire visual field are all components of *optic flow* information.

Motion selective neurons can show adaptation, leading to motion after effects that are similar in principle to the color after effects described above. For example, after viewing a waterfall for at least 30 seconds, those neurons that best respond to downward motion become fatigued. If a person then looks at a non-moving, textured stimulus, it will appear to float upward. Adaptation to an expanding stimulus will lead to the perception of the contraction of the visual field. This latter type of adaptation can occur if flying through the dense clouds.

Motion-detecting neurons can also be selective for speed/velocity, that is, the rate of change in the motion signal. Neurons in the MT area are sensitive to object speed and respond well to moving edges (Perrone and Thiele, 2001). The ability to perceive speed is important to provide feedback regarding the rate of observer movement, as well as to support the visual pursuit tracking of moving objects in the environment. Velocity information combined with depth information is used to determine time to contact (see below).

Visual Cue Integration and Applications in Aviation

The above section outlined several types of visual processing as if they were independent from each other. However, in real environments color, depth, and motion cues are all processed simultaneously and they are integrated to aid our understanding of the visual scene. This section will highlight a few processing examples especially pertinent to aviation. Top-down processes (e.g., memories, expectations, attentional influences) also play an important role in many of these more complex examples.

Size, depth and visual angle (size constancy) Size and depth will be explored together in this section because the processing of those stimulus characteristics is interrelated; assumptions (top-down processes) about size will influence the perception of depth, and vice versa. The following equation summarizes the relationship and introduces retinal image size:

$$S = R * D, \tag{1}$$

where S is the perceived size (which may or may not match reality), R is the retinal image size, and D is the perceived distance (which also may or may not match reality). The retinal image size is the size of the image (proximal image) created by a real world object (distal image) on the 2D retina.

As described in the above equation, as distance increases for an object of fixed size, the retinal image size will decrease. If either size or distance is accurately perceived, then the visual system can accurately determine the other based on the size of the retinal image, which is directly perceived by the system. Thus, in an environment rich with cues, our perceptions of distance and size can be highly accurate, which leads to good size-constancy (the fact that our perception of an object's real size remains stable even though the retinal image size corresponding to that object systematically changes as a function of distance).

However, with impoverished conditions, as may happen when flying at night or in fog, errors can be made. Collett and Parker (1998) stated, "in an impoverished scene, the accuracy of our distance judgments is low: We are very bad at locating isolated points of light" (p. 410). McKee and Smallman (1998) went a step further stating that, "in reduced viewing conditions, inaccuracies in estimating distance should necessarily lead to systematic biases in size constancy" (p. 378). Consequently, maintaining size, shape, or depth constancy in a visually impaired approach environment is nearly impossible, yet pilots confidently fly night visual approaches to landing. Pilots also continue to fly into environments quickly deteriorating due to weather or nightfall during low-level or cross-country navigation flights.

A common way that the relationship between size and distance is summarized using a single value is through the visual angle, which can be calculated with the following equation:

$$\theta = 2*(\text{arc tan})[(\text{size}/2)/(\text{distance})] \tag{2}$$

As an object of fixed size is moved to a greater distance, the visual angle will decrease, and vice versa as it is moved closer. Larger objects at a fixed distance will have larger visual angles than smaller objects. Based on acuity measures, we can predict the detectability of objects that have different visual angles and predict when an object will become too small in the distance to be detected (see Chapter 3 for a discussion of acuity measures).

The above visual angle formula assumes that the object of interest is viewed in the frontal plane. In reality, objects may vary in orientation with respect to the viewing angle, which increases the complexity of the relationship between size and distance. Baird and Wagner (1991) proposed the following equation which accounts for various angles when viewing objects:

$$\cos\theta = d^2 + ds\cos\varphi - hs\sin\varphi + h^2 \qquad (3)$$

$$[\mathrm{sqrt}((d + s\cos\varphi)^2 + (h - s\sin\varphi)^2)] * [\mathrm{sqrt}(h^2 + d^2)]$$

For an object that is flat on the ground observed in the medial direction, like a runway, the above equation can be simplified because cos θ when θ = 0 is 1, and sin θ when θ = 0 is 0:

$$\cos\theta = (h^2 + d^2 + ds) \qquad (4)$$

$$[\mathrm{sqrt}(h^2 + d^2)] * [\mathrm{sqrt}(h^2 + (d + s)^2)]$$

Thus, the visual angle equation becomes a function of the height (h) of a pilot, the distance (d) of the pilot from the runway, and the length of runway. See Figure 2.6 for a schematic of the variables in these equations.

Table 2.3 shows visual angle calculations for a fixed runway size at different distances and altitudes as might be experienced when landing using a 3-degree glide-path. Note the small visual angle in the medial extent regarding a long 2,743 m runway. Even when a pilot is 185 m from landing, the length of the runway

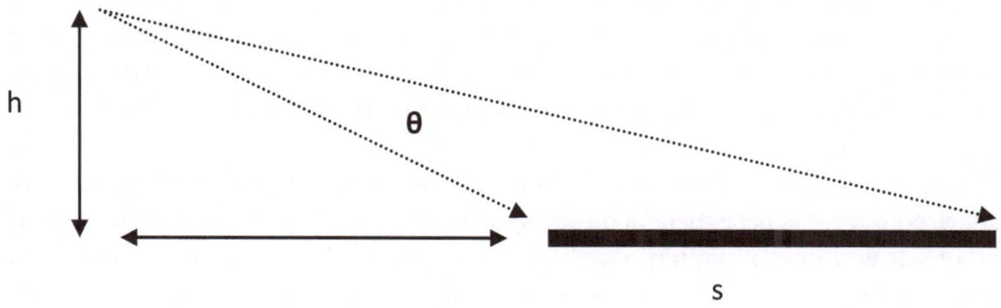

Figure 2.6 Visual angle geometry from an elevated perspective

Source: Adapted from Baird and Wagner (1991).

Table 2.3 Visual angle of runway width and length at different altitudes and distances along a 3-degree glide-path for landing

Distance in nm	Distance in m	Altitude in m	Width visual angle	Length visual angle
7	13,878	679	0.27	0.52
5	9,260	485	0.37	0.68
3	5,556	291	0.62	0.99
1	1,852	96	1.87	1.78
0.5	926	48	3.75	2.23
0.25	463	24	7.48	2.57
0.10	185	9.7	18.6	2.79

does not take up much of the pilot's retinal area, having only a visual angle of 2.79 degrees.

To help ascertain distance estimates it has been suggested that cues are pooled or data fusion occurs and cross-talk happens between differing depth modules (Collett and Parker, 1998). Although the process is unknown, numerous researchers agree that the process carries a statistical weighting of dominant cues which are situationaly dependent (Collett and Parker, 1998; Galanis, Jennings and Beckett, 1998). Those cues, weighted dependent upon the situation, must be consistent to formulate accurate size constancy (McKee and Smallman, 1998). McKee and Smallman proposed a model of dual-processing for the judgment of angular size and objective size. In this model two separate processes occur when viewing a scene: (1) judgment of angular size based on retinal image and (2) judgment of objective size based on the perceived distances of global and local features for comparison. Most important in this process is the calculation of the relative size information because it validates the crosstalk between the two outcomes. The model also contains "noise", which represents the lack of accuracy in estimating angular dimensions and angular depths given the inputs. This model will be discussed more thoroughly with respect to aviation misperception in Chapter 6.

Shape and depth (shape constancy) Similar to size constancy, our ability to maintain shape constancy is dependent upon accurate depth perception because closer parts of an object will have a proportionately larger retinal image than a farther part of an object. Thus, with the door example at the beginning of the chapter, the retinal image of a rectangular door is only truly rectangular when it is viewed in a perfectly frontal plane; otherwise the retinal image will really be trapezoidal. In processing shape, our visual systems use both top-down and bottom-up processing to interpret the incoming retinal image based on available depth information and prior knowledge of objects.

This intertwined process is especially relevant to aviation with respect to the perception of runways on approach to landings. When perceiving runway slant and depth/altitude, the primary monocular visual cues are linear perspective, foreshortening, and texture density (of surrounding trees, buildings, etc., as the runway itself has little texture), with the strongest cue being linear perspective (Previc, 2004). As a pilot comes in for a landing, the rate of change in these cues can be used to accurately guide the plane. Linear perspective is a function of altitude, in that more splay is perceived if at a lower altitude. Splay is the widening of the front end of the runway and the narrowing of the far end of the runway. Thus, the lower the point of perspective, the more splay and the larger visual angle the runway will consume. The change in splay can inform a pilot of changing altitude. At a constant altitude, the splay angle is constant, but as a pilot descends the splay angle increases and, if descending at a constant rate (ft per minute descent), it is a constant splay rate of change (Flach and Warren, 1995). Splay is discussed again in Chapters 4 and 5.

A major real-life difficulty in accurately using the above cues, however, is that the retinal image size of the runway is very small until the pilot is at a very close distance (see Table 2.3 above). Further, the pilot is not always viewing the runway from a line of sight looking straight down the length of the runway. We know from perceptual illusions such as the "Turning the Tables" illusion that when our systems try to interpret the shape of objects from different perspectives, we may make errors due to assumptions that have been internalized based on previous experiences. In this text (see Figure 2.7), the shape and size of the tabletop surface is exactly the same (but rotated approximately 90 degrees) for both tables. However, because our visual systems try to interpret them as 3D figures, we do not perceive them to be the same size and shape. Instead the one on the left looks longer/narrower and the one on the right seems more square-like.

More specifically, Crassini, Best, and Day (2003) proposed that, when viewing a 3D object in a 2D setting (or in a 3D setting with reduced depth cues), the visual system assumes reduced visual angles which in turn leads to the interpretation of

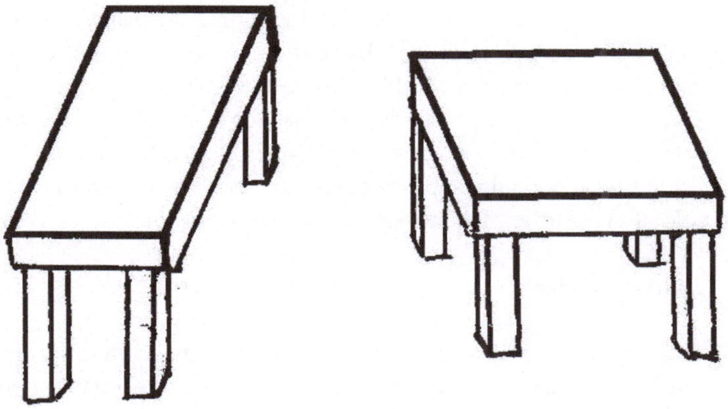

Figure 2.7 Turning tables illusion

greater medial extents than really exist (an overcompensation process). Figure 2.8 illustrates the reduced visual angle of an object when it is viewed on the medial plane rather than the in the plane perpendicular to the frontal line of sight. For an object that is 270 cm in length, if it is viewed from a Transverse perspective (left figure), it will have a retinal visual angle of 19.9 degrees, while if it is viewed from a medial perspective (right figure), it will have a retinal visual angle of 3.7 degrees. Using Crassini et al.'s logic, a pilot in a reduced depth cue situation (e.g., nighttime landings), who was also dealing with a small visual angle due to both the distance and the medial plane of the runway, will have a strong likelihood to perceptually elongate the runway in depth-orientation. If the runway looks long, then for the given position from the runway the pilot may perceive himself as being higher and farther than he really is. In turn, the pilot will be motivated incorrectly to initiate a descent—as in a black-hole approach discussed in Chapters 1, 6 and 7.

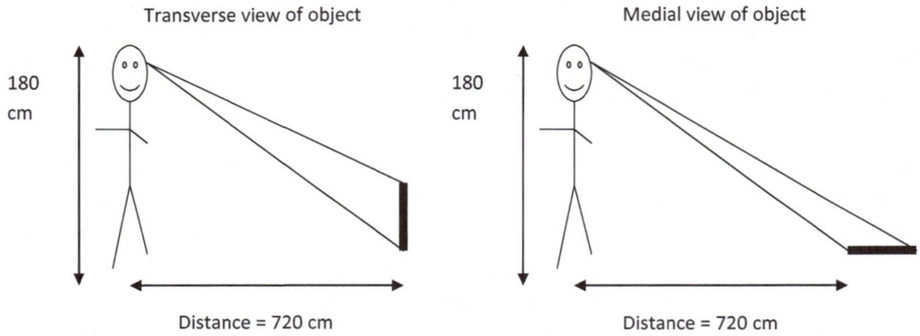

Figure 2.8 Transverse and medial visual angle schematics

Source: Adapted from Crassini, Best, and Day, 2003.

Depth, motion and time to contact With respect to aviation, pilots will use motion and velocity perception to determine two things for successful collision avoidance: (1) will I collide with the object if I continue to travel in the same direction? and (2) how much time do I have before collision will occur? Consider the case of flying towards a tree. The lateral distance at which the tree will cross the fronto-parallel plane that contains the eyes (the 'crossing distance') can be estimated using both monocular and binocular sources of visual information. If the approach velocity is constant the crossing distance (X_c) of the tree is given by:

$$X_c = [2W(d\alpha/dt)]/(d\theta/dt) \quad (5)$$

where $d\alpha/dt$ is angular lateral speed of the tree's retinal image, $d\theta/dt$ is the rate of change of the tree's visual angular subtense and W is the tree's width (Bootsma, 1991; Regan and Kaushal, 1994).

The number of seconds remaining before the tree crosses the fronto-parallel plane containing the eyes [the 'time-to-passage' or (TTP)] is also specified by both monocular and binocular sources of visual information. As illustrated in Figure 5.6A, for a tree approaching point P some distance from the pilot's eyes:

$$TTP \approx \frac{\theta}{d\theta / dt} - \frac{\gamma}{d\gamma / dt} \tag{6}$$

where γ is the optical angle at the eye subtended by the current location of the tree and the point at which the tree will cross the fronto-parallel plane containing the eyes (P) and $d\gamma/dt$ is the rate of constriction of this angle (Bootsma and Oudejans, 1993). In the special case that the tree is directly approaching the catcher's eye, $\gamma=0$ and the above equation reduces to the correlate of the time to collision (TTC) commonly called 'tau' (Lee, 1976):

$$TTC \approx \frac{\theta}{d\theta / dt} \tag{7}$$

For direct approaches it has been demonstrated that humans can accurately estimate TTC with estimation errors ranging from 2 to12 percent of the actual TTC.

Binocular information about TTP relies on relative disparity information (Regan, 2002) as described by:

$$TTP \approx \frac{2d\delta / dt}{d^2\delta / dt^2} - \frac{\gamma}{d\gamma / dt} \tag{8}$$

where $d\delta/dt$ is the rate of change of retinal disparity relative to a fixed reference point (F). Estimation of TTP based on binocular information alone has not been previously investigated except in the special case of an object directly approaching the midpoint between the eyes where estimation errors range from 2 to 10 percent of the actual TTC (Gray and Regan, 1998).

At this point a reader familiar with the topic of visual perception may be asking the question: but wouldn't stereopsis be ineffective in flying because the objects are too far away? While it is true that the relative retinal disparity (i.e., δ) for a given depth separation between two objects is inversely proportional to the square of the viewing distance, this limitation only applies to judgments of static depth. When judging the direction of motion in depth and TTP or TTC, the relevant retinal image variable is the rate of change of disparity ($d\delta/dt$). Because the magnitude of $d\delta/dt$ is proportional to the approach velocity, the high speeds involved in flying would ensure the value of $d\delta/dt$ is well above perceptual threshold in most situations.

Optic flow, speed and altitude Height/altitude and optic flow (the systematic change in the retinal image as a person scans or moves through the environment) interact in our interpretation of speed. A general principle that becomes hardwired in the system is that a slower rate of optic flow across the retina indicates a slower speed of movement. As described above with respect to motion parallax, the further a person is from objects in the environment, the slower the rate of change of those objects across the retina as the person (or object) moves. Our systems accurately use this latter relationship to interpret depth (motion parallax).

The interaction between height/distance and optic flow was not a problem for the perception of speed until humans created vehicles that quickly shifted the driver's height (e.g., normal standing height when walking, versus height when driving a low sports car, versus height when driving a large pick-up truck, versus height when in a plane). Experience with one particular height causes the system to learn the optic flow and speed relationships for that height, that is, those relationships become "hardwired" into the system. Thus, people who drive mid-sized cars regularly report that they feel like they are driving faster when they are in the low sports car (due to the rapid optic flow of the nearer ground texture) and slower when in they are in the truck (due to the less rapid optic flow from the greater height). These mis-perceptions often lead to the maintenance of inappropriate speeds as they match the optic flow rate to whichever height they are most accustomed.

Pilots can experience similar speed mis-interpretation when taxiing, taking off and landing; different planes place the pilots at different heights from the ground. For pilots such mis-interpretations can be deadly, as speed will influence takeoff and landing safety. On a more positive note, often-practiced landings in a specific plane will allow the optic flow rate to become learned and become an influential part of ambient (non-focal) vision, which can help guide behavior. For example, the Hudson River landing of Flight 1549 mentioned at the end of Chapter 1 was possible due to the rich optic flow cues available from buildings along the riverbanks, even though the river itself provided few texture gradient cues to determine altitude and speed.

A more extreme example of the mis-perception of speed based on optic flow is that of the speed of an airplane up in the sky (whether from the perspective of an observer on the ground looking at the plane or from the pilot's perspective based on optic flow of the ground from the plane). Warren (1988) compared aviation to driving and pointed out that a car on a freeway and a jet flying at the speed of sound 15 m (50 ft) above the ground have comparable values of global optic flow rate. Thus, even at an extremely low flying altitude, optic flow is not of much use to a pilot to determine speed. At normal flying altitudes optic flow is of no use at all for the perception of plane speed.

An additional consideration about the optic flow/speed relationships is that they can be "rewired" due to different consistent experiences, which will lead to a shift in the mis-perceptions. Although these mis-perceptions are somewhat hardwired, knowledge of these relationships can lead a driver/pilot to pay more attention to external equipment measures of speed, and thus reduce the likelihood of accidents.

The Vestibular System and Body Perception

Several sensory systems support our ability to perceive how our body is positioned and its movement through the environment: vision, touch (pressure and skin stretching sensations), kinesthesia (body position awareness from muscle and joint feedback), and vestibular processes. The vestibular system supports the ability to perceive body accelerations and decelerations due to translation and rotation. While touch, vision, and kinesthesia can provide useful input to a pilot (e.g., the sensation of gravitational pressure from the seat when sitting, and the ability to develop good hand-eye coordination based on vision and kinesthetic feedback, respectively), the vestibular system has traditionally received much focus in aviation due to its role in spatial disorientation.

The vestibular system is part of the inner ear, which has three main parts, the cochlea, the otolith organs, and the semicircular canals (see Figure 2.9), all of which are full of fluid called endolymph. The cochlea is part of our auditory system and converts acoustic energy into neural information. The two otolith sensory organs, the utricle and the saccule, are used to sense gravity and linear acceleration, and the semi-circular canals are used to sense rotation. The otolith organs have sensory hairs with small calcium carbonate crystals attached to the ends that give the hairs more mass to detect movement. When our bodies move, the fluid in the chambers

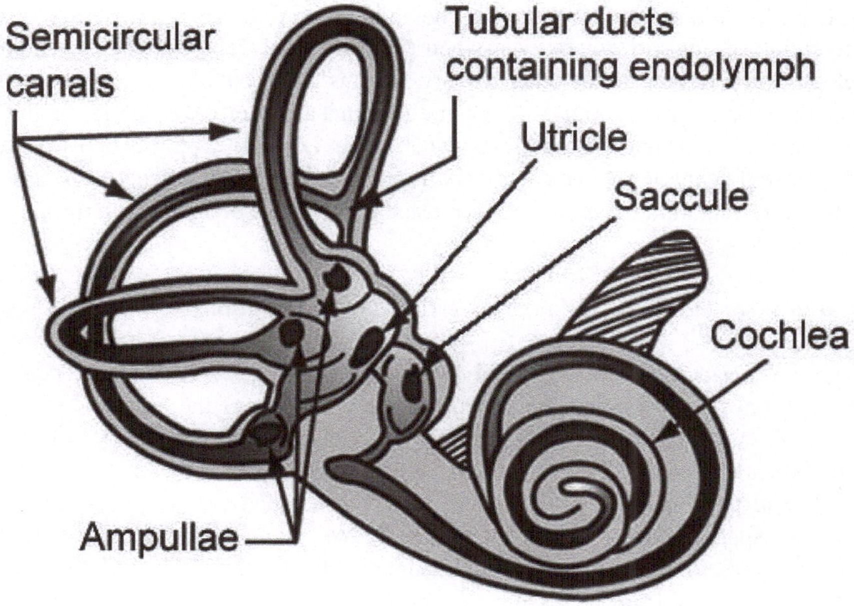

Figure 2.9 Vestibular system components

Source: Courtesy of NASA Educational Brief

also moves, and causes the hair cells to bend. This bending initiates a neural response (sensation) and, depending upon the direction of the bending of the hair cells, movement is perceived in a particular direction.

The semicircular canals consist of three distinct canals at right-angles to each other. They allow us to sense acceleration/deceleration in three angular dimensions, roll, pitch, and yaw. As in the otolith organs, sensation (the initiation of a neural signal) occurs when the fluid movement in the canals causes the hair cells to bend. Sensations from the canals in both ears are combined to more accurately determine the direction of rotation.

As with the other senses, there is a minimum amount of stimulus change that is necessary for detection, that is, a threshold. According to Lee (2005), the amount of translational acceleration that is necessary for a pilot to detect the motion is 10 cm/sec^2 for oscillating motion, which is equivalent to approximately 0.01 G of acceleration or deceleration. For rotation, between .44 and .80 deg/sec^2 is necessary for a pilot to report the first detection of angular movement.

Because the vestibular system requires the movement of fluid to cause the hair cells to bend in order to influence neural activity, the system is somewhat slow to respond (due to inertia of the fluid), and slow to cease responding (e.g., the fluid can continue to swirl inside the chambers after that body has ceased movement), which can lead to motion after effects. Further, movements occurring with constant velocity and direction will cease to be detected by the vestibular system (the fluid motion catches up to the body motion), leading to the erroneous perception of non-movement even when the body is actually moving at a high velocity. Finally, as with the other sensory systems, the vestibular system evolved to function well within a range of body motions that are often exceeded in aviation (e.g., accelerations and G-forces encountered in aviation are well beyond what a person would experience in a natural environment). All of these effects can lead to spatial disorientation in a pilot and increase the likelihood of an aviation mishap.

Spatial Disorientation: Vision and Vestibular Contributions

Spatial Disorientation was defined by Gillingham (1992) as, "an erroneous sense of one's position and motion relative to the plane of the earth's surface" (p. 297). As mentioned above, a person's awareness of his or her body position and movement comes from several sensory inputs: visual inputs, kinesthetic inputs and vestibular inputs. Newman (2007) concluded that visual processes dominate spatial orientation by contributing 80 percent of the inputs. Visual cues can also correct for inaccurate vestibular signals (due to adaptation, etc.). In fact, if the vestibular system is absent and the only remaining senses are vision and kinesthesia, balance and equilibrium can be maintained. However, when vision is absent, spatial orientation deteriorates quickly (for example, try to stand on one leg with your eyes closed). Given that spatial disorientation can occur even in rich visual environments, it is easy to understand how much more likely spatial disorientation would be in visually limited aviation environments such as at nighttime or in cloudy weather.

The means by which vision, kinesthesia, and vestibular processes are combined with respect to spatial orientation have been reviewed in detail by Parmet and Ercoline (2008). They stated that, in general, vision is processed at higher cortical levels than vestibular inputs. Within the visual areas, the *where* pathway (ambient vision responsible for orienting actions) receives significant vestibular inputs, while the *what* pathway (focal vision responsible for identifying objects) receives no inputs. More specifically, recent human neuro-imaging evidence has shown vestibular inputs to the anterior areas of MST (the medial superior temporal area), which processes complex patterns of motion usually associated with optic flow (Smith, Wall, and Thilo, 2009). Other vestibular processing is accomplished within the more primitive brain structures such as the cerebellum, medulla, pons and reticular formation. Based on this physiological information, Parmet and Ercoline concluded that the, "integration of visual and vestibular information in the cerebellum and brainstem appears to allow visual control of basic equilibratory reflexes of vestibular origin" (p. 149). The fact that a majority of vestibular processing occurs in areas not directly accessible to conscious interpretation highlights the ease by which vestibular inputs can lead to spatial disorientation, especially in extreme conditions such as when flying inverted, flying at excessive G-forces or flying in restricted visual environments.

An unfortunate example of how a restricted visual environment can lead to spatial disorientation is the highly publicized aircraft accident of John F. Kennedy, Jr. (detailed in Ch 7). This accident brought to light the connection between a pilot flying "visually" and then losing a discernable horizon (the visual 80 percent of orientation inputs no longer accessible). The resulting loss of spatial orientation was due to the unreliable and un-capable vestibular and kinesthetic systems (the remaining 20 percent input for spatial orientation) being unable to provide credible spatial orientation cues to the pilot. The accident investigation of this particular mishap determined the probable cause as, "the pilot's failure to maintain control of the airplane during a descent over water at night, which was a result of spatial disorientation … Factors in the accident were haze, and the dark night" (National Transportation Safety Board, 1999). Spatial disorientation is often assumed to refer to a loss of vestibular orientation, but as stated in the above "probable cause," a more complete explanation was provided in terms of the restricted/limited visual environment.

Conclusion

This chapter has overviewed some of the basic visual and vestibular system parts, functions, and processes that are especially relevant to aviation. These processes are complex and interact in ways that are not yet completely understood. However, many decades of research have led to solid understanding of the fundamentals of these processes. When taking all of these studies into account, it is clear that we do not see the world exactly as it is (naïve realism); it is also apparent that the majority of perception does not come about from pure bottom-up processing of the external stimuli. A pilot's previous experiences, expectations, motivations, and desires can greatly influence interpretation of his/her environment. Smallman and St. John (2005) aptly described perceptual processing as, "assumptions [that] distort interpretation and result in imperfect, just-in-time, just-good-enough approximations of reality" (p. 8).

However, by learning about our sensory and perceptual systems and taking that knowledge into account both when actually flying and when designing aviation systems, aviation mishaps due to visual and vestibular misperceptions can hopefully be substantially decreased. Some of the following chapters further illustrate how visual perception influences flying and detail examples of how misperceptions can lead to mishaps. On a more positive note, Chapter 3 covers how knowledge of visual perception is used in the assessment of pilot vision, and Chapter 8 includes examples of how it relates to technical advancements in aviation.

References

Baird, J. C. and Wagner, M. (1991). Transformation theory of size judgment. *Journal of Experimental Psychology: Human Perception and Performance, 17*, 852–864.

Benjamin, W. (2006). *Borish's Clinical Refraction*. Oxford: Butterworth-Heinemann.

Benson, A. J. (1988). Spatial disorientation—general aspects. In J. Ernsting and P. King (Eds.), *Aviation Medicine* (pp. 277–296). London: Butterworths.

Bootsma, R. (1991). Predictive information and the control of action: what you see is what you get. *International Journal of Sports Psychology*, *22*, 271–278.

Bootsma, R. J. and Oudejans, R. R. (1993). Visual information about time-to-collision between two objects. *J Exp Psychol Hum Percept Perform*, *19*(5), 1041–1052.

Bruce, V., Green, P., and Georgeson, M. (1996). *Visual Perception: Physiology, Psychology, and Ecology*, *3rd Edition*. East Sussex, UK: Psychology Press.

Calvert, E. S. (1950). Visual aids for landing in bad visibility with particular reference to the transition from instrument to visual flight. *Transactions of the Illumination Engineering Society of London*, *6*, 183–219.

Collett, T. S. and Parker, A. J. (1998). Depth constancy. In V. Walsh and J. Kulikowski (Eds.), *Perceptual Constancy: Why Things Look as They Do* (pp. 409–435). New York: Cambridge University Press.

Crassini, B., Best, C. J., and Day, R. H. (2003). Misperceiving extents in the medial plane. In J. Andre, D. A. Owens, and L O. Harvey, Jr. (Eds.), *Visual Perception: The Influence of H. W. Leibowitz* (pp. 97–23). Washington, DC: American Psychological Association.

Conway, B. (2003). Colour Vision: A Clue to Hue in V2, Current Biology, Vol. 13, R308–R310, DOI 10.1016/S0960–9822(03)00233–1.

Curcio, C., Sloan, K., and Packer, O. (1987). Distribution of cones in human and monkey retina: Individual variability and radial asymmetry. *Science 236*, 579–582.

Cutting, J. E., and Vishton, P. M. (1995). Perceiving layout and knowing distances: The integration, relative potency, and contextual use of different information about depth. In W. Epstein and S. Rogers (Eds.), *Handbook of Perception and Cognition, Vol 5: Perception of Space and Motion* (pp. 69–117). San Diego, CA: Academic Press.

Department of Defense (2005). Human factors accident classification system.

Flach, J. M. and Warren, R. (1995). Low-altitude flight. In P. Hancock, J. Flach, J. Caird, and K. Vicente (Eds.), *Local Applications of the Ecological Approach to Human-Machine Systems* (pp. 65–103). Hillsdale, NJ: Lawrence Erlbaum Associates.

Galanis, G., Jennings, A., and Beckett, P. (1998). A mathematical model of glide-slope perception in the visual approach to landing. *The International Journal of Aviation Psychology*, *8*, 83–101.

Gibson, J.J. (1979). The Ecological Approach to Visual Perception. Boston: Houghton.

Gibson, J. J. (2002). A theory of direct visual perception. In A. Noe and E. Thompson (Eds.), *Vision and Mind: Selected Readings in the Philosophy of Perception* (pp. 77–89). London: The MIT Press.

Gillingham, K. K. (1992). The spatial disorientation problem in the United States Air Force. *Journal of Vestibluar Research*, *2*, 297–306.

Gray, R. and Regan, D. (1998). Accuracy of estimating time to collision using binocular and monocular information. *Vision Research*, *38*(4), 499–512.

Harris, J. L. (1977). What makes a visual approach "nonvisual?" *Air Line Pilot*, *46*, 14–19.

Hasbrook, A. H. (1971). Anatomy of a landing cue by cue. *Business Commercial*.

Huang, J., Wu, S., and Chen, C. (2008). Enhancing color representation for the color vision impaired. Proceedings from the *Dans Workshop on Computer Vision Applications for the Visually Impaired*.

Irani, P., Slonowsky, D. and Shajahan, P. (2006). Human perception of structure in shaded space-filling visualizations. *Information Visualization*, (5), 47–61.

Kastner, S., Schneider, K., and Wunderlich, K. (2006). Beyond a relay nucleus: Neuroimaging views on the human LGN. *Progress in Brain Research, (155)*, 125–143.

Kern, T. (2002). *Controlling Pilot Error: Approach and Landing*. New York: McGraw-Hill.

Lee, A. (2005). *Flight Simulation: Virtual Environments in Aviation*. Farnham, UK: Ashgate Publishing

Lee, D. N. (1976). A theory of visual control of braking based on information about time-to-collision. *Perception*, *5*(4), 437–459.

McKee, S. P and Smallman, H. S. (1998). Size and speed constancy. In V. Walsh and J. Kulikowski (Eds.), *Perceptual Constancy: Why Things Look as They Do* (pp. 373–408). New York: Cambridge University Press.

Mertens, H. W. (1981). Perception of runway image shape and approach angle magnitude by pilots in simulated night landing approaches. *Aviation, Space, and Environmental Medicine*, *52*, 373–386.

Mosier, K. L., Sethi, N., McCauley, S., Khoo, L., and Oransanu, J. M. (2007). What you don't know can hurt you: Factors impacting diagnosis in the automated cockpit. *Human Factors*, *49*, 300–310.

Murray, S. and Wojciulik, E. (2003). Attention increases neural selectivity in the human lateral occipital complex, *Nature Neuroscience* (7), 70–74.

NASA Educational Brief. The Effects of Space Flight in the Human Vestibular System. EB-2002–09–011-KSC. Retrieved May 22, 2009. http://weboflife.nasa.gov/learningResources/vestibularbrief.htm.

National Transportation Safety Board (1999). Accident occurred Friday, July 16, 1999 in Vineyard Haven, MA. NTSB Identification: NYC99MA178.

National Transportation Safety Board (2002). Collision with trees on final approach: Federal Express Flight 1478, Tallahassee, FL.

Newman, D. G. (2007). An overview of spatial disorientation as a factor in aviation accidents and incidents. Australian Government, Australian Transport Safety Bureau, Aviation research and analysis report—B2007/0063.

Owens, D. A. (2003). Twilight vision and road safety. In J. Andre, D. A. Owens, and L. W. Harvey, Jr (Eds.), *Visual Perception: The Influence of H. W. Leibowitz* (pp. 157–180). Washington, DC: American Psychological Association.

Parmet, A. J. and Ercoline, W. R. (2008). Spatial orientation in flight. In J. R. Davis, R. Johnson, J. Stepanek, and J. A. Fogarty (Eds.), *Fundamentals of Aerospace Medicine*, (pp. 142–205). Philadelphia: Lippincott, Williams, and Wilkins.

Parmet, A. J. and Gillingham, K. K. (2002). Spatial orientation. In R. L. Dehart and J. R. Davis (Eds.), *Fundamentals of Aerospace Medicine* (pp. 184–244). Philadelphia: Lippincott, Williams, and Wilkins.

Perrone, J. and Thiele, A. (2001). Speed skills: Measuring the visual speed analyzing properties of primate MT neurons. Nature Neuroscience, (4), 526–532.

Previc, F. H. (2004). Visual orientation mechanisms. In F. H. Previc and W. R. Ercoline (Eds.), *Spatial Disorientation in Aviation, Progress in Astronautics and Aeronautics*, Vol. 203 (pp. 94–144). Reston, VA: AIAA.

Proffitt, D. and Caudek, C. (2003). Depth Perception and the perception of events. In A. Healy, R. Proctor, I. Weiner (Eds.), *Handbook of Psychology: Experimental psychology* (pp. 214–231). Hoboken, New Jersey: John Wiley and Sons.

Proctor, R. W. and Proctor, J. D. (1997). Sensation and perception. In G. Salvendy (Ed.), *Handbook of Human Factors and Ergonomics* (pp. 43–88). New York: John Wiley and Sons.

Regan, D., and Kaushal, S. (1994). Monocular discrimination of the direction of motion in depth. *Vision Research*, *34*(2), 163–177.

Rock, I. (1983). The Logic Of Perception. Cambridge: MIT Press.

Scharff, L. (2008). Perception. In S. Davis and W. Buskist (Eds.), *21st Century Psychology: A Reference Handbook.* (pp. 245–255). Thousand Oaks, CA: Sage Publications, Inc.,

Sharpe, L., Stockman, A., Jagle, H., and Nathans, J. (1999). Opsin genes, cone photopigments, color vision, and color blindness. In K. Gegenfurtner, and L. Sharpe (Eds.), *Color Vision: From Genes to Perception*, (pp. 3–52). Cambridge: Cambridge University Press.

Smallman, H. S. and St. John, M. (2005). Naïve realism: Misplaced faith in realistic displays. *Ergonomics in Design*, Summer, 6–13.

Smith, A. T., Wall, M. B., and Thilo, K. V. (2009). Vestibular input to human MST but not MT [Abstract]. *Journal of Vision*, 9(8):679, 679a, http://journalofvision.org/9/8/679/, doi:10.1167/9.8.679.

Smith, V. C. and Pokorny, J. (1975). Spectral sensitivity of the foveal cone photopigments between 400 and 500 nm. *Vision Research*, *15*, 161–171.

Soto, D., Hodsoll, J., Rotshtein, P., and Humphreys, G. W. (2008). Automatic guidance of attention from working memory. *Trends in Cognitive Sciences*, *12*, 342–348.

Spillman, L. and Werner, J. (1990). Visual Perception: The Neurophysiological Foundations. Boston: Academic Press, Inc.

Swanson, W., Birch, D. and Anderson, J. (1993). S-cone function in patients with retinitis pigmentosa. *Investigative Ophthalmology and Visual Science*, *(34)*, 3045–3055.

Tredici, T. J., and Ivan, D. J. (2008). Ophthalmology in aerospace medicine. In: *Fundamentals of Aerospace Medicine*, Davis, Johnson, Tepanek and Fogarty, (Eds.). New York: Lippincott Williams and Wilkins.

Turner, P. and Mainster, M. (2008). Circadian photoreception: ageing and the eye's important role in systemic health. *Br. J. Ophthalmol. (92)*, 1439–1444.

Johnson, J. Stepanek, and J. A. Fogarty (Eds.), *Fundamentals of Aerospace Medicine*, (pp. 349–379). Philadelphia: Lippincott, Williams, and Wilkins.

Wandell, B. (1995). *Foundations of Vision*. Sunderland, Massachusetts: Sinauer Associates, Inc.

Warren, R. (1988). Visual perception in high-speed low-altitude flight. *Aviation, Space, and Environmental Medicine*, *59*, A116–124.

Wickens, C. D., Lee, J. D., Liu, Y. and Becker S. G. (2004). *An Introduction to Human Factors Engineering*, 2nd edn. Upper Saddle River, New Jersey: Pearson Education, Inc.

Chapter 3
The Role of Basic Visual Functions in Aviation

There is no doubt that flying is visually demanding, but how well does a pilot need to be able to "see" to fly safely and effectively? In both civilian and military aviation, considerable effort has been put forth to answering this question with the hope of establishing hard-and-fast visual standards for pilot recruiting and licensing. Overall, vision problems are the single most common reason for denial of a medical certificate and they account for more than 90 percent of all medical restrictions assigned to civilian pilots (Nakagawara, Montgomery and Wood, 2002). Until the last decade, the visual standards for aviation were largely borrowed from other occupations with little experimental or medical justification (Kumagai, Williams and Kline, 2005). In recent years, changes in human rights legislation and court challenges (e.g., Holt, 2002) have placed increasing pressure on aviation's governing bodies to establish standards that are occupationally relevant and supported by aviation-specific research. One of the major challenges in establishing visual standards for pilots is that, as discussed in detail in Chapter 2, visual perception actually comprises several different functions (e.g., acuity, motion perception, and depth perception) that operate largely independently and in parallel. Therefore, understanding how well a pilot needs to be able to "see" to fly actually requires several different studies, each examining the relationship between a specific visual function and different flight tasks. In this chapter we review research on the relationship between different basic visual abilities and flight performance.

Visual Acuity

The visual function most strongly associated with our everyday conception of what it means to "see well" is visual acuity. Visual acuity refers to the ability to see high levels of detail in a visual image and is most commonly measured using a Snellen eye chart (although there are numerous other methods). Visual acuity is often further subdivided into *near acuity*, referring to the ability to see details of objects that are less than 1 metre away, and *far acuity*, referring to the ability to see details of objects that are more distant. In addition, visual acuity can be assessed both with low and high contrast stimuli. Aviation subject matter experts (SMEs) have identified reading maps and navigation charts, and monitoring cockpit controls as flight tasks that rely on a high near acuity while identifying objects or hazards (e.g., other aircraft, runway hazards) are tasks that rely on high far acuity (Kumagai et al., 2005).

The most common method used for measuring visual acuity is the Snellen eye chart. Snellen charts test an individual's ability to identify static, high-contrast letters from a distance of 20 ft. An individual's performance compared with a standard observer creates a *Snellen* fraction, where the numerator is the distance from the chart at which the observer can accurately identify the smallest letters, and the denominator is the distance at which a normal person with average vision can identify the smallest letters. Therefore, a Snellen acuity of 20/100 indicates that an observer needs to be 20 feet from the chart to read letters that an average person could read from 100 feet. Snellen acuity can also be expressed in terms of visual angle to allow comparison with other visual performance measures. Normal, focal visual acuity is considered to be 1.0 (a resolution of 1 minute of arc) and is equivalent to the Snellen 20/20.

Internationally, visual acuity standards for recruitment of military pilots vary substantially. For example, the UK visual acuity standard is 6/6 m (20/20 ft) uncorrected in each eye while the standard in the United States is only 6/21 uncorrected (as long as it can be corrected to 6/6). Whether optical correction (i.e., glasses or contacts) is allowed during acuity measurement also varies from nation to nation and is a highly controversial issue in aviation (see below). In U.S. civilian aviation, 6/6 or better (corrected) is required for a pilot's license with the stipulation that if corrective lenses are necessary for 6/6 vision, the person may be eligible only on the condition that corrective lenses are worn while exercising the privileges of an airman certificate. In most cases these standards were borrowed from other occupations (e.g., heavy trucking) and have not been validated specifically for aviation. There have been relatively few studies that have specifically examined the relationship between acuity and flight performance.

Mann and Hovis (1996) investigated the effect of optically degrading acuity (i.e., requiring pilots to wear lenses with differing levels of blur) on flight simulator performance. Instrument Flight Rules (IFR) approaches were simulated, therefore near acuity was assessed. The blurring lenses created equivalent levels of acuity ranging from 6/6 to 6/30 (20/20 to 20/100). The most surprising result of this study was that 86 percent of participants could complete a successful IFR approach to decision height even under the worst levels of acuity. Furthermore, control errors were rare even at the highest blur levels. There were also no significant effects of blur on either glide-slope or altitude maintenance. The only significant effects of blur found in the study were that at the highest blur levels pilots could not read the approach plates or set communication and navigation frequencies. Why does acuity have larger effects on reading and setting dials when it seems to have little or no impact on more complex flight tasks such as glide-slope control? This effect seems to be due to the nature of the instruments and displays involved. Under IFR, control of altitude and glide-slope involves maintaining alignment with an indicator (bar or dots) in an instrument, while setting a communication frequency requires the pilot to pick up detailed visual information. Whether pilots can achieve a similar level of approach performance under degraded acuity and Visual Flight Rules (VFR) has not been investigated. From the results of this study, the authors recommended that the near acuity minimum standard for pilots should be no worse than between 6/10–6/12 (20/32–20/40).

Tanzer et al. (1999) correlated low- and high-contrast acuity measures with the aircraft carrier landing grades of 446 deployed naval aviators. They found a significant positive correlation between high contrast acuity (100 percent contrast letter charts) and landing grades. There were no significant relationships found between low-contrast acuity and day, night, or overall landing grades.

Kruk and Regan (1983) compared several measures of flying performance (including bombing accuracy in low-level runs and missile hits in air-to-air combat) with basic measures of visual performance including Snellen acuity. No significant correlation between flight performance and visual acuity was found. This was somewhat surprising given that the acquisition distance in the flight tasks ranged from 4,300 to 10,600 metres and therefore presumably required a high level of far acuity. Similar findings have been found in driving, as it has been shown that steering performance is relatively unaffected by the introduction of optical blur (Owens and Tyrrell, 1999).

To summarize, research evaluating the relationship between Snellen acuity and flying performance is mixed. While some studies have found that pilots with higher acuity perform better, others have found no significant effects. How the importance of high acuity varies as a function of pilot experience has not been determined. Furthermore, it seems that aircraft control can be reasonably successful when low acuity is simulated with blurring lenses. Together these results indicate that screening applicants with Snellen acuity tests will provide little insight into how successful they will be as pilots, although more research is needed. This lack of relationship between high-contrast acuity and performance has also been found in related domains such as driving (e.g., Wood and Owens, 2005). We are unaware of any experiments that have examined the relationship between low contrast acuity and flight performance.

Contrast Sensitivity

Contrast sensitivity is the ability to distinguish an object from its background and is typically measured at several different object spatial frequencies (i.e., levels of detail). Contrast sensitivity measures provide a much more detailed and complete description of visual performance than visual acuity. Visual perception decrements indicated using a Contrast Sensitivity Function (CSF) measurement are often not detectable in Snellen acuity measures (Goldstein, 1999), because visual acuity measures such as Snellen acuity provide a single ratio to quantify performance (e.g., 20/20 ft), but CSF measures provide a performance values over a range of size and contrast conditions. Taking measures across a range of conditions is particularly important when trying to assess vision in adverse conditions, such as conditions of high glare (discussed below). Furthermore, contrast sensitivity testing has proven to be highly effective in detecting visual disorders such as glaucoma and early cataract. SMEs have identified tasks such as detecting approaching aircraft, identifying targets in search and rescue missions, and locating objects in bad weather as tasks that require high contrast acuity (Kumagai et al., 2005). Despite the high level of effectiveness of contrast sensitivity measures demonstrated in basic and clinical research, to date no contrast sensitivity standards have been developed for aviation. However, several studies have investigated the relationship between contrast sensitivity and flight performance.

The primary tool used for measuring *spatial contrast sensitivity* is the sinewave grating (shown in Figure 3.1). A sinewave grating is an image that has an intensity variation following a sinewave function, giving the appearance of light and darks bars. The commonly manipulated properties of the sinewave grating are its *contrast* (the difference in brightness between the dark and bright bars) and its *spatial frequency* (the number of dark and light bars within a given area). Spatial frequency is measured in cycles per degree of visual angle, where a cycle includes one dark bar plus one light bar. The top and bottom panels in Figure 3.1 show examples of high and low spatial frequency sinewave grating patterns, respectively. When the contrast of a sinewave grating becomes too low it becomes impossible for the observer to distinguish the grating from an image with uniform average luminance. The middle panel of Figure 3.1 shows a sinewave grating with the same spatial frequency as the top panel but with reduced contrast.

Quantifying an individual's *contrast sensitivity* involves finding the minimum contrasts at which sinewave gratings of different spatial frequencies are visible (i.e., the observer can reliably distinguish between the grating and a uniform

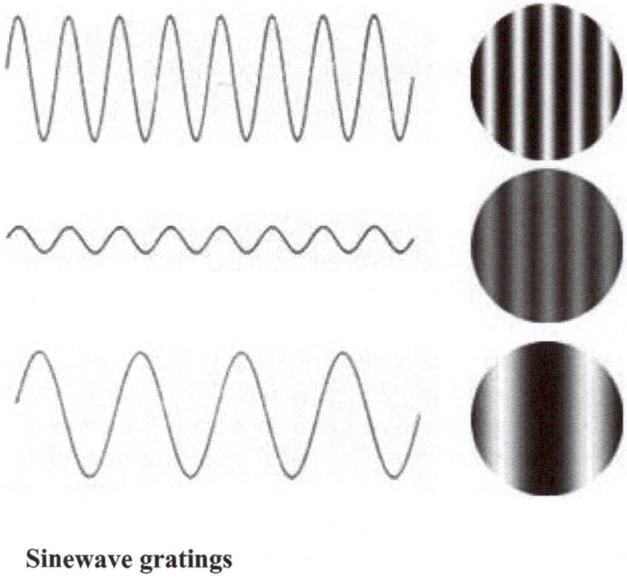

Figure 3.1 Sinewave gratings

patch). These contrast thresholds are then used to create a *contrast sensitivity function* (CSF) for the observer as illustrated in Figure 3.2. One can immediately see the advantage of the CSF over a Snellen acuity measurement. Whereas the latter only provides a single measure of performance using only high-contrast letters, a CSF quantifies acuity over a wide range of stimuli. The typical adult is most sensitive (i.e., has the lowest contrast thresholds) for spatial frequencies of about 3–5 cycles per degree of visual angle, and then performance drops off rapidly at higher or lower values. The 20/20 Snellen acuity rating is equivalent to 30 cycles per degree in contrast sensitivity, explaining the high contrast required to see at the 20/20 standard. The maximum possible human acuity is 60 cycles per degree (also requiring very high contrast to be detected), which corresponds to 20/10 vision.

Ginsburg, Evans, Sekuler and Harp (1981) investigated the relationship between contrast sensitivity and performance in an air-to-ground detection task in an F-16 video-based simulator. The task involved detecting the presence of aircraft on a runway during randomly selected landings. Eleven active-duty instructor pilots participated. Sinewave gratings of varying contrast were used to measure a CSF for each pilot. The range at which the obstacle aircraft was detected was significantly correlated with contrast sensitivity ($r = 0.83$) but not with visual acuity measured with a Snellen eye chart ($r < 0.15$).

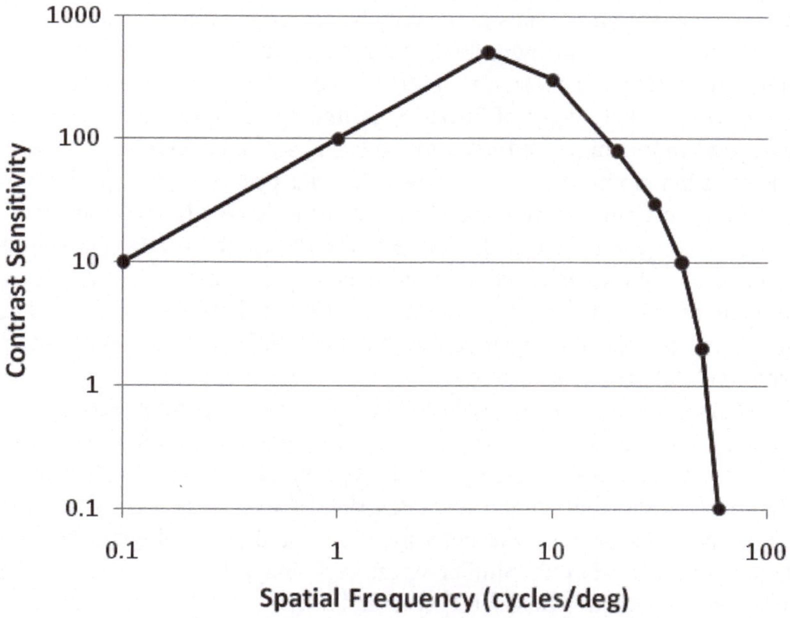

Figure 3.2 Contrast sensitivity function

Stager and Hameluck (1986) examined whether the detection of crash sites in a simulated search and rescue task was related to contrast sensitivity. The task required subjects (55 Search and Rescue specialists) to view slide presentations of aerial photographs and to identify the locations of crash sites. Detection accuracy was used as the main measure of performance. CS measured over the range of 12–18 c/deg was significantly related to detection accuracy while there was no relationship for CS measures at lower spatial frequencies. Eighteen c/deg was the highest spatial frequency used in the study.

Shinar and Gilead (1987) measured the relationship between CS measures and non-pilots' ability to perform complex object discrimination tasks. These tasks involved detecting tanks or human forms presented against a background of mountainous terrain. Subjects in the study were split into high and low CS groups. Subjects with high CS had target detection times that were more than twice as fast as low CS subjects. There were no differences in detection time when subject were divided into groups based on low and high Snellen acuity.

Grimson, Schallhorn and Kaupp (2002) argued that the development of contrast sensitivity standards for military pilots would be valuable for screening applicants and for evaluating deficits caused by factors such as laser surgery and age. They proposed that a necessary first step would be to establish baseline, normative data for a large population of military personnel. Thus, contrast sensitivity was measured using the Small Letter Contrast Test on 107 student naval pilots, 35 experienced pilots and 360 non-pilot military personnel matched for age. Both sets of pilots had significantly better mean contrast sensitivity than the non-pilots, even when individuals were matched for visual acuity. So, for example, pilots with a visual acuity of 20/16 had better contrast sensitivity than non-pilots with visual acuity of 20/16. This finding stresses the importance of going beyond simple high-contrast acuity measures when evaluating vision, and demonstrates that contrast sensitivity tests detect aspects of visual performance that are not discriminable by standard acuity tests. Overall, visual acuity was only moderately correlated with contrast sensitivity ($r = 0.45$). Given that contrast sensitivity measures are not used in screening potential pilots, why did the pilots in this study have better contrast sensitivity? This is most likely due to self-selection (individuals who are aware that their vision is weak do not attempt to become pilots) and/or motivation.

The studies reviewed so far reported significant relationships between pilot performance and contrast sensitivity (measured by finding the minimum contrast threshold for a range of different spatial frequencies). However, other studies have found no relationship. In the study described above, Kruk and Regan (1983) found that contrast sensitivity was not a significant predictor of flight performance in telemetry-tracked aircraft. Similar negative findings have been reported for detection and recognition tasks (O'Neal and Miller, 1987; Task and Pinkus, 1987).

McFadden (1994) compared the contrast sensitivity of 50 aircrew candidates. The main goal of this study was to determine whether the addition of contrast sensitivity measures would add any additional information above and beyond the basic pilot-screening battery. Using two different CS tests, it was found that there was a relatively small variance in CSF scores across candidates, indicating that CSF measures did not provide additional information for aircrew recruitment. As suggested by the authors, such measures may be useful for quantifying changes in pilot visual capabilities over time (since many CS changes are observed with ageing, see below) and/or how visual capabilities are affected by visual enhancement and protection devices. However, these benefits must be weighed against the cost and administration time—the best CSF test in McFadden's study took over 30 min to complete.

In summary, contrast sensitivity measures appear to capture some aspects of visual function that cannot be identified with visual acuity measures and in some cases CS has been shown to predict performance in flying and flight-related tasks. But should a CS measure be adopted as part of the visual test battery for pilots? As discussed above, there is currently no such standard and, given the inconsistent findings, clearly more research is needed to determine how well it can predict flight performance. However, it does seem to have some clear benefits. For flight tasks that involve detection, recognition and identification of highly detailed targets (e.g., reading complex instruments, search and rescue, etc.) previous research suggests that CS may be useful for predicting performance. For tasks that rely on motion processing (e.g., landing, formation flight) it is unlikely that CS will significantly predict performance because it does not measure an aspect of vision that is crucial for these tasks. Thus, there may be benefits for assessing CS within specific aircraft types, such as those with missions that tap into aspects of CS more so than others (e.g., air-to-air and air-to-ground military jet fighters, and those used by search-and-rescue crews). Finally, any potential benefits of CS measures must be weighed against the potential impracticality due to cost and measurement time.

Optical Correction: Lenses and Surgery

As presented in detail in Chapter 2, there are four common refractive problems that require optical correction: myopia (nearsightedness), hyperopia (farsightedness), astigmatism (uneven image focus), and presbyopia (loss of accommodative power). A significant proportion of pilots in both civilian and military aviation require visual correction for one or more of these problems. Miller et al. (1990) sampled 5000 USAF pilots and found that 27 percent of pilots and 52 percent of navigators/weapons systems operators were required to wear corrective lenses while flying. Myopia and astigmatism were the most common refractive problems. Nakagawara et al. (2004) sampled FAA records and found that roughly 51 percent of civilian pilots had visual restrictions in 2001. More recently, pilots have turned to refractive surgery, specifically photorefractive keratectomy (PRK) and laser in situ keratomileusis (LASIK), to correct refractive errors. Additionally, the number of pilots requiring visual correction appears to be increasing as the age of our population increases; Nakagawara et al. (2004) found roughly a 9 percent increase in the percentage of FAA certified pilots with visual restrictions between the years 1976 and 2001.

Given the high prevalence of optical correction in aviation it is critical to determine what effect wearing lenses and/or laser eye surgery has on flying performance. In general, research has found that corrective lenses can be used safely and effectively in aviation even for critical and hazardous missions. Subjectively, extended-wear contact lenses favorably affected job performance, but empirical studies vary in their conclusions regarding the impact of wearing corrective lenses. Bachman (1990) conducted a field study of 44 helicopter pilots wearing extended-wear soft and rigid contact lenses. No pilot was grounded due to contact lens-related problems. Still and Temme (1992) investigated night carrier-landing performance of 122 US Navy fighter pilots (16 wore corrective lenses, 106 had no correction). There was no significant difference in night carrier-landing scores for pilots who had refractive error and were required to wear an eyeglass correction and pilots with uncorrected vision. Mittelman et al. (1993) evaluated contact lens use in 90 Marine Corps aviation personnel and found that safety and health were not compromised, and subjective ratings of job performance were favorably affected.

But there are also contradictory findings. Still and Temme (1992) compared the air-to-air target detection distances for US Navy pilots wearing corrective lens with pilots who did not require optical correction. They found that the pilots who did not require visual correction could detect the target at a distance 20 percent further than the pilots who wore corrective lenses. In an analysis of civilian pilots using data from the FAA and NTSB, Nakagawara et al. (2002) reported that contact lenses and eyeglasses were contributing factors for 19 different mishaps over the period 1983–2001. These mishaps included difficulties with broken glasses, sunglasses, incompatibility with breathing equipment, or inappropriate prescriptions. In an examination of aircraft ejection events among US Navy pilots, O'Connell and Markovits (1995) reported that in 37/46 of the events corrective eyewear was lost during ejection. Finally, Partner et al. (2005) conducted research suggesting there may be safety differences between eye glass and contact lens use in pilots. Seven hundred active Royal Air Force pilots were surveyed about vision correction. Over a 12-month period, 5 percent of the corrective-spectacle users reported flight-safety incidents (including misting and conflict with HMDs), while there were no flight safety incidents reported by contact lens wearers.

Another issue raised is the use of mono-vision contact lenses by pilots. Mono-vision contact lenses provide the correct focus for far objects to one eye and the correct focus for near objects to the other eye, thus eliminating the need for bifocals. However, this differential focus in the two eyes effectively impairs the use of any binocular/stereo visual information. As discussed in detail in Chapter 7, mono-vision contact lenses have been cited as a potential cause in a 1996 mishap at LaGuardia International airport, NY; however, in general there is a large amount of disagreement as to whether or not they hurt flying performance.

The use of LASIK or PRK amongst aviators has been even more controversial. In the US Navy and USAF, pilots who have had have corrective eye surgery are prohibited from flying in combat jets. Markovits (1993) raised many concerns regarding the appropriateness of corrective vision surgery for pilots, including structural stability in high G-force conditions and visual functioning post surgery. Schallhorn et al. (1996) reported that glare susceptibility in 30 Navy/Marine personnel was abnormally high for the first month following PRK; however, susceptibility returned to normal levels after 3 months. Two of their participants also reported prolonged problems with night vision following their surgery. More recent studies have suggested that corrective eye surgery may be less of a problem for aviators than was first anticipated. Levy et al. (2003) performed a case study of one jet pilot who had LASIK correction for myopia. The pilot resumed flight 2 weeks after surgery and had no safety events in several dozen flights. Van de Pol et al. (2004) investigated 20 Army Black Hawk helicopter pilots before and after corrective surgery. Testing involved pilots flying a visually demanding flight profile under night, unaided conditions in a research–instrumented Black Hawk flight simulator and in a research–instrumented Black Hawk aircraft. Corrective surgery only produced transient effects on flight performance; both performance in the simulator and real-aircraft returned to pre-surgery, baseline levels at one month post-op.

Finally, in a recent report the FAA concluded that:

> It is unknown at this time how the long-term effects of refractive surgery may affect the performance of civil airman and if the known refractive surgery complications summarized in this paper may be exacerbated by age. It is important that pilots be aware of possible problems that may result from having refractive surgery that may affect their ability to safely perform aviation tasks. (Nakagawara, Wood, and Montgomery, 2006)

Visual Fields and Useful Field of View

An observer's visual field is the spatial extent of his or her vision, that is, how far an object can be placed from the center of their visual field and still be detected. Visual fields are typically measured using a technique called perimetry, where small lights are presented in a concave dome. The observer is required to maintain fixation on a target in the center of the dome and press a button each time a light is detected. For a person with normal vision, the horizontal extent of the visual field for binocular viewing is about 180 deg. In the vertical dimension, the inferior field (below the line of sight) extends down approximately 75 deg and the superior field (above the line of sight) extends up approximately 60 deg. Research has shown that visual field size can be reduced by diseases such as glaucoma and retinitis, and disease or injury can lead to "holes" (i.e., locations within an observer's field of vision where objects are not detected).

Because peripheral vision is crucial for visually-guided behavior and for search/surveillance tasks, it not surprising that all Air Standardization Coordination Committee (ASCC) nations have visual field standards for both military and civilian flying. Loss of peripheral vision (not specifically defined in terms of visual field size) or the presence of any scotomas (blind spots within the useable field) are both cause for candidate rejection. Subject matter experts (SMEs) have identified peripheral tasks such as detection of some cockpit warning signals, detection of obstacles (birds, other aircraft), stability and glide-slope maintenance relative to horizon as requiring good peripheral vision (Kumagai et al., 2005).

How does the extent of a pilot's visual field effect flying performance? Large reductions in visual field extent have been shown to dramatically reduce performance in flying and other visual guidance tasks. Using a helicopter simulator, Kruk and Regan (1996) measured pilots' ability to estimate the time to collision (TTC) with another helicopter. Field of view (FOV) was varied using a head-mounted display. For the two largest FOV's (66 vertical ×127 deg, 19 × 25 deg), pilots underestimated the TTC (i.e., they thought they would have collided earlier than the actual TTC) by roughly 150 ms. The smallest FOV (3 × 3 deg) produced dangerous overestimations of TTC of roughly 100 ms. Similar results have been found in accident analyses of drivers. Binocular field losses have been shown to be strongly related to driving accidents (Johnson and Keltner, 1983): field reductions ranging between 3 and 13 percent were found to be associated with accident and violation rates more than double those of age-matched control subjects.

Despite these findings, perimetry does not appear to be a useful tool for predicting pilot performance. In subjects with otherwise normal vision, it has been shown that visual field extent is not correlated with performance on visual-motor tasks such as driving (Myers et al, 2000); therefore, it is unlikely it will be related to flight performance. In other words, small variations in visual-field size would not be expected to affect flying. Instead, perimetry seems best suited for evaluating the effects of large reductions in visual-field size that result from ocular diseases. A measure that seems more suitable for screening/evaluating flying performance is the Useful Field of View (UFOV).

Whereas perimetry measures simple detection performance, UFOV measures observers' ability to selectively direct their attention to and identify briefly-presented targets. As shown in Figure 3.3, UFOV tests vary in difficulty from (1) identifying an object that appears near the center of the visual field (where the subject is instructed to look), (2) identifying an object that appears near the center of the visual field while at the same time judging the location of an object that appears in the periphery, and (3) identifying a central object while simultaneously localizing a peripheral object that is embedded in distracter objects. It has been argued that UFOV is a more ecologically valid measure of performance than perimetry because it captures some of the key task characteristics of visually-guided actions in the real world, namely segregating

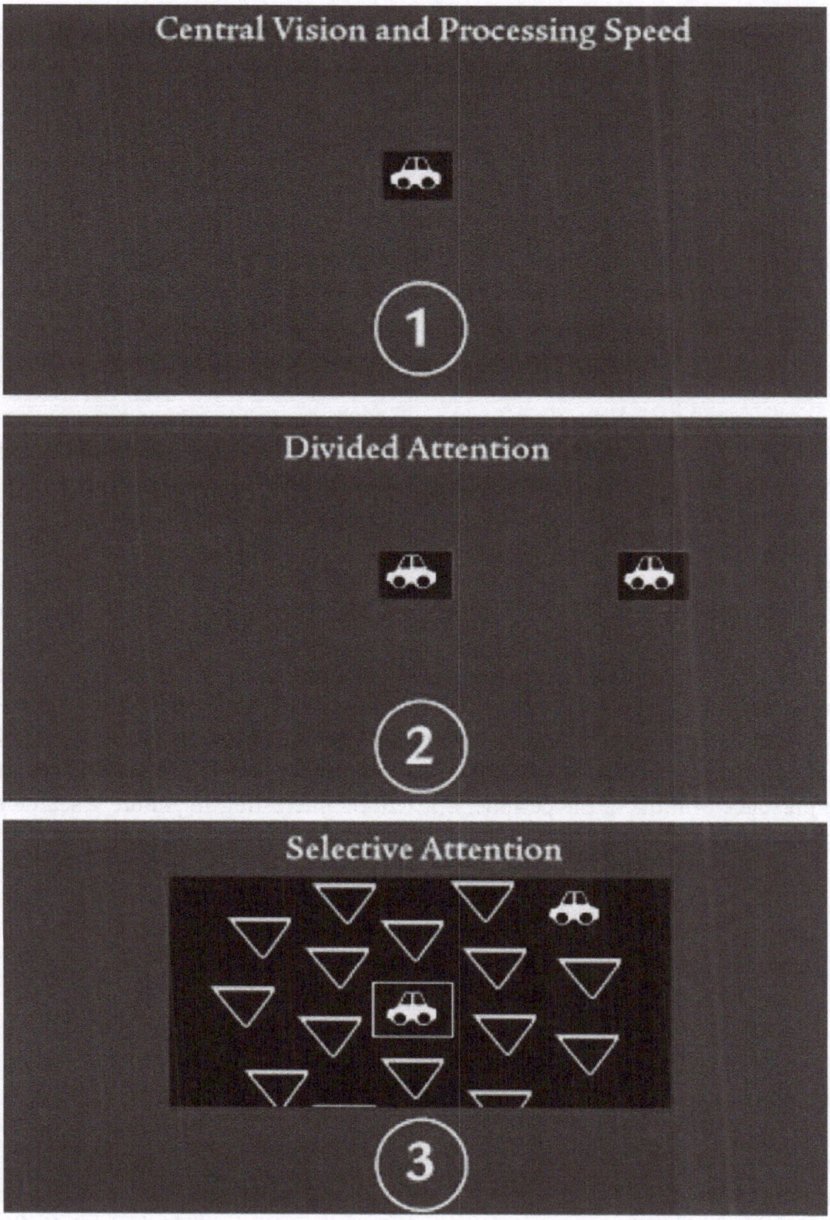

Figure 3.3 Useful field of view

Source: Reprinted with permission from Crabb, D. P., Fitzke, F. W., and Hitchings, R. A. (2004).

an object from its background, selective attention, and performing dual tasks. It provides a measure of the size of the visual field that is *useful* for extracting complex visual information.

Performance on UFOV tasks has been shown to decline with age much more dramatically than perimetry measures (Ball et al., 1990). It is also affected by the level of difficulty of the task the observer is performing in central vision. For example, if the object presented in the center of the display in Figure 3.3 (middle panel) is presented for a very short duration, observers are less able to localize objects presented in the far periphery. This effect is often referred to as "cognitive capture" or inattentional blindess (Mack and Rock, 1998); focusing on a highly demanding task causes an observer to develop "tunnel vision" and miss the object presented in the visual periphery. Similar tunneling effects have been reported for pilots using head-up displays (Prinzel et al, 2004), as discussed in Chapter 8.

While there has been no research on the relationship between UFOV and flying, research on driving suggests that UFOV may be a useful measure for predicting performance in visually-guided tasks. Among older drivers, UFOV size accounts for nearly 20 percent of the variability in accident rates (Ball et al., 1990). The UFOV test consistently predicts on-road and simulated driving performance (e.g., Wood and Troutbeck, 1995; Rizzo et al., 1996) and at-fault crash involvement (Ball et al, 1993; Owsley et al., 1998) in older drivers.

The UFOV test appears to have great potential for use in aviation; however, to date it has been largely unexplored. SMEs have identified many flight tasks that appear on the surface to be similar to the UFOV test (Kumagai et al., 2005). These include avoiding an object on a runway while taxiing, hovering, and identifying the presence of birds in the periphery. It will be interesting for future research to explore the effectiveness of the UFOV test as screening tool and as a predictor of performance in different flight tasks.

Glare Susceptibility

Even for a pilot with high acuity and a large visual field, environmental conditions can greatly hinder the ability to perform visual tasks. One of the most serious conditions in aviation is glare. Although the eye can operate over a wide range of illumination (from roughly 10^{-6} to 10^{6} lx), it only performs optimally over a very narrow range (10^{3} to 10^{4} lx). When flying in the presence of a strong external light source, light from the source is scattered within the eye onto the retina (called *veiling glare*), resulting in a reduction in the contrast of the retinal image (Miller and Benede, 1973; Van den Berg, 1986). This reduction of object contrast is called *disability glare*. In flying, there are a variety of external light sources that can produce disability glare including sunlight, fires, flares, explosions, and onboard camera flashes. Glare can also be produced by reflected light such as sunlight reflecting off a snowy terrain or reflection of the pilot's own face in the visor. Glare can be a problem both during daylight hours (e.g., flying into a setting sun) and night flight (e.g., bright city lights during an approach and landing). The problems associated with disability glare generally increase with age, especially for individuals with early cataract i.e., protein build-up on the lens (Miller and Benede, 1973; Allen and Vos, 1967; AAO, 1990; Regan et al., 1993), although there is large individual variation in glare susceptibility at all ages (Regan, 1990). Receptors also tend to become more splayed with age, which increases the negative impact of glare (they will respond to the scattered light more easily than receptors lined up to the line-of-sight).

The problems associated with disability glare become more severe as the distance between the glare source and the visual target of interest becomes smaller. To take advantage of this effect, German fighter pilots in World War II would commonly position their aircraft between the sun and their target (Brennan, 1989). Using glare to reduce their visibility in this manner gave rise to the maxim: "Beware the Hun in the sun'. Disability glare is also exacerbated when objects are being viewed through media or atmospheres that scatter the light above and beyond what occurs in the eyeball (Brennan, 1989; Hughes and Vingrys, 1991). Such conditions include scratched or dirty windscreens or visors and flying in haze, fog or mist. SMEs have identified taking off or landing directly into the sun, flying over water in sunshine, and formation flight as situations where glare can inhibit performance (Kumagai et al., 2005).

Nakagawara et al. (2004) used the NTSB Aviation Accident/Incident Database to determine the prevalence of glare-related events in civilian flight. Of the total of 25,226 accidents investigated, glare from the sun was identified as a contributing factor in 130 (0.5 percent). The most common types of accidents associated with glare were mid-air collisions with other aircraft and collision with objects/terrain during approach and landing. Because military aviation involves more extreme flight conditions and more demanding tasks than civilian aviation it is expected that the incidence of glare-related events is even higher in military flight.

Despite the well-recognized problems associated with disability glare there has been very little research quantifying the effects of disability glare in aviation and no glare resistance standards have been set. One study, however, did examine mishaps due to bright lights. Using NTSB reports between the years 1982 and 2005, and Federal Aviation Administration reports between the years 1978 and 2005, Nakagawara, Montgomery, and Wood (2006) examined accidents and incidents that involved bright lights distracting pilots. The data query was performed using search words associated with night vision problems. The authors found 58 documented vision-related accidents and incidents, with 17 accidents (57 percent) and 10 incidents (36 percent) occurring during the approach and landing phase. Nakagawara et al. 2006, reported that pilots had difficulty in perceiving distances and depth due to the glare of some approach/airfield lights. One pilot reported glare from contact lenses as being the problem, while other investigative reports cited reflection of aircraft lights back at the pilots by dust, fog, snow, rain, and ice as causing difficulties in perceiving the landing environment.

Bright light sources also led to documented visual misperceptions of height and distance, and a momentary flash blindness/glare was reported to distract pilot veridical perception. Takeoff accidents/incidents were less common; however, one Boeing 737 pilot was visually incapacitated and became spatially disoriented from a ground-based laser.

Nakagawara et al.'s (2006) article also summarized findings from the Aviation Safety Reporting System, a confidential and voluntary program to encourage pilots to share their safety incidents with others. That data base produced 153 pilot reports of vision problems resulting in unsafe flying situations between the years 1988 and 2004. Taken together, these reported data sources clearly present the dangers of bright lights to night-adapted vision, rendering pilots momentarily physiologically incapable of sensation and perception.

A recent driving study (Gray and Regan, 2007) suggests that it may be possible to measure the effects of disability glare on specific flight tasks and to predict which pilots will be most susceptible to its effects. In this study, a driving simulator was used to quantify the effects of driving into a low, setting sun. Drivers made left-turns in front of oncoming traffic in both glare and no-glare conditions. The contrast (high/low), approach speed, and initial distance of the oncoming vehicle were varied. As shown in Figure 3.4, the presence of glare resulted in a significant reduction in the safety margin used by drivers (by 0.65 sec on average) that would have resulted in many collisions in everyday driving. The effect of glare was larger for low-contrast than for high-contrast oncoming vehicles. Older drivers (45–60 years) had a significantly greater reduction in safety margin than younger drivers (19–29) years), though there was a large inter-individual variability in both age groups. Prior to completing the driving experiments, glare susceptability was measured for each participant using eye charts and a "Brightness Acuity Tester" (BAT) glare source (Mentor™). As shown in Figure 3.5, glare susceptibility measured using this simple acuity test was significantly correlated with the effects of glare on driving performance, suggesting that adding a glare susceptibility test to the visual tests taken when granting a driving license might well cause many of those with high susceptibility to avoid driving in high-glare conditions.

How can a pilot reduce the effects of disability glare? Nakagawara et al. (2004) provide the following recommendations:

- Enlist the assistance of a copilot to help perform instrument tasks so attention can be fully devoted to viewing out-of-cockpit objects.
- Deploy the aircraft's sun visor and/or use sunglasses.
- Make sure all transparent surfaces in the line of sight are clean including the windscreen, sunglasses, NVGs, etc.
- Avoid wearing light-colored clothing that can create a reflection on the windscreen or instrument panel.
- Be cautious when using photosensitizing medications such as certain antibiotics.

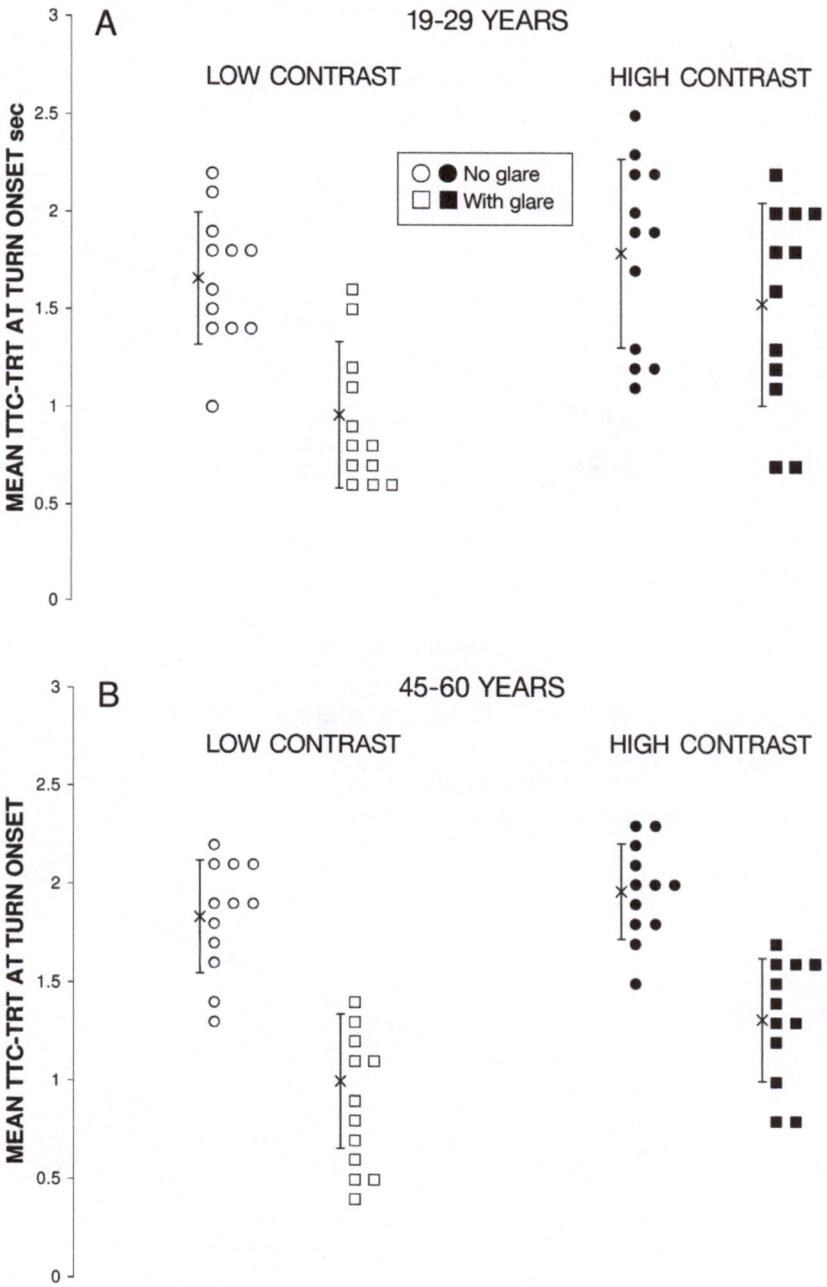

Figure 3.4 Mean left-turn safety margin study data

Source: Reprinted with permission from Gray, R. and Regan, D. (2007).

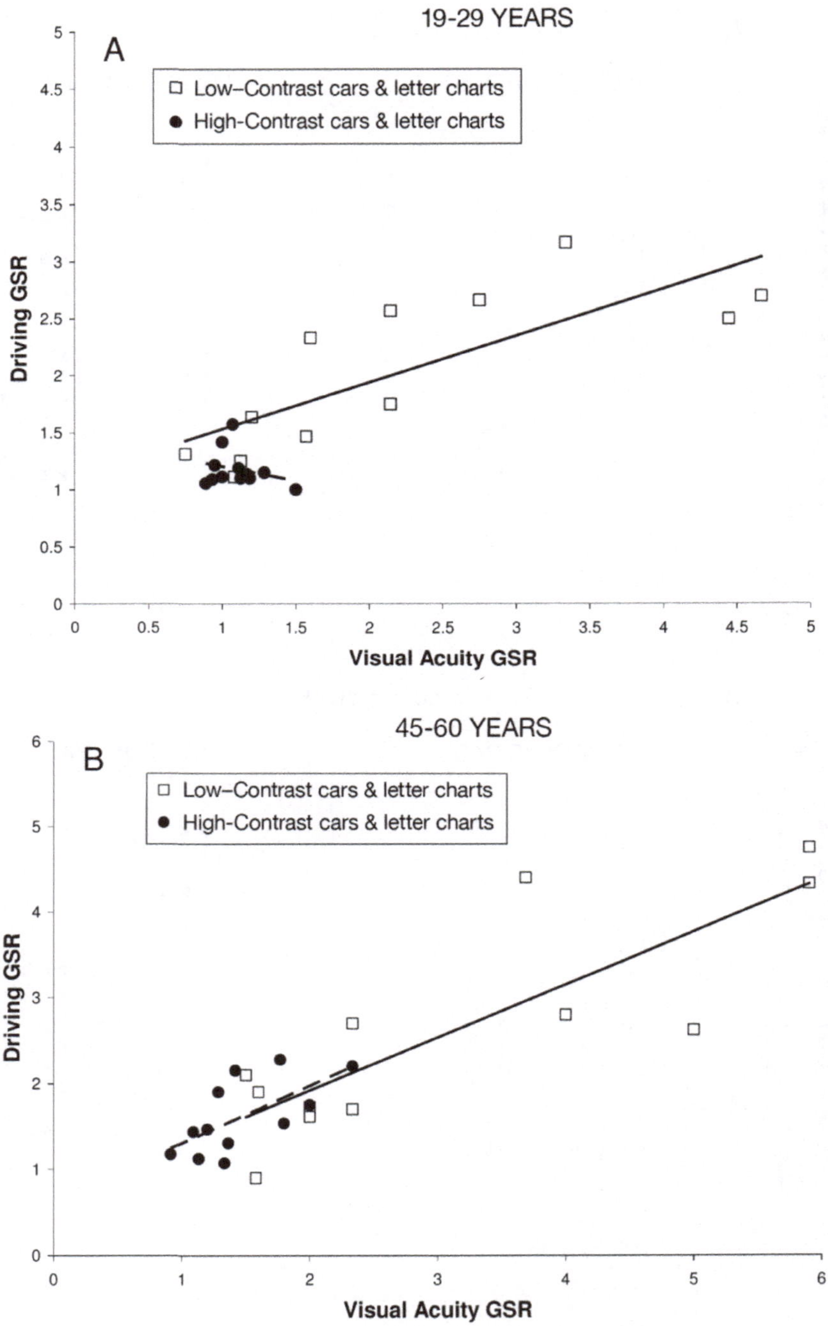

Figure 3.5 Relationships between visual acuity and glare susceptibility

Source: Reprinted with permission from Gray, R. and Regan, D. (2007).

A less serious problem associated with glare is when light sources are annoying to a pilot but do not degrade visual performance. These effects, called *discomfort glare*, occur more commonly for flashing or scintillating light sources. A wide variety of sources can produce discomfort glare, including headlights, reflections from snow, windows, CRT screens, windshields and aircraft canopies.

SMEs reported some discomfort glare is experienced when approaching bright runway lights from a dark environment (Kumagai et al., 2005). Working toward a solution to this problem, Schmidt (1999) examined physiological compatible lighting for airfields, suggesting that with the "right" lights pilots will experience fewer glare problems. Schmidt proposed that spatial disorientation results from lights that have a "point light source" and cause an after-image. It was suggested that a cold cathode light at approximately 512 nm (a blue-green light) would be the most physiologically compatible airfield lighting as it does not result in an after-image.

In summary, disability glare appears to be a problem that can seriously affect vision in flying. Although some initial research has been done, much more work is needed to determine the extent to which glare causes accidents/incidents and which flight tasks are most affected. Research on driving (Gray and Regan, 2007) suggests that it may be possible to use a simple visual screening test to determine which pilots will be the most susceptible to glare effects.

Color Vision

A pilot's ability to process color information is important in many flying tasks because color contrast (i.e., the difference in color between an object and its background) aids in the detection, recognition and identification of objects in natural images, displays and instruments. SMEs have identified several information sources in aviation that involve color coding including VFR maps, approach plates, contour maps, CRT instrumentation, weather radar, ground proximity systems, flight management systems, runways approach lighting and runway lights (Kumagai et al., 2005). Color discrimination is also important for SAR, search-and-rescue operations, as flares are color-coded to indicate different meanings. As discussed in detail in Chapter 2, color-vision deficiencies (i.e., "color blindness") are among the most common types of visual deficiencies—it has been estimated that 2–8 percent of men (depending on ethnicity) and 0.4 percent of women have some form of color-vision deficiency (Delpero et al., 2005). These problems can be both congenital and acquired through disease, trauma or toxicity.

Given the importance of color perception in aviation it is not surprising that there are elaborate color-vision tests and standards in both military and civilian aviation (e.g., International Civil Aviation Organization, 2001). Pilots are required to have normal color vision, or if they have defective color vision they are required to demonstrate that they have the ability to recognize the colors used in air navigation. Until recently, this was assessed using the Farnsworth Lantern test

(FALANT). This test involves presenting pairs of lights to subjects and having them identify the colors. By design, the FALANT allows persons with mild color blindness to pass the test, while still excluding the more serious cases of color blindness. The USAF discontinued use of the FALANT in 1993 due to its frequent failure to identify serious cases of cases of color blindness. The use of this test by the FAA has recently been challenged in light of the 2002 crash at Tallahassee airport (described in detail in Chapter 1). In this accident it was found that a contributing factor was the inability of the pilot to distinguish the PAPI signals because of his deficient color vision (even though he had passed the Farnsworth Latern test). The NTSB report concluded that:

> … existing aviation medical certification standards for color vision and use of related screening tests may not ensure detection of color vision deficiencies that can be detrimental to safety; it is possible that in some emergency situations, the speed of color recognition may assume an importance that is not currently reflected in the standards … Therefore, the Safety Board concludes that one or more of the color vision screening tests currently approved for use in the aviation industry (for example, the FALANT) are not adequate and that these tests should be identified and their use discontinued. (pp. 62–63)

As a follow up to this investigation, Cole and Maddocks (2008) investigated the ability of subjects with color-vision deficiency (CVD) who pass the FALANT to distinguish PAPI signal colors. Fifty-two subjects with CVD and 52 subjects with normal color vision (NCV) were required to name the colors of simulated PAPI signals (red, white, or no light) observed in the dark. Ten of the 52 subjects with color-vision deficiency had previously passed the FALANT. Color-identification error rates were 0.2 percent for NCV subjects and 8.2 percent for CVD subjects that failed the FALANT. Most importantly, the error rate for CVD subjects who had passed the FALANT was 4 percent (i.e., 20 times higher than for NCV subjects). This finding supports the NTSB conclusion that passing the FALANT does not ensure that pilots can distinguish PAPI signal colors.

Recent research has investigated other possible color vision tests that could be used for aviation standards. Squire et al., (2005) examined different color-vision tests used in aviation. They pointed out that technology has significantly changed over the years, yet the manner in which color vision is assessed in aviation has not. Their research compared four different secondary color-vision tests: Nagel anomaloscope and three different lantern tests, Holmes Wright Type A, Beyne, and Spectrolux. Prior to testing on the four secondary tests, each of the 79 participants was screened with the primary testing method used by the Joint Aviation Authorities, the Ishihara PIP. The Ishihara PIP has a 38-plate and a 24-plate version. The 38-plate test has the first 25 plates for regular testing as well as alternate plates, and the 24-plate version uses the first 15 plates for a screening test, per the Joint Aviation Requirements, and they are included in the 38-plate version as well.

If participants made any mistakes at all during the Ishihara PIP assessment, then they were assessed with the following: American Optical Hardy, Rand, and Rittler plates, Farnsworth D15, City University test, and the Nagel anomaloscope. If the Ishihara PIP was passed, then only the Nagel anomaloscope test was used to confirm color-vision efficacy. Squire et al. (2005) found that of their 79 participants they had 36 deuteranomalous trichromats of various levels of deficiency (weak in their green cone), 5 deuteranopes (missing green cones), 9 protanomalous trichromats of two levels (weak in their red cones), 5 protanopes (missing red cones), and 24 normal trichromats.

According to Joint Aviation Requirements, failing the Ishihara PIP then would require them to pass one of the four secondary tests. All of the dichromats (those missing a red or green cone) failed all the secondary tests. All nine protanomalous trichromats (weak in the red cone) failed all 3 lantern tests, but 3 of the 9 passed the nagel anomaloscope and 14 of the 36 deuteranomalous trichromats (weak in the green cone) passed at least 1 of the secondary tests. Of the 24 normal trichromats, 7 made between 1 and 3 errors with the Ishihara PIP test and thus had to continue to take the secondary tests. Of those 7 normal trichromats, they all failed at least one of the secondary tests except for the Holmes-Wright lantern assessment. It is interesting that some normal trichromats did not pass a color-vision test.

Taking into account all of the results, the research by Squire et al. (2005) clearly depicts that the color vision assessments used in their study (and by the Joint Aviation Requirements) yield inconsistent findings. Some normal trichromats failed assessments and some color-deficient participants passed assessments. On a positive note, none those with the most severe deficiencies passed any of the tests. The authors did present some compelling arguments regarding whether or not Protans (red weak) ought to be declared medically able to fly. For instance, the inability to correctly perceive red still may allow for discriminating between a perceived dark color compared to white and green. This, however, is not supported due to the many aviation displays that incorporate red and green colors into their warning systems. For instance, terrain, weather, and collision avoidance displays use color as a means to communicate to the pilots the level of impending threats.

Current color vision screening procedures now typically involve multiple phases employing different tests. For example, in Europe color vision is tested first with the Ishihara PseudoIsochromatic plates (PIP). If the applicant fails that color test, then a Nagel anomaloscope is used or a lantern test (Squire et al., 2005). Therefore, screening procedures have begun to take into account the findings that a single color vision test does not give a full picture of a pilot's color vision performance.

Night Vision

As discussed in detail in Chapter 2, as light level in the environment decreases, a transition from photopic (bright light) to scoptic (very dim light) vision occurs in a process called dark adaptation. SMEs have identified landing approaches at night (especially in the rain) and flying from bright conditions to very dim conditions as situations in which night vision is important (Kumagai et al., 2005). Assessing the rate of dark adaptation and effectiveness of an individual's night vision is a time consuming process that requires specialized facilities and highly-trained personnel.

Glovinsky, Belkin and Hammer (1992) compared the effectiveness of 20 different night vision tests in predicting pilot performance. Dark adaptation rate (DAR) and mesopic (low light such as dawn or dusk) contrast sensitivity (CS) were both found to be predictive of performance in detecting military targets at night. The authors concluded that CS measurements under mesopic illumination may be the most effective for aviation screening purposes.

Night vision is again addressed in Chapter 6 regarding hazards to pilots and is part of numerous mishaps presented in Chapter 7. In Chapter 8, many of the problems associated with nighttime aviation can be overcome with the use of night-vision goggles (NVGs); however, NVGs come with their own set of problems.

Stereoscopic Depth Perception

Stereoscopic vision involves combining the images from the two eyes in order to judge the depth of objects in one's environment. The difference between the two eyes' images (the binocular disparity) is proportional to the depth separation between objects. It is generally accepted that the effectiveness of stereo vision falls off rapidly as the viewing distance is increased beyond several meters (Goldstein, 1999). However, the specific upper limit of distance for binocular cues is not well established, with estimates ranging from (out to 100 ft) stereopsis (Proctor and Proctor, 1997; McKee and Smallman, 1998). Binocular disparity (δ) is related to distance (D) and interocular separation (I) as follows:

$$\delta \approx \frac{I\Delta D}{D^2} \qquad (1)$$

In this situation the effectiveness of δ as a cue to relative depth (ΔD) is substantially affected by distance (i.e., it decreases with the square of distance).

Given the large distances involved in aviation, it is not surprising that it has been concluded that there is not a sufficient reason to have stereopsis standard for pilots (Diepgen, 1993). Instead it is assumed that pilots will rely primarily on monocular cues (e.g., height in field, relative size, etc.) for judgments of depth. The FAA does not use stereo vision tests and will, in fact, allow pilots who have lost an eye to fly after a period of training to consciously use monocular cues. And of course, Wiley Post is a well-known example of a highly successful monocular pilot. In contrast, the USAF has used stereoacuity tests (e.g., the Verhoeff test) since World War I for screening pilot candidates; however, there is no specific standard.

Snyder and Lezotte (1993) conducted a prospective study that provided support for the conclusion that good stereo vision is not important for pilots. Using pilot records, this study compared the attrition rate for pilots who had passed the stereoacuity screening test with those who had not. There was no significant difference between these groups, suggesting that whether pilots have good or bad stereopsis is not a determining factor in their success. Similar findings have also been reported for driving (which involves shorter distances than aviation and would therefore be expected to be more reliant on good stereopsis). McKnight et al. (1991) compared the performance of 40 monocular and 40 binocular tractor-trailer drivers matched for age and driving experience. Assessment of driving performance (performed on a close test course) revealed no significant differences between the two groups on measures of visual search, lane keeping, gap judgments, and hazard detection. Null findings for monocular versus binocular drivers have also been reported in several other studies (e.g., Johnson and Ketner, 1983); however, there have also been a few reports that monocular drivers are involved in significantly more accidents than drivers with normal vision (e.g., Lovsund et al., 1991; Rogers and Janke, 1992).

It is important to note that these distance limits of stereovision apply only to stationary objects. When an object is moving (or the observer is moving towards a stationary object), the difference between the two eyes' images (binocular disparity) changes and it has been shown in numerous studies that the rate of this change can be used to estimate the time to collision and direction of travel of the object (reviewed in Regan and Gray, 2000). Furthermore, this rate of change of binocular disparity can potentially be used for large distances. The rate of change of disparity ($d\delta/dt$) is related to distance as follows:

$$\frac{d\delta}{dt} \approx \frac{IV}{D^2} \qquad (2)$$

where V is the speed of the approaching object along a line that passes midway between the eyes. Note that in equation [2] the effect of the square of distance on the rate of change of disparity is pitted against the object speed. Therefore, for fast moving objects the effectiveness of binocular cues could extend well beyond the distances reported for static stereopsis and could be useful for pilots (e.g., in judging the time to collision and direction of travel of an approaching aircraft). This possibility is considered in more detail in Chapter 5.

Motion Perception

The ability to process the motion of objects in the environment with a high degree of accuracy and sensitivity is obviously of critical importance to a pilot. SMEs have identified landing on an aircraft carrier and helicopter hovering as tasks that particularly rely on acute motion perception (Kumagai et al., 2005). Furthermore, a high degree of sensitivity to motion is also important for detecting the presence of other aircraft because motion is one of the most powerful cues for drawing attention to a new object (Goldstein, 1999). In Chapter 5 we discuss how specific motion cues are important for tasks such as landing and low-altitude flight. In low-altitude flight there can be situations where the cues that normally allow a pilot to distinguish different terrain features (i.e., color, luminance, and texture) can be rendered ineffective, leaving motion as the only reliable information (Regan, 1995). For example, a small hill that is the same color, brightness and texture as the surrounding terrain can be difficult to see when the pilot is stationary. It is the relative motion between the hill and its background that allows the pilot to perceive it and avoid a collision.

Despite its obvious importance, there is no required motion-sensitivity threshold for pilot entry in either military or civilian aviation. This is largely due to the fact that there is no clinically accepted test for its measurement. One promising test for use in aviation is motion-defined form (MDF), a simple laboratory test developed by Regan and Hong (1990). This test resembles a Snellen acuity test in that subjects are required to read letters, but differs in that the letters are made visible only by their motion relative to the background. The letters and background are presented on a computer monitor and made up of randomly positioned small dots. To make the letter visible, the dots inside the letter boundary move in one direction (e.g., left) while the dots outside the letter boundary move in the opposite direction. Research has demonstrated that this MDF test measures visual abilities not assessed by other tests because some patients with multiple sclerosis cannot identify MD letters while having perfectly normally Snellen acuity and contrast sensitivity. It would be interesting for future research to investigate whether performance on the MDF test can predict pilot performance in such tasks as low altitude flight.

Vision and Ageing

The changes in visual functioning that occur over a lifetime pose a challenging problem for civilian and military aviation regulators. Meeting initial entry standards for vision does not mean that visual functioning will not later decline below this level due to the effects of ageing or the acquisition of a disease. Refractive errors requiring corrective lenses become increasingly prevalent with increasing age. Furthermore problems with cataracts, acuity at low light-levels, and glare susceptibility become more common with increasing age. It has been estimated that by 65 years of age 1 person in 3 will suffer from some form visually debilitating eye disease (Quillen, 1999).

Rebok et al. (2007) investigated the incidence of age-related visual problems in commuter pilots. A total of 3,019 male pilots were studied over a period from 1987 to 1997 using FAA data files. In this sample there were 419 cases of vision problems (i.e., roughly 14 percent). Pilots in the oldest age group in the sample (60–64 years) were more than twice as likely to have a vision problem as compared with the youngest group (45–49 years). The three most common types of problems were corneal problems (16 percent), glaucoma (15 percent), and cataracts (7 percent). These findings are important because they demonstrate that even though the incidence of vision problems in pilots is lower that the general population (due to the rigorous visual testing that excludes a person with early onset problems from becoming a pilot), they still occur in a significant number of pilots. These problems could greatly interfere with a pilot's ability to perform the SME identified tasks related to acuity, contrast sensitivity, and color vision discussed above. Even more dramatically, a vision readiness screening of 4,825 active duty personnel at 13 Department of Defense sites Buckingham et al. (2003) found that 83 percent of the personnel surveyed were not vision-ready and 74 percent had eye-related health deficiencies.

Clearly age-related visual problems are relatively common in both active civilian and military pilots. Furthermore, many of these problems may go undetected and untreated because in many cases pilots' vision is not regularly screened after passing the initial licensing evaluation. It is crucial for future research to understand how pilots' visual abilities change as they age, how these changes affect flight, and to develop protocols for detecting these changes.

Summary

Taken together, Chapters 2 and 3 have laid out the foundation for aviation visual perception to include the physiology of the vestibular system as well. As stated previously, the visual perception process has many independent but parallel working systems that capture and interpret environmental cues to guide pilot's aircraft inputs. Similar to these last two chapters, Chapters 4 and 5 are highly related in that they examine the visual cues in the environment for a pilot's perception. Chapter 4 is presented from the perspective of a pilot and it uses more "pilot speak," whereas Chapter 5 presents the human-factors visual-perception research that has attempted to make sense of the visual cues used by pilots.

References

Allen M. J. and Vos J. J. (1967). Ocular scattered light and visual performance as a function of age. *Amer J Optom Arch Amer Acad Optom. 44*, 717–727.

American Academy of Ophthalmology report (1990): Contrast sensitivity and glare testing in the evaluation of anterior segment disease. *Ophthalmology. 97*, 1233–1237.

Bachman, W. G. (1990). Evaluation of extended wear soft and rigid contact lens use by Army aviators. *Journal of the American Optometric Association*, *61*(3), 203–210.

Ball, K., Owsley, C. and Beard, B. (1990). Clinical visual perimetry underestimates peripheral field problems in older adults. *Clinical Vision Sciences 5*, 113–125.

Ball, K., Owsley, C., Sloane, M. E., Roenker, D. L. and Bruni, J. R. (1993). Visual-attention problems as a predictor of vehicle crashes in older drivers. *Investigative Ophthalmology and Visual Science 34*, 3110–3123.

Brennan, D. H. (1989). Glare in Aviation. *Health Physics*, *56*, 665–669.

Buckingham, R. S., McDuffie, D., et al. (2003). Follow-up study on vision health readiness in the military. *Military Medicine*, *168*, 789–791.

Cole, B. L. and Maddocks, J. D. (2008). Color vision testing by farnsworth lantern and ability to identify approach-path signal colors. *Aviation Space and Environmental Medicine*, *79*, 585–590.

Crabb, D. P., Fitzke, F. W., and Hitchings, R. A. (2004). A practical approach to measuring the visual field component of fitness to drive. *British Journal of Opthalmology*, *88*, 1191–1196.

Delpero, W. T., O'Neill, H., et al. (2005). Aviation-relevent epidemiology of color vision deficiency. *Aviation Space and Environmental Medicine*, *76*, 127–133.

Diepgen, R. (1993). Do Pilots of Airplanes Need Stereopsis—a Survey of the Literature of This Topic. *Zeitschrift Fur Flugwissenschaften Und Weltraumforschung*, *17*, 331–342.

Ginsburg, A. P., Evans, D. W., Sekuler, R., Harp, S. A. (1982). Contrast sensitivity predicts pilots' performance in aircraft simulators. *American Journal of Optometry and Physiological Optics*, *59*, 105–109.

Glovinsky, Y., Belkin, M. et al. (1992). Night-vision in young individuals—correlation between laboratory findings and performance in the field. *Investigative Ophthalmology and Visual Science*, *33*, 1416–1416.

Goldstein, E. B. (1999). *Sensation and Perception*. New York: Brooks/Cole Publishing.

Gray, R. and Regan, D. (2007). Glare susceptibility test results correlate with temporal safety margin when executing turns across approaching vehicles in simulated low-sun conditions. *Ophthalmic and Physiological Optics*, *27*, 440–450.

Grimson, J. M., Schallhorn, S. C., and Kaupp, S. E. (2002). Contrast sensitivity: establishing normative data for use in screening prospective naval pilots. *Aviation Space and Environmental Medicine*, *73*, 28–35.

Holt, J. L. (2002). Arizona verdict for pilot provides lessens on Lasik. (Post v.University of Arizona). *Trial*, *38*(8), 70–72.

Hughes, P. K. and Vingrys, A. J. (1991). Reduced contrast sensitivity when viewing through an aircraft windscreen. *Aviation Space and Environmental Medicine*, *62*, 254–257.

Johnson, C. A. and Keltner, J. L. (1983). Incidence of visual-field loss in 20,000 eyes and its relationship to driving performance. *Archives of Ophthalmology*, *101*, 371–375.

Kruk, R. and. Regan, D. (1983). Visual test results compared with flying performance in telemetry-tracked aircraft. *Aviat Space Environ Med. 54*, 906–11.

Kruk, R. and Regan, D. (1996). Collision avoidance: a helicopter simulator study. *Aviat Space Environ Med. 67*, 111–14.

Kumagai, J.K., Williams, S. and Kline, D. (2005). Vision Standards for Aircrew: Visual Acuity for Pilots. *Defence Research and Development Canada, W7711047921/001/TOR*.

Levy, Y., Zadok, D., and Barenboim, E. (2003). Laser in situ keratomileusis in a combat jet aircraft pilot. *Journal Of Cataract And Refractive Surgery, 29*, 1239–1241.

Lovsund, P., Hedin, A. and Tornros, J. (1991). Effects on driving performance of visual-field defects—a driving simulator study. *Accident Analysis and Prevention 23*, 331–342.

Mack, A. and Rock, I. (1998). *Inattentional Blindness.* Cambridge, MA: MIT Press.

Mann, J. and Hovis, J. (1996). The effect of optical blur on simulated IFR approaches. *Aviation Space and Environmental Medicine*, *67*, 739–745.

Markovits, A. S. (1993). Photorefractive Keratectomy (Prk)—Threat or Millennium for Military Pilots. *Aviation Space and Environmental Medicine*, *64*, 409–411.

McFadden, S. M. (1994). A Comparison of 2 Contrast Sensitivity Tests and Their Usefulness as a Screener for Aircrew. *Aviation Space and Environmental Medicine*, *65*, 710–717.

McKee, S. P, and Smallman, H. S. (1998). Size and speed constancy. In V. Walsh and J. Kulikowski (Eds.) *Perceptual Constancy: Why Things Look as They Do* (pp. 373–408). New York: Cambridge University Press.

McKnight, A. J., Shinar, D. and Hilburn, B. (1991). The visual and driving performance of monocular and binocular heavy-duty truck drivers. *Accident Analysis and Prevention*, *23*, 225–237.

Miller, D. and Benede, K. G. (1973). *Intraocular Light Scattering*. Thomas, Illinois.

Miller, R. E., 2nd, Woessner, W. M., Dennis, R. J., O'Neal, M. R., and Green, R. P., Jr. (1990). Survey of spectacle wear and refractive error prevalence in USAF pilots and navigators. *Optometry and Vision Science*, *67*(11), 833–839.

Mittelman, M. H., Siegel, B., et al. (1993). Contact-Lenses in Aviation—the Marine Corps Experience. *Aviation Space and Environmental Medicine*, *64*, 538–540.

Myers, R. S., K. Ball, K., et al. (2000). Relation of useful field of view and other screening tests to on-road driving performance. *Perceptual and Motor Skills*, *91*, 279–290.

Nakagawara, V. B., Montgomery, R. W., et al. (2002). Aviation accidents and incidents associated with the use of ophthalmic devices by civilian airmen. *Aviation Space and Environmental Medicine*, *73*, 1109–1113.

Nakagawara, V. B., Montgomery, R. W., and Wood, K. J. (2004). Changing demographics and vision restrictions in civilian pilots and their clinical implications. *Aviation Space And Environmental Medicine*, *75*, 785–790.

Nakagawara, V. B., Montgomery, R. W., and Wood, K. J. (2006). Aircraft accidents and incidents associated with visual disturbances from bright lights during nighttime operations. DOT/FAA/AM-06/28, Office of Aerospace Medicine.

Nakagawara, V. B., Wood, K. J. and Montgomery, R. W. (2006). New refractive surgery procedures and their implications for aviation safety. DOT/FAA/AM-06/9, Office of Aerospace Medicine.

O'Connell, S. R. and Markovits, A. S. (1995). The Fate of Eyewear in Aircraft Ejections. *Aviation Space and Environmental Medicine*, *66*, 104–107.

O'Neal, M. R. and Miller, R. E. (1987). Further investigation of visual acuity and contrast sensitivity in pilot detection of aircraft. *Proceedings of the Human Factors Society, 31st Annual Meeting, Santa Monica, CA: Human Factors Society, 1987: 1189–93.*

Owens, D. A. and Tyrrell, R. A. (1999). Effects of luminance, blur, and age on nighttime visual guidance: A test of the selective degradation hypothesis. *Journal of Experimental Psychology-Applied*, *5*, 115–128.

Owsley, C., Ball, K., McGwin, G., Sloane, M. E., Roenker, D. L., White, M. F. and Overley, E. T. (1998). Visual processing impairment and risk of motor vehicle crash among older adults. *Jama-Journal of the American Medical Association* 279, 1083–1088.

Partner, A. M., Scott, R. A. H., et al. (2005). Contact lenses and corrective flying spectacles in military aircrew—Implications for flight safety. *Aviation Space and Environmental Medicine*, *76*, 661–665.

Prinzel, L. J., Comstock, J. R., et al. (2004). The efficacy of head-down and head-up synthetic vision display concepts for retro- and forward-fit of commercial aircraft. *International Journal of Aviation Psychology*, *14*, 53–77.

Proctor, R. W. and Proctor, J. D. (1997). Sensation and perception. In G. Salvendy (Ed.) *Handbook of Human Factors and Ergonomics* (pp. 43–88). New York: John Wiley and Sons.

Quillen, D. A. (1999). Common causes of vision loss in elderly patients. *American Family Physician*, *60*, 99–108.

Rebok, G. W., Qiang, Y. D. et al. (2007). Age-related vision problems in commuter and air taxi pilots: A study of 3019 pilots, 1987–1997. *Aviation Space and Environmental Medicine*, *78*, 706–711.

Regan, D. (1990). The Charles F. Prentice Award Lecture 1990: Specific tests and specific blindness: keys, locks, and parallel processing. *Optom Vis Sci. 68*, 489–512.

Regan, D. (1995). Spatial orientation in aviation: visual contributions. *Journal of Vestibular Research*, *5*, 455–471.

Regan D., Giaschi D. E. and Fresco B. B. (1993). Measurement of glare sensitivity cataract patients using low-contrast letter charts. *Ophthalmic Physiol Opt. 13*, 115–123.

Regan, D. and Gray, R. (2000).Visually-guided collision avoidance and collision achievement. *Trends in Cognitive Neuroscience*, *4*, 99–107.

Regan, D. and Hong, X. H. (1990). Visual acuity for optotypes made visible by relative motion. *Optom Vis Sci.* 67, 49–55.

Rizzo, M., Reinach, S., McGehee, D. and Dingus, T. (1996). Simulated car crashes and the useful field of view. *Investigative Ophthalmology and Visual Science 37*, 4406–4406.

Rogers, P. N. and Janke, M. K. (1992). Performance of visually-impaired heavy vehicle operators. *Journal of Safety Research 23*, 159–170.

Schallhorn, S. C., Blanton, C. L., et al. (1996). Preliminary results of photorefractive keratectomy in active-duty United States Navy personnel. *Ophthalmology*, *3*, 5–21.

Schmidt, R. T. (1999, June). Reduce risk of inducing spatial disorientation using physiologically compatible ground lighting. *Aviation, Space, and Environmental Medicine*, *70*.

Shinar, D. and Gilead, E. (1987). Contrast sensitivity as a predictor of complex target detection. Proceedings of the Human Factors Society, 31st Annual Meeting, Santa Monica, CA: Human Factors Society, 1987: 1194–6.

Stager, P. and Hameluck, D. (1986). Contrast sensitivity and visual detection in search and rescue. *Defence and Civil Institute of Environmental Medicine #95454 86–04582.*

Still, D. L. and Temme, L. A. (1992). Eyeglass Use by United-States-Navy Jet Pilots—Effects on Night Carrier Landing Performance. *Aviation Space and Environmental Medicine, 63*, 273–275.

Squire, T. J., Rodriguez-Carmona, M., et al. (2005). Color vision tests for aviation: Comparison of the anomaloscope and three lantern types. *Aviation Space and Environmenta Medicine*, *76*, 421–429.

Snyder, Q. C. and Lezotte, D. C. (1993). Prospective Assessment of Stereoscopic Visual Status and USAF Pilot Training Attrition. *Aviation Space and Environmental Medicine*, *64*, 14–19.

Tanzer, D. J., Rossi, W. A., Schallhorn, S. C., Kaupp, S. E., van de Pol, C., and Brown, M. C. (1999). High and low contrast visual acuity and optical aberration correlations with naval aviation flight performance. *Investigate Ophthalomolgy and Visual Science*, *46*.

Task, H. L. and Pinkus, A. R. (1987). Contrast sensitivity and target recognition performance: a lack of correlation. Society for Information Display 87 Digest of Technical Papers. New York, NY: Palisades Institute for Research Services Incorporated, 1987: 127–9.

Van den Berg, T. J. T. P. (1986). Importance of pathological light scatter for visual disability. *Docum Ophthal. 61*, 327–333.

Van de Pol, C., J. L. Greig, et al. (2004). Helicopter night flight operations following refractive surgery. *Investigative Ophthalmology and Visual Science*, *45*, 2823.

Wood, J. M. and Owens, D. A. (2005). Standard measures of visual acuity do not predict drivers' recognition performance under day or night conditions. *Optometry and Vision Science*, *82*, 698–705.

Wood, J. M. and Troutbeck, R. (1995). Elderly Drivers and Simulated Visual Impairment. *Optometry and Vision Science 72*, 115–124.

Chapter 4
Pilot Perspective of Cues Used for Visual Flying

> What makes it so hard? There is first of all the fact that the modern airplane has a very shallow gliding angle and thus must approach its field along so shallow slanting a line. It is as if you had to shoot at a target, with the target not facing you but set almost edgewise to you; the slightest error would make you miss not only the bull's eye but the whole target! (Langewiesche, 1944, p. 263)

Flying an airplane is a perceptually learned process in terms of the pilot understanding invariants for the ecological control of the aircraft (Benson, 1999; Lintern, 2000). Salient visual cues used by a pilot are above sensory threshold and provide perceptual information for accurate aircraft control. Therefore, for a visual cue to be perceptually effective, it must convey information to the pilot. If an aircraft control input is made that alters the spatial position of the aircraft, cues within the visual scene need to reflect that change. Thus, the best visual cues for a pilot to perceive are those that are sensitive to even the slightest aircraft positional change. A visual cue that does not change given gross aircraft control input is probably not an effective cue (Lintern). However, some visual cues, by remaining constant, provide information to the pilot; these are called invariants.

There are many sources of information pilots have access to regarding their aircraft's status. Those sources of information range from throttle(s), yoke/stick, and rudders as well as the engine sound and the feel of the aircraft in terms of vibration and pull of gravitational forces. The dominant and most important perceptual cue, however, is vision (Leibowitz, 1988; Smith, 1999). Aviation is a visually-controlled behavior, hence vision's dominance in aviation regardless of the type of flying or maneuvering. As Langewiesche (1944) stated above, visual perception is not easy. Recall in Chapter 1, Smallman and St. John (2005) coined the term *naïve realism*, defined as a "misplaced faith in people's ability to extract information from realistic displays" (p. 7). The authors also explained the challenges of understanding visual perception and stated that visual perception is hard, flawed, and spartan. So it is with this challenge that this book and specifically this chapter attempt to describe the visual perceptual experiences pilots encounter.

This chapter briefly introduces the visual aspects of aviation. The intent is to enhance understanding of the operational maneuvers pilots accomplish as well as the complex environmental perception requirements needed to safely operate an aircraft. This information should help the reader better understand the visual illusions shared in Chapter 6 and the mishap descriptions presented in Chapter 7.

This chapter is written from a pilot's perspective regarding the cues that a pilot uses when flying. The following chapter, Chapter 5, addresses many of the same visual aspects of flying but from a completely different perspective. Chapter 5's approach is not from a pilot's perspective, but more from a vision research or human factor's perspective of visual perception cues. Consequently, some of the terms, concepts, and studies introduced in this chapter are repeated in the following chapter but the discussion is couched in research rather than "pilot-speak" operational use.

Environmental Perception

The phrase "environmental perception" is introduced to assist the reader in its use throughout this chapter and beyond. We use environmental perception to refer to a pilot's perception of cues from his/her aviation viewing environment, to include all prior experience, training, and expectations a pilot has of that viewed environment. Our use of environmental perception acknowledges both direct external perception of the cues perceived as well as the pilot's cognition of those cues. This is contrasted with the strict definition of "ecological perception." Ecological perception refers to purely direct perception of one's environment and does not include the cognitive components. Gibson, the father of ecological perception, advocated no cognitive aspects to this form of perception; he believed that everything an observer needed to perceive in the environment was directly available within their visual scene. Gibson's perspective takes perception to its extreme, but fails to account for prior experience, learning, expectations, and mental models an observer has when interacting with an environment. Many of the illusions and disorientations experienced by pilots are the result of a disconnect between what a pilot directly sees (sensation) and what a pilot expects to see (perception of the scene). These illusions are presented in Chapter 6.

Environmental perception of the aviation visual scene is what allows a pilot to create and maintain spatial orientation. This is contrasted with spatial orientation created and maintained via instrument flying … the use of displays to indirectly form aircraft spatial position. Because this book and specific chapter pertain to "visual flying," environmental perception refers to the act of pilots using vision of their external visual scene to directly guide their actions, without the use of instruments.

One final term needs clarification prior to exploring visual flying cues. In the opening paragraph the concept of an "invariant" was introduced. In environmental perception an invariant is a cue that does not change within the optical array across time and can serve as a macro-level cue used for the pilot's perception. For instance, if a pilot is flying an approach to landing and the horizon is visible, the horizon is an invariant. The horizon's general location relative the runway does not change as the pilot descends. Another invariant during an approach to landing in terms of environmental perception is the location of the runway in the pilot's environment, remaining relatively stable in the windscreen. This is contrasted with changing texture of the foliage beneath the pilot that increases in detail as the pilot descends toward the terrain. Therefore, some visual cues provide information to the pilot by not changing, whereas other visual cues that do change provide information to the pilot. The pilot's perceptual challenge is determining which of many environmental cues are the most pertinent toward a safe landing or for maintaining safe altitude clearance.

Approach to Landing

The most challenging portion of learning to fly any aircraft is the approach-and-landing phase because of the small margin for error (Benson, 1999). Complicating the visual interpretation of flying an appropriate visual glide-path down to landing are the highly interdependent and dynamic actions required to maintain the desired glide-path. A glide-path is the angle to aimpoint line along which a pilot maneuvers the aircraft to landing. It is described in more detail later in this chapter. Figure 4.1 depicts a 3-degree glide-path. If flying an instrument approach to landing, this exact glide-path is displayed via aircraft instruments and navigation displays. Flying a visual glide-path to landing, without the aid of instrumentation, is more of an approximation via a pilot's "mental picture." This mental picture of the appropriate glide-path is practiced over and over and engrained into a pilot. The intent of the training is that a pilot will know what the appropriate visual picture to a runway looks like regardless of the surrounding environment and different runways. Bressey (1976) graphically articulated the experience a commercial airline pilot feels when flying an instrument approach to landing as one of "sliding down an electronic banister in the sky with some 200 tons of aircraft strapped to

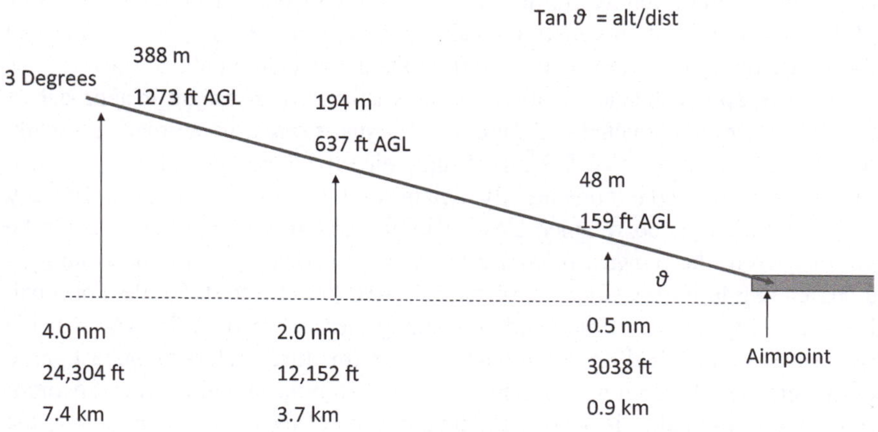

Figure 4.1 Altitude and distance of a 3-degree glide-path

his posterior while the ground is coming up to meet him at an approximate speed of 230 ft per second" (p. 16).

Figure 4.1 demonstrates that to maintain the appropriate 3 degree glide-path at certain distances from the runway the aircraft must be at a particular height above the ground. This perceptual task requires constant updating of one's distance from the runway and altitude above the ground (surface plane of the runway). This task, however, is prone to many perceptual errors and these errors are presented in detail in Chapter 6. For example, perception of runway shape and size based on previous training and experience make a pilot prone to perceptual constancy problems, also addressed in Chapter 6.

There are as many techniques available to use during an approach to landing as there are the number of instructor pilots teaching people how to fly (Pitts, 1967). Hodgson detailed the basic information a pilot must have for a visual approach to landing: (1) pitch and roll attitude information, (2) horizontal and vertical velocity information, (3) position information relative to the desired landing runway, and (4) vertical flight path information in terms of distance from landing point, height above touchdown point, and height above terrain (1971, p. 205). Hasbrook (1971) described the necessary control of certain variables for a safe and successful landing. He listed aircraft vertical and horizontal speed as well as heading. Hasbrook stressed speed control because it incorporates pitch angle and thrust setting dynamics. He also mentioned runway size and shape as cues for a pilot conducting a visual approach to landing.

Bressy (1976) in his description of a visual approach to landing stressed two reference planes a pilot must monitor: a vertical plane through the center of the runway and a glide-path. A pilot must continuously monitor the intersection and displacement of the two planes, as well as heading and rate of closure and rate of change of heading relative to each plane. Bressy acknowledged that it is easy to monitor heading relative to the runway, but the displacement above or below glide-path is difficult because it results in extension or compression (termed foreshortening) of the runway picture from the desired glide-path.

A visual approach to a runway can be flown either via a straight-in approach or from an overhead, depicted in Figure 4.2. A visual straight-in approach normally occurs from a distance of 9.3 km (5 nm) directly lined-up with the intended runway. Visibility conditions usually dictate when the visual straight-in formally begins, which in good weather, clear visibility, and appropriate terrain could be 10 miles from the runway. In contrast, a visual overhead pattern and landing is a descending turn to position the aircraft 2.4–0.8 km (1.5 to 0. 5 miles) on final. Imagine a 180-degree turning and descending merge lane on a freeway—this is what the overhead final turn is similar to. The term "final" refers to the last phase of the approach to landing common to both the straight-in and overhead pattern. If a plane is on final, it is nearing the final portion of the approach, almost to the runway. This is contrasted with the final approach fix, the latest point in space at which a pilot would normally configure the aircraft with gear and flaps. This final approach fix is normally associated with an instrument approach and is comparable

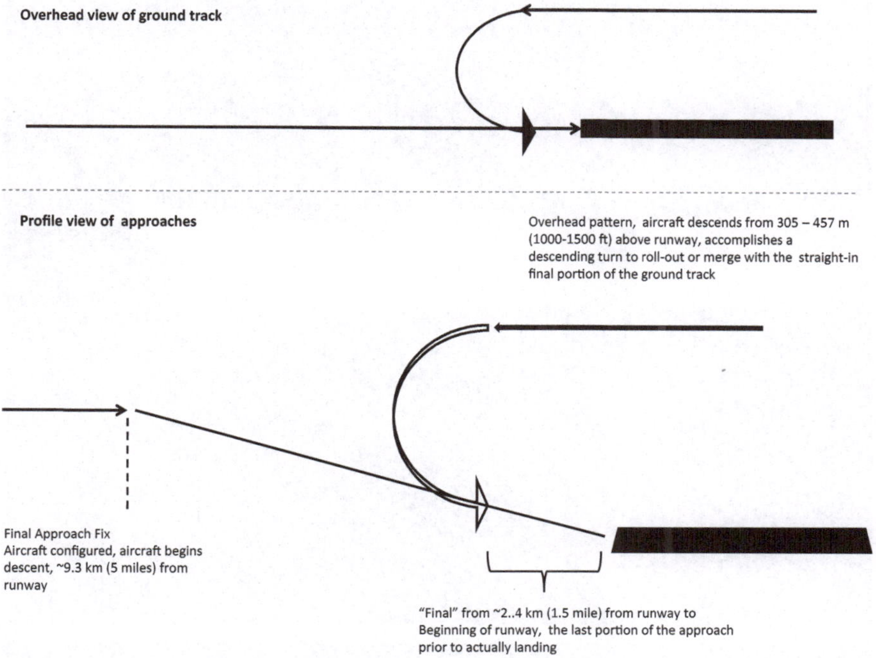

Figure 4.2 Visual overhead and visual straight-in approach depictions

to a 9.3 km (5 mile) point on a visual approach. Often it is at this point that the aircraft no longer maintains level altitude and begins the descent to the landing runway on the straight-in.

The straight-in visual approach and landing is broken down into six parts: (1) runway 2) airspeed and power, (3) horizon, (4) glide-path, (5) aimpoint, and (6) field of view references. In the following chapter, the approach to landing is only discussed in terms of runway alignment, glide-path control, and the actual landing phase referred to as the "flare." Because this chapter presents a pilot's perspective of visual flying maneuvers, for completeness the six aspects of an approach and landing are detailed. Figures 4.3 to 4.10 depict photographs of final approach to landings in a variety of locations representing terrain differences, distances to the runway, weather conditions, and day/night conditions. As the discussion continues on the six aspects of a landing, reference to the photographs in those figures will aid in the description.

Runway

The starting point for a visual straight-in for landing begins with runway alignment. The aircraft vector is controlled directly toward the intended runway. In no-wind conditions this equates to pointing the nose of the aircraft at the runway. A cross-

Figure 4.3 Three photographs of an approach to landing

Source: With permission from Steven Kaplan photographer, http://www.airliners.net/photo/Ansett-Australia-Airlines/Airbus-A320-211/0084956/&sid=8947f6d638361fe73446e8bd6a1a2980.

Figure 4.3 ***Concluded***

wind, however, may require the pilot to "crab into the wind," so that the nose may not point at the runway (it points into the wind), but the direction of flight (vector) is still towards the runway.

Riordan (1974) surveyed 360 commercial airline pilots and found the pilots rated the runway image and shape/size as their most used visual reference for an approach to landing. The concept of runway alignment is the easiest for inexperienced pilots to grasp as the runway also serves as the aimpoint. Initially it is a distant, non-descript visual cue, but as the pilot closes the distance, the runway systematically increases in size and detail, allowing for more specific aimpoint targeting. Figures 4.3 to 4.10 all depict very unique landing environments that pilots must correctly interpret for a successful landing. Figure 4.3 includes three photographs depicting a plane as it approaches the runway; note the change in the shape/size of the runway as the pilot gets closer.

Related to the runway is the concept of splay, described as the angle from a line perpendicular to the horizon. For a given runway, splay increases as altitude decreases and decreases as altitude increases. Only for extreme departures from desired glide-path at close distances to the runway will splay be a salient visual cue for landing (Lintern, 2000). A more effective way to understand splay is to consider the size and shape of the runway. If a pilot is shallow or low while on approach, the runway will appear short and wide, with the front of the runway very wide relative to the far end. This is how the runway appears in the third photograph in Figure 4.3. Splay also can be described in terms of linear perspective because the runway may appear to emanate from a distant point near the horizon. In contrast, if a pilot is high on glide-path, well above the runway, or simply farther out from the runway, it will appear more orthogonal, thinner with the near and far end more equal in width, that is, less splay. The photograph, Figure 4.3, showing the view furthest from the runway illustrates an image of the runway that is more orthogonal. This concept of splay is discussed in Chapter 5 from a more visual perception research perspective.

The size of the runway may also influence perception of runway distance. This idea will be discussed again later in Chapter 6 but is worthy of introduction here as well. For instance, Figure 4.4 depicts a runway that is long and relatively thin, with a length-to-width ratio of 91 (13,793 × 150 ft). Runways with high ratios appear more orthogonal and have less splay than runways that have smaller ratios. High ratio runways may be interpreted as being farther away than runways with low ratios. As is more completely described in Chapter 6, the interpretation of runway ratio will be affected by a pilot's previous experiences with runways of other ratios, and the pilot's interpretation can influence his choice of glide-path.

One last point needs mentioning regarding the runway. As stated above, pilots use the runway as their intended target or aimpoint (the visual goal); however, most pilots fail to appreciate visual cues other than the runway itself. The runway serves as the egocentric objective, but the surrounding terrain provides global environmental perspective. It is the environmental orientation via ambient vision that allows for accurate, visually controlled actions to safely and successfully land an aircraft.

Figure 4.4 Landing at Albuquerque International Airport

Source: With permission from Dan Jordan, photographer, Airliners.net, from http://www.airliners.net/photo/-/-/1010744/.

Airspeed and Power

Flying a straight-in approach requires the pilot to maintain a constant airspeed, which varies greatly depending upon type of aircraft, type of flap setting, and aircraft weight/fuel load (approach speed is a function of stall speed). Appropriate throttle settings, thrust, for a given aircraft-configuration greatly stabilize the airspeed control. But even having the correct power setting does not always result in a good approach. For instance, nose-low attitudes may result in an increase in airspeed unrelated to power setting, just as an increase in pitch attitude, nose up, will slowly deplete the airspeed. Also, adding or reducing power may not only increase/decrease airspeed but may impact glide-path. A common mantra in flying an approach and landing is "aimpoint, airspeed." Consequently, maintaining proper airspeed is an important aspect for an approach and landing (Hasbrook, 1971).

Horizon

The horizon serves multiple purposes for a pilot during an approach to landing. For the most part, it is not appreciated by a pilot as his/her gaze is fixated upon the runway and the intended spot of landing, (i.e., the aimpoint); however, the horizon provides essential ambient and global environmental feedback to the pilot regarding wings-level status as well as pitch information without having to reference the attitude indicator. Thus, while focal vision is consciously focused on the aimpoint, the horizon provides feedback to pilots regarding their orientation relative to an earth-fixed reference. When the horizon is absent (as in Figures 4.5 and 4.8), the pilot must rely on his/her attitude indicator to ensure proper pitch and bank are maintained during an approach.

The horizon may either be explicitly or implicitly defined for use by a pilot. Figures 4.3 and 4.4 have well-defined, explicit horizons. Thus, a pilot has a world-sized attitude indicator that lacks no ambiguity. But sometimes meteorological conditions do not allow for an explicit horizon and the implicit horizon is all that is available. An implicit horizon is simply one that is not clearly defined. Figure 4.8 with the fog in the Azores does not provide reliable cues for the horizon's location. However, there are times on a clear night when the horizon can easily be seen given moon illumination and the appropriate cultural lighting and terrain features.

From a pilot's perspective, the horizon is also used for establishing a gap, which is a distance of sky versus ground within the field of view of the pilot (Lintern and Liu, 1991; Benson, 1999). Later discussion presents the horizon as a visual invariant for horizon-angle (or H-angle), an angle generated between horizon and aimpoint. For a normal landing within the pilot's field of view, the horizon perspective is half ground and half sky. Thus, the horizon is the dividing line. The horizon also helps establish the runway surface plane for ground orientation. Langewiesche (1944) described the horizon as an invariant for landing because the horizon never moves relative to eye-level.

The horizon also plays a role during the landing flare. The "flare" is when the pilot is just about to land and thus decreases the aircraft's speed and sink rate to allow for a smooth touchdown onto the runway. A technique in landing an airplane involves looking to the horizon during the flare to help gauge height above the ground. This also helps avoid ground-rush, the "optically violent" global optic flow rate (Warren, 1988, p. A121) that occurs when looking directly at the ground while traveling at a high speed in the downward vertical vector. As the pilot looks towards the horizon during the landing flare, peripheral vision "picks-up" external cues to the side of the airplane that provide "sink-rate" information. This environmental perception is accomplished by ambient vision, which is physiologically able to detect movement and rate of change of movement. These environmental ambient cues along the Hudson River may have made the safe water landing possible for Captain Sullenberger. Landing at night or in impoverished visual conditions, however, often produces less smooth landings because of the limited ambient vision cues.

Glide-Path

The biggest challenge in accomplishing a straight-in approach and landing is maintaining a stabilized glide-path to the runway. The normal glide-path is 3 degrees. There are variations ranging from 2.5 degrees to 4 degrees, but 3 degrees is the standard (see Figure 4.1). In Figure 4.1, note the glide-path arrow points at a spot just a bit past the beginning of the runway; this is the aimpoint. This spot is normally 152–305 m (500 to 1,000 ft) down the runway, although some airplanes do aim for "brick-one" of the runway, (i.e., the very beginning), but this is the exception. A pilot may aim at the beginning of the runway if the braking ability of the aircraft is in question or the surface condition of the runway may prohibit stopping within the available runway length (e.g., wet, snow, ice). By aiming just past the beginning of the runway the pilot allows for a margin of error in case they do land shorter than their desired aimpoint. A pilot does not want to aim too far beyond brick-one and subsequently land too far down the runway either, for a pilot never knows when they may have needed an extra 152 m to ensure being able to stop on the runway as opposed to departing a prepared surface and running off the end of the runway.

Glide-path control involves keeping the aircraft's vertical velocity within desired range given the target airspeed. For an approach speed of 232 km/h (125 knots), a vertical velocity of 203 m/minute (665 feet/minute) descent maintains a 3 degree glide-path to the runway. The logic behind the glide-path is that if it is too low (less than 2.5 degrees), it brings the aircraft dangerously close to possible terrain hazards. Too steep a glide-path (greater than 3.5 to 4.0 degrees) brings the aircraft onto the runway at high rates of descent that present three problems. One, higher rates of descent result in harder landings. Consequently, structural fatigue problems may occur with the aircraft because the landing gear is absorbing excessive forces upon landing. Second, pilot and passenger comfort come into play with hard landings. Finally, steep approaches involve low throttle settings. To maintain airspeed in a steep descent requires retarding throttles possibly to idle. This becomes unsafe in the event of a go-around or missed approach; the aircraft is in an unsafe position to land and the pilot must immediately increase power to gain enough thrust to change the aircraft's vector, up and away from the runway. Therefore, the 3-degree approach is the right mix of obstacle clearance towards the runway, smooth vertical descent for transition to a comfortable and safe landing, and shallow enough to require the pilot to carry power toward the runway in the event that a go-around (an aborted landing) is necessary.

Maintaining glide-path involves a perceptual learning process of a "mental picture" of the appropriate angle of descent towards the runway. Mertens (1979) described it when a pilot's visual scene looks "correct" based on appropriate runway slant, size, and shape. Much of flying an approach and landing is based on direct perception; however, as mentioned at the beginning of this chapter, the cognitive aspect of prior experience is also involved in any visually-guided action (Haber, 1981). Figure 4.5 portrays time-elapsed photography of an aircraft landing at Madeira Airport, Azores, Portugal. This photograph depicts the glide-path the pilots took as they guided their aircraft toward the runway using a combination of approach instrumentation as well as stored knowledge of the appropriate visual picture of a 3-degree glide-path to landing. Additionally, Figure 4.3 and its series of three photographs show from the cockpit view the glide-path flown toward landing—the mental picture guiding a pilot's visually controlled actions. The

Figure 4.5 Night time-lapsed photograph at Funchal, Portugal in the Azores

Source: With permission from Rui Sousa, photographer, Airliners.net, http://www.airliners.net/photo/TAP-Air-Portugal/Airbus.../0646366/M/.

scene in Figure 4.6 may challenge pilots to maintain their glide-path because of the close proximity of final approach to the hill and houses on the pilot's left. And finally, Figure 4.9, the landing in Estonia, shows the pilot "on glide-path" with the two white lights on the left and the 2 red lights on the right as depicted by the VASIs (visual approach slope indicator lights). The visual mental picture is reinforced with the VASIs confirming the visual cues.

The proper mental glide-path picture must be fairly robust to overcome countless changes in the runway environment, surrounding terrain differences, and varying levels of visibility as pilots attempt to maintain "the picture." At greater distances to the runway a pilot is able to safely make some extreme deviations above or below glide-path to determine what the proper glide-path is given a particular runway. The first author has done this numerous times at unfamiliar locations in an attempt to determine just where the current visual scene relates to a stored mental model of the "visual picture" from memory. For instance, if unsure of the glide-path, a pilot can alter his/her altitude to go above and below until it "looks right" given the distance from the runway and the altitude above the ground. This maneuver by a pilot on a visual approach is a good example of environmental perception. Galanis, Jennings, and Beckett (2001) discussed this idea of a build-up or lag in glide-path correction whereas Lintern and Koonce (1991) called this a build-up of "visual evidence" (p. 69).

Thus far the primary discussion of risk involved with landing an airplane has centered on impacting the terrain prior to the runway. When a perfectly worthy

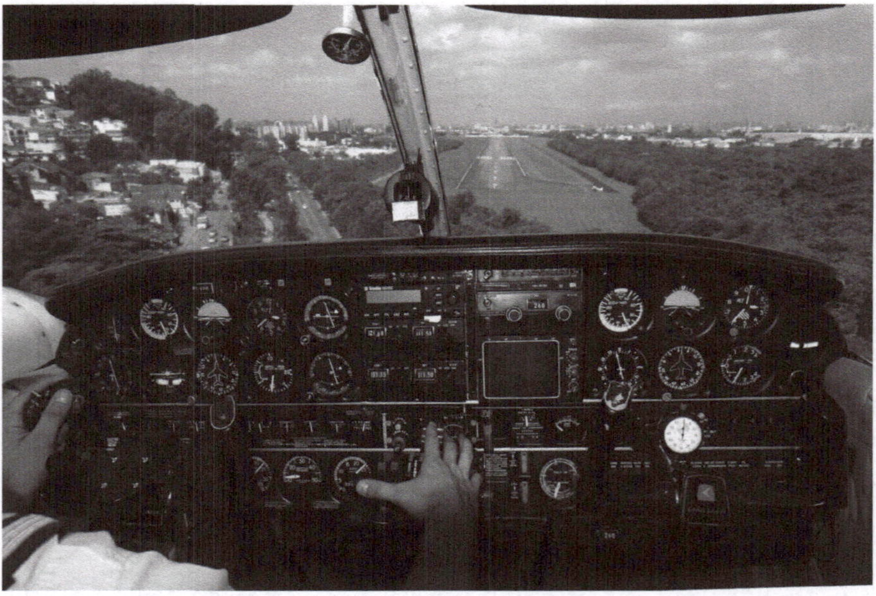

Figure 4.6 Landing at Matre Airport, Sao Paulo, Brazil

Source: With permission from Stephan Klos Pugatch, photographer, airliners.net, http://www.airliners.net/photo/Blue-Air-Taxi/Embraer-EMB-810C-Seneca/1413563/M/.

aircraft unintentionally impacts the ground it is called "controlled flight into terrain -- CFIT." It is fitting within the discussion of glide-path to discuss poor glide-path maintenance and too steep of an approach. The consequences of being too steep on approach are not as severe unless the landing runway is unusually short. If steep on an approach the pilot has a few options:

1. Accomplish an aggressive descent and re-establish the desired glide-path to the original aimpoint. It is difficult to intercept the glide-path from above if there is limited distance to the runway. Normally a pilot does not want to have an excessive vertical descent rate toward the ground with limited altitude, which is the case if trying to re-establish glide-path prior to the runway.
2. Accept the current glide-path and simply establish a new aimpoint further down the runway. This is probably the safest option if the runway is long enough to still allow normal braking action and stopping within the given runway length.
3. If the runway is short, accept the steep glide-path angle to the current aimpoint and prepare for a faster and harder (less flare) landing.

In this short runway case, which is the most dangerous, if the aimpoint is shifted down the runway the pilot is landing well beyond usable landing surface and greatly increases the chance of running out of runway prior to stopping the aircraft. If the runway is wet or the braking of the aircraft is less than optimal this could result in departing the prepared surface of the runway.

The discussion above went into detail to clearly present to the reader that being steep on approach is a dangerous place to find oneself. Impacting the terrain is not the main concern if a visual illusion induces a pilot to fly higher and away from the ground; however, it has its own dangers in that there are hazards while attempting to land. An accident that demonstrates this relates to Figure 4.5, Madeira Airport. In 1977 an aircraft attempting to land in a heavy storm landed too far down the runway, failing to stop due to the wet runway and high speed in the available runway remaining. The runway at the time was only 1,600 m (5,250 ft). Two construction projects, one in the 1980s and one in 2000, extended the runway to what it is today, 2,781 m (9,124 ft) making it much safer for large aircraft to take off and land.

According to Pitts (1967) the judgment of height is the most important perceptual skill for a pilot. This concept of height estimation, vertical awareness, plays a crucial role in glide-path maintenance to determine rate of change of height perception. Known pitch and power settings ensure the aircraft is at the appropriate height given the distance to the runway for ideal conditions but even those parameters change as a result of aircraft weight and winds from one approach to the next.

Aimpoint

The aimpoint cue refers back to the first half of the "aimpoint, airspeed" mantra and is closely related to glide-path, as the angle to aimpoint is simply the desired glide-path. Once a pilot starts down glide-path, shown at the "final approach fix" in Figure 4.1, a common practice is to put the runway in the center of the windscreen (Figures 4.6, 4.8, and 4.9). The runway then becomes a perceptual invariant, a source of visual information that should not change while flying the approach. This "gun-sight" technique is questioned, however, by some (e.g., Hasbrook, 1971; Mertens, 1979) because turbulence and head movements may appear small and insignificant but could result in excessive glide-path deviations. This common technique, which must be continuously monitored, is one of many visual cues to establish glide-path towards the appropriate spot on the runway.

As the aircraft gets closer to landing the pilot can see more detail in the runway environmental cues. This availability of detail signals the point in time when the pilot picks a spot on the runway as their specific aimpoint. Often the chosen aimpoint is 152 m (500 ft) or 304 m (1,000 ft) down from the beginning of the runway (from "brick-one"). It must be stressed this is the aimpoint, not the eventual landing point. The aimpoint is a spot on the runway that the pilot directs the aircraft down towards the runway until 15 and 2 meters (50 and 6 ft) above the surface. At that time, the pilot initiates the "flare" by completely reducing power to idle (if not already) and breaking the descent rate to smoothly touch down on the runway surface. Depending upon the aircraft, this could be a reduction of 15.8 km/h (10 knots) and cutting the 183 m/minute (600 ft/minute) descent in half to 91 m/minute

Figure 4.7 Final approach to landing at Great Barrier Reef Airport, Hamilton Island, Australia

Source: With permission from Darren Howie, photographer, airliners.net, http://www.airliners.net/photo/-/-/0978838/L/.

Figure 4.8 Short final for landing at Sao Jorge, Portugal in the Azores

Source: With permission from Joao Resendes, photographer, www.positiveclimb.com and airliners.net, from http://www.airliners.net/photo/SATA-Air-Acores/British-Aerospace-ATP/0596916.

Figure 4.9 Landing at Tallinn Airport, Estonia

Source: With permission from T. Phelps.

(300 ft/minute). The aimpoint shift results in a landing 152–304 m down the runway beyond the aimpoint, thus landing the aircraft 610 m (2,000 ft) down the runway beyond the threshold. Due to the pitch of the aircraft and the limited downward viewing angle of a pilot out the windscreen, the aimpoint and touchdown landing point are not actually viewed by the pilot in the last phases of the approach and landing. This would only be possible if the airplane had a glass bottom for the pilot to see through, similar to some commercial helicopters.

Another aspect of glide-path and aimpoint is pitch control. Pitch is the vertical movement of the aircraft's nose. Pitch is related to power and airspeed in that, for a given throttle setting, too low of a pitch will greatly increase airspeed and too high of a pitch will decrease airspeed and neither condition is acceptable for a stabilized 3-degree visual glide-path. Pitch control is how a pilot directs the aircraft towards the aimpoint. Improper pitch in the nose down direction will cause the aimpoint to move up in the windscreen from its center position, giving visual feedback to the pilot regarding an aimpoint short of intended landing. Pitch up of the aircraft in turn may cause the aimpoint to move down from center in the windscreen, resulting in the aimpoint further down the runway.

The aimpoint is important to maintain throughout the approach. Often an aircraft may deviate slightly from desired 3-degree glide-path, and the pilot may elect to re-establish the 3-degree glide-path. This must be done relative to the original aimpoint. To correctly accomplish glide-path correction, a pilot must momentarily fly either more steeply or more shallow to re-establish the 3-degree glide-path for the same, consistent aimpoint. For example, if a pilot deviates from the 3-degree glide-path and lowers the nose, thereby increasing pitch in the down direction, the aimpoint will move up in his/her windscreen and if continued on this path, the vector would put the aircraft to land short of the runway (flying well below the desired glide-path). To correct this, the pilot must raise the nose of the aircraft, pitch up, and fly shallow-to-level momentarily until a re-established 3-degree glide-path is intercepted for the original aimpoint, as shown in the time-lapsed photograph of Figure 4.5.

The ideal pitch control is to maintain the aimpoint in the center of the windscreen aligned with the center of the runway. As the aircraft gets closer to the runway, the detail of the aimpoint can be seen and the aimpoint "fills the windscreen." It is at that time (15–3 m above the runway), the pilot shifts the aimpoint to transition to the flare and landing. This interdependence of pitch and power for glide-path control towards an aimpoint is what makes this a challenging maneuver to learn. The expansion of the aimpoint is the concept of "focus of expansion" and the flow of the optic array from the center point of "ego-motion" (Gibson, 1950; Gibson, 1979; Gibson, Olum, and Rosenblatt, 1955; Hasbrook, 1971). As a visual cue, the expansion of the aimpoint occurs just prior to landing as previously described, therefore it is not used during the majority of the approach. The flare is discussed in more detail in the following chapter in terms of visual perception research.

A paved runway is not a requisite to maintain an aimpoint as pilots land successfully on open fields or snow/ice, as shown in Figure 4.10. This ability demonstrates the importance of an aimpoint and landing surface—rather than the runway as a

Figure 4.10 Short final for landing at McMurdo Station, Antarctica

Source: With permission from T. Phelps.

constructed man-made object. If landing on an open field, there are normally some effective environmental cues to assist in height and distance estimation. The change of texture and color between fields can help define where the "runway" begins; possibly a row of trees can serve as a point to attempt to land abeam (an aimpoint). These types of environmental cues are similar to those that allowed the pilot of Flight 1549 presented in Chapter 1 to safely "land" upon the Hudson River. One advantage the pilot had in that case was an extremely long "runway" in terms of the river not constraining the aimpoint and touchdown point. Thus, even without an actual runway (the preferred and most used visual cue according to Riordan's 1974 research), ample visual cues were available to perceive height and distance. Figure 4.10, however, shows a runway carved out of snow, and the entire area is featureless for the most part; it is an axample of an extremely challenging environment for purely visual height and distance estimation.

Field of View References

It has been explained that pilots use pitch control to maintain the runway and aimpoint in the center of their windscreen (Benson, 1999). The windscreen is part of the pilot's field-of-view reference for aimpoint stability. Other parts of the aircraft used are the glare-shield and canopy bow or top of the windscreen. These field-of-view references can be used in reference to the horizon. The horizon may be in the top third of the windscreen and become an invariant for glide-path control. The gap or distance between the glare-shield and the aimpoint provides feedback regarding stability of the glide-path. Field of view references are depicted in Figures 4.3, 4.6, 4.8, and 4.9 in terms of where the runway and aimpoint fall relative to the pilot's out-the-window perspective.

Overhead Pattern

The overhead pattern is a much shortened version of the straight-in approach, but has similar characteristics. The runway serves as the primary visual reference throughout the overhead pattern. The final 2.4–0.8 km (1.5 to 0.5 mile) portion is the same; however, to get to that point, the pilot must maneuver the aircraft in a 180-degree descending turn (shown in Figure 4.2). It is much more difficult than a straight-in because the pilot must visually control the aircraft's bank to guide the changing heading as well as deal with the previously mentioned aspects of runway variables. As mentioned earlier in the chapter, the overhead pattern requires the pilot to maneuver the aircraft around the final turn; analogous to a 180 degree downhill merge lane of a freeway. The two most challenging aspects of the overhead pattern are runway displacement upon roll-out for lateral control and intercepting the desired glide-path for vertical control. This second challenge combines rolling out on final with the proper vertical displacement given the horizontal displacement … equating to the appropriate glide-path.

Another aspect of the final turn to landing is that the pilot must account for wind in terms spacing from the runway, vertical descent during the final turn, rolling out on final, and how to position the aircraft to counter any drift due to winds; the aircraft's vector must continually point toward the aimpoint. Environmental perception also includes clues that inform the pilot of wind direction and how that wind directly influences the control inputs used by the pilot to fly the aircraft to final.

Beall and Loomis (1997) investigated optic flow and visual analysis of the base-to-final turn in their research. They colorfully suggested how pilots accomplish such a maneuver: "the pilot perceives the 3D spatial layout of the terrain (including the runway), perceives the 3D motion of the aircraft prior to the turn, and then plans the turn through some process tantamount to solving differential equations" (p. 206). Their description was a bit tongue-in-cheek as they doubted this process was required for pilots and instead offered a theory consisting of optic flow variables and splay rates. Beall and Loomis had pilots fly the same approaches at both night and day and explored the role of ground texture on pilot's performance. They hypothesized that performance would not be different between day and night and that is what they found. They attributed performance to the pilots' perceptual assessment of the runway's splay as they turned to final.

Beall and Loomis' (1997) finding suggests that in calm winds at a standard altitude for an overhead pattern, 305–457 m (1,000–1,500 ft), minimal visual perception is needed to get close (i.e., complete the final turn). Thus, experienced pilots who know the pitch and power settings as well as bank angle are able to fly the maneuver with minimal environmental cues. For instance, the first author's experience in five different but similar aircraft had approximately the same parameters: 30 degrees of bank and a pitch setting of $^{2}/_{3}$ ground and $^{1}/_{3}$ sky. Beall and Loomis' (1997) research is again examined in Chapter 5 and expands the discussion in terms of visual perception of splay.

Stabilized Approach Criteria

The idea of established parameters or criteria, known as "windows of acceptable performance" for pilots to continue their approach to landing is a lesson often learned the hard way in aviation circles. According to Turner (2007), a stabilized approach is the key to workload management and a good approach leads to a good/safe landing. Unfortunately, many pilots have the wrong mindset while flying, and that is to "salvage an approach" no matter how unsafe because "pilots are overwhelmingly optimistic and the culture of aviation places great value on innovating to recover from nearly hopeless situations … the result is a mindset that we can salvage any situation, no matter how bad" (p. 2). Far too often pilots have continued approaches despite airspeed, vertical velocity, and vertical/lateral approach limits being out of acceptable safety margins. Commercial aviation was the first truly to embrace the use of parameters to help with the decision either to continue the approach or execute a go-around/missed approach.

When evaluating the criteria, it is important to remember that all of the aspects of an approach to landing are interrelated. The current section on visual approaches began by stating flying and maintaining precise parameters were very challenging because of the highly interdependent and dynamic factors involved. It is due to this interaction between airspeed, pitch, power, glide-path, and aimpoint that the correction of one directly influences other parameters and may lead to unstabilized approaches. The following is a scenario often encountered by pilots.

A pilot determines that the aircraft is above the desired 3-degree visual glide-path. To correct the glide-path the pilot reduces power and pushes forward on the controls to lower the nose (and momentarily moving the aimpoint extremely short of the runway). Also during this correction, the vertical velocity increases due to the descent. Then upon re-intercepting the glide-path the pilot brings the nose of the aircraft back up and establishes the original aimpoint. While correcting the glide-path, however, the pilot fails to re-establish the proper power setting (having left it reduced to account for the original descent) and now with the power too low, the airspeed of the aircraft decelerates. To correct the slow airspeed, the pilot pushes the power up and accelerates. But this also has the inadvertent effect of generating thrust and subsequent lift. And although the aircraft's speed is no longer an issue and the descent rate reduced, the pilot has shifted the aimpoint down the runway, and is now climbing above the desired 3-degree glide-path.

This scenario has happened to all pilots when they were learning how to fly and may have also happened during a task-saturating approach if juggling other distracting tasks, such as an emergency procedure. The above example described an unstabilized approach … the pilot failed to maintain aircraft parameters within expectable limits. Experienced pilots know that accurate and precise flying requires numerous micro-corrections. Pilots need to avoid over-correcting or becoming impatient by making large corrections.

Khatwa and Helmreich (1998–9) produced an impressive report for the Flight Safety Foundation on controlled flight into terrain. Recall, this is when an airworthy aircraft unintentionally impacts terrain. One of their recommendations specifically addressed stabilized approach criteria as a change for standard operating procedures to minimize the risk of controlled flight into terrain. A "no fault go-around policy" would also encourage pilots to cognitively prepare themselves to "go missed approach" and not worry about any employment repercussions. The primary changes advocated by the authors were (p. 52):

1. Acceptable stabilized-approach criteria that would require necessary visual cues to continue descent below MDA/DH (minimum descent altitude/decision-height) as well as flight deck alerts (e.g., GPWS [ground proximity warning system]) requiring timely action.
2. An approach-ban policy that prohibits the continuation of an approach beyond a point not less than 1,000 ft [305 m] above the threshold of the landing runway, unless minimum visibility or runway visual range requirements as appropriate for that particular approach type are met or exceeded.

Turner (2007) in an *Aviation Safety* magazine article presented industry accepted norms for stabilized approach criteria, listing specific criteria for both instrument and visual approaches. Basically a pilot is to execute an immediate missed-approach or go-around if deviations are observed in terms of (1) more than ¾ scale lateral deflection on Course Deviation Indicator if instrument or not aligned with the landing runway if visual, (2) airspeed: 5 knots slow or 10 knots fast, (3) rate of descent exceeds 1,000 ft/minute, and/or (4) pitch is lower than 10 degrees below or 5 degrees above horizon.

The point of a stabilized approach is that pitch, power, and attitude parameters are met and minimal changes are made to the aircraft to allow for the safest possible approach to landing. Given the aircraft's configuration, it is of the utmost importance to have the aircraft properly trimmed (aerodynamic controls pressures are equaled out) and power set. In many ways a stabilized approach is similar to aerial refueling, which is very common in military aviation. Figure 4.11 depicts aerial refueling. Attempting to rejoin (two or more aircraft coming together during flight) on a tanker aircraft is exceptionally difficult due to the lack of ambient visual cues as one approaches another aircraft. To make it manageable the airspeed of the tanker needs to be known to the adjoining aircraft and known visual cues of the tanker's aircraft can provide approximate distance information. For instance,

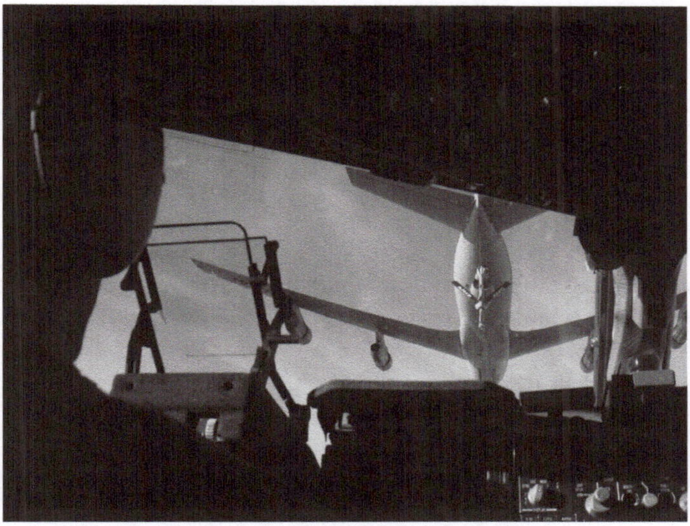

Figure 4.11 Aerial refueling photograph

Source: With permission from T. Phelps.

if certain details on the tanker's aircraft can be seen, these visual cues provide feedback to the approaching pilot regarding the distance from the tanker aircraft. Also, calculated distance information between the two aircraft is available (using internal specific identification codes) and is necessary to confirm distance as well as the known overtake speed relative to the tanker. The visual cues are not reliable enough on their own. Also, a purely visual rejoin is very difficult and often results in a stagnant rejoin (not generating any overtake due to fear of too much speed) or an overshoot because of too much overtake speed.

This refueling example depicts the difficulty in visually-guided actions and perceiving environmental cues. Attempting to gracefully contact a tanker aircraft when it is the only object in the sky is as perceptually challenging as trying to gracefully land on a runway when it is the only object on a dark night. The key to success in any landing environment is a stabilized approach to ensure a safe landing. Given impoverished visual conditions, a stabilized approach will reduce cognitive workload requirements and allow a pilot to better focus on other factors and not be saturated with erratic aircraft control inputs. According to Turner (2007), unstabilized approaches are correlated with approach-and-landing accidents. Correlation is not equated with causation; however, a predictive relationship exists that must be respected by pilots.

According to Captain Tarnowski, an Airbus test pilot, "stabilized, constant descent angle final approaches significantly raise the safety level of this flight phase" (2007, p. 21). Tarnowski suggested that pilots can fly every approach similar to a precision approach using stabilized approach criteria with the help of cockpit technology—eliminating the unstabilized manner in which non-precision approaches had been flown in the past.

Go Around or Missed Approach

In order to better understand much of the discussion in this book, reference is often made to a pilot accomplishing a go-around or missed approach. This maneuver occurs when a pilot makes the decision to not land and immediately increases power and pitches the aircraft up to climb away from the ground or runway. The reason to initiate a go-around is primarily due to safety of flight. Many different situations may cause safety of flight concerns: a pilot not in a safe position to land, being either too high or too low relative to the runway; a pilot with excessive airspeed on final or approaching a stall (airspeed too low); a pilot seeing another aircraft or vehicle on the landing runway; or possibly, the pilot loses sight of the runway due to environmental conditions such as fog.

Pilots must have a sufficient amount of self-confidence to fly. This self-confidence, however, may prevent pilots from admitting that the aircraft is in a situation that may be beyond their capability to safely recover. The aviation community has come a long way in terms of improving cockpit decision-making. Many pilots, however, have successfully "salvaged" an approach (found a way to not have to go-around and landed the airplane). This history of success, combined with ego-driven decisions, makes it difficult for some pilots to "admit defeat" and perform the go-around—it may be seen as a sign of an inferior pilot. Of course, that is not the case; environmental conditions may arise that make a safe approach near impossible. Windshear conditions, visibility limitations, or runway hazards often make go-arounds a simple decision.

In summary, the runway, airspeed and power, horizon, aimpoint, and field of view references are used to maintain glide-path during the approach and landing. The paradox is that the visual approach-and-landing cues taught and used by pilots are those that are consciously accessed and described (runway size, shape, and perspective) but in actuality they may not always serve as the most useful. In reality, the ambient cues may provide the most functionally invariant sources of information and they are processed with little cognitive effort. Thus, the runway environment, the terrain, the surrounding man-made or natural features that define the location, orientation, and position of the runway within the airport's environment are not specifically instructed landing cues, but they may prove to be the most important.

Low Level Flying

Low level flying is a thrilling but unforgiving aviation activity. The element of risk is greatly increased due to the close proximity to the ground and other hazards (birds, obstacles, other aircraft) that exist within the low-level environment. In low level flying the focus of expansion is the point on the terrain where a pilot does not want to collide. While maneuvering, the expansion point communicates to the pilot the possible point of terrain impact if the pilot fails to make a correction. This simple, yet valuable piece of information regarding the low altitude is taught to young and experienced pilots alike.

Haber and Haber (2003) described military low-altitude flight as a type of flying that, "comprises a constellation of tasks that represent the very edge of human perceptual and attentional capabilities" (p. 21). Flach and Warren (1995) described low-altitude flight as, "the unforgiving character of the dynamic ecology that imparts an element of ever present, if not always clear, danger to each moment of action or inaction and makes prescience desirable" (p. 66). Throughout this chapter and others, the topic of low-level or low-altitude flight is alluded to in demonstrating visual-perceptual limitations.

Compared with higher-altitude flight, flying low to the ground requires the additional task of terrain clearance. Regardless of the current crew activity internal to the cockpit and stimuli external to the aircraft, the number one priority must be vertical awareness with the terrain, not only at the present, but along a future vector. The future aspect of vertical awareness involves the aircraft's flight path if left unaltered. If pilots are distracted and momentarily drop vertical awareness from their cross-check, where will the aircraft be? This is a question the pilot must continuously assess and answer. The mental-processing capacity metaphor that is commonly used in aviation is a "bucket." Only so many tasks can be accomplished and/or handled before the pilot's bucket becomes full. Once full, something is not going to be accomplished or handled properly. The aviation bucket of cognition, or conscious working memory, directly relates to the pilot's ability to juggle the limited resources of attention.

Haber and Haber (2003) shared the term "time to die" in their writing and described it in terms of particular flight maneuvers and the resultant time remaining, if no action is taken, prior to ground impact. This is an operational aviation phrase, (i.e., pilot term), that certainly doesn't hold back the consequence of failing to act. It is driven by G-forces, altitude above the terrain, and velocity vector (speed and flight path). For instance, if a pilot enters a level 5-G turn at 30 m (100 ft) above the ground, the pilot can maintain a level turn at those parameters if a bank angle of 75 degrees is maintained; however, if the pilot overbanks by 5 degrees (to 80 degrees of bank) the nose will slice toward the ground if the G is not increased. In this case, the time to die is 3.7 seconds. According to the experts in the flying unit, 2.5 seconds is provided for pilot reaction time (Haber and Haber). Consequently, 1.2 seconds is available for the pilot to sense the problem … if not perceived within that time frame, ground impact is unavoidable.

Low-level turn accidents are deadly. Due to the minimal time to react, pilots rarely have the time to eject. It has been reported that turning maneuvers only encompasses 1 percent of a US Air Force pilots' mission time while flying low levels, but unfortunately account for 6 percent of Class A mishaps and they are fatal every time (Lyons, Gillingham, Thomae, and Albery, 1990). The cause of low-level turn mishaps can be due to many factors, such as cognitive-task saturation resulting from mission-oriented tasks competing with aircraft-control tasks. Lyons et al. stated that these types of accidents come down to two major factors: one, the physics and aerodynamics of high-speed low-altitude flight; and two, the "frailty of our orientation senses which predisposes us to develop undetected overbanks and descent rates" (p. 3). In other words, aviation provides sensation and perception dilemmas with which our body is unable to accurately cope.

Another dangerous aspect of low-level flying is ridge crossings. While flying low, often a pilot needs to maneuver over and around a ridgeline or some higher obstacle. The problem with flying too high over the ridgeline or obstacle is that the aircraft may momentarily be exposed to an enemy or, sometimes—and just as dangerously—it may fly into instrument-required weather conditions. Consequently, aggressive maneuvering is required. This aggressive maneuvering momentarily places the aircraft's vector below the horizon, toward the ground. Thus, in terms of global optic flow, the point of expansion is a point on the ground. Recall that aviation is a visually guided behavior in which impact on the runway is desired while impact with other terrain spots is a tragic accident. Thus, pilots fly a thin line in terms of hitting what they want to hit—the runway—and not hitting what they want to avoid—the terrain.

Especially when flying low, any misperception or any delay in aircraft-control input results in minimal if not unavoidable time to die. Following an examination of one of the low-altitude accidents, Haber (1987) created a list of five characteristics of terrain that a pilot uses to formulate terrain clearance during low-altitude flight. Those five were: (1) terrain's texture, (2) irregularity of ground elevation, (3) visible surface detail, (4) known size references, and (5) pilot's familiarity. Especially important is the fourth characteristic, known size references. These references ensure that size-distance constancy can be kept reliable for retinal image interpretation by helping a pilot estimate height and distance. For example, Figure 4.9 shows an aircraft of known size waiting to take off; another pilot looking at this scene would immediately have a good sense of how far away he was from the other aircraft and the other objects in the surrounding runway scene. In terms of flying a low-level sortie, any object of known dimensions along the route of flight will help a pilot "calibrate his eyeballs" regarding his or her current altitude above the ground. Finally, worth noting is the fifth characteristic, pilot's familiarity with the terrain and route of flight. Because cognition plays a strong role in visual perception, any route-planning and map-study can greatly assist the pilot with interpretation of the environment's visual cues.

Haber and Haber (2003) further studied the perceptual and attentional factors and consolidated their findings into four categories of terrain visual information that are especially useful during low-altitude flight:

1. *Terrain contour irregularity*: The irregularity of terrain geography is one of the main reasons why optical concepts presented earlier such as edge

rate fail to provide adequate perceptual cues. Terrain contour irregularity can make it difficult for pilot to predict altitude changes and maintain a constant altitude clearance over the terrain; it will be addressed again in Chapter 6.

2. *Terrain texture*: this is also a highly variable environmental cue based upon the ground cover across the ground that the pilot is flying over.
3. *Linear perspective*: the dominant visual cue for aviation depth perception, it is dependent upon sufficient environmental cues to provide the necessary depth perspective.
4. *Resolution of fine details of objects on the ground*: Calvert (1950, 1954) presented concepts for the use of global and local cues to altitude. For instance, when flying over a field of cows, if the pilot can easily distinguish the cows' four legs then the aircraft is below 152 m (500 ft). Focal details can later serve as ambient cues when they pass and leave the pilot's field of view.

According to Haber and Haber (2003), global optic flow is usually the most important source of visual information for maintaining altitude and ground clearance. They stated that, "patterns of optic flow at the retinas geometrically reflect properties of the terrain, both its variation in contour and its distance from the observer" (p. 46). They discussed different hypotheses regarding how ambient cues result in visually guided behavior and concluded that, however the processing occurs, it is automatic and highly informative. The authors presented limitations for efficacious perception of global optic flow for terrain clearance. One limitation included pilot physiology and G-forces restricting peripheral vision. Another optic flow perception issue concerns optic flow rates changing either due to terrain texture or aircraft ground speed. And finally, optic flow rates may be perceptually influenced by the pilot's prior experience and perceptual adaptation.

Haber and Haber (2003) summarized the pilot's perceptual and cognitive limitations while flying in the low level environment:

> Pilots cannot depend on their automatic perceptual processes during low-altitude flight because misperceptions are likely to occur. As a result, the pilot must engage in three processing steps, all of which require focused attention. First, he has to consciously override automatic processes when they potentially provide him with incorrect information; second, he has to consciously remember to refer to his instrument or other sources for that information; and third, he has to process the alternative sources of information, using focused attention. (p. 46)

As a pilot, the first author can attest to the cognitive load it takes to disregard perceptual information, seek-out and process displayed aircraft status information, and then generate a mental model of the situation. Referring back to the "bucket"—the bucket quickly gets full in the above-mentioned scenario.

The key for a pilot to safely fly low-level sorties is to have a constant cross-check of the aircraft's altimeter and recalibrate his or her eyes, given the terrain, to that elevation above the ground. Pre-flight map studies can inform the pilot of any drastic terrain elevation changes during the course of the flight. Of course, the best instrument for a pilot is a radar altimeter. This device provides the pilot with instantaneous height information, eliminating the need for mental computation of the difference between current terrain elevation and sea-level elevation. Also, the radar altimeter provides a more accurate value for pilots to calibrate their eyeballs to during environmental perception.

Visual Integration and Cognitive Challenges

The topic of cognitive attentional demands in aviation is worthy of its own book; here it needs mentioning in order for the reader to appreciate one more aspect of aviation and the pilot's visual requirements. While there are many possible scenarios that might lead to an increase in cognitive load, two occur frequently in non-emergency situations: instrument flying and the transition from instrument flying to visual flying (or vice-versa). In both cases, accurate perception is crucial, but the lack of ambient cues or the shift in perspective can lead to misperceptions and error.

Reading instrument displays and aircraft status information is completely different from flying visually using environmental perception, where much is done via unconscious ambient processing with a low cognitive footprint, (although focal visual perception is still required for internal cockpit perception). While flying an approach in impoverished visual conditions, it is extremely difficult to perceive environmental cues accurately. It might seem that cognition and attention allocation would be less taxed because of the reduction of information. However, while visual perception of internal cockpit displays is perceptually more accurate, it is also cognitively more challenging. The pilot no longer is making probabilistic and intuitive perceptual visually guided control inputs, but he is making rather deterministic and accurate decisions because of the detail provided by the instrumentation. There is no impoverished viewing environment within the cockpit, but a pilot has high attentional allocation demand for the visual processing of each data source viewed with focal vision, all at the conscious level. A pilot must individually attend to, process, assimilate, and synthesize the aircraft's navigational and system data. Similar to what occurs when flying visually, all these incoming data sources must be compared with other short-term and long-term memories for the creation, maintenance, and projection of situational awareness—a highly cognitive effort.

Cognitive challenges in the interpretation of perceptual information also arise when pilots flying an instrument approach reach their decision point and they must transition to a visual landing using visual references. At this point, ambient visual information must be attended and rapidly integrated into the pilot's situational awareness. The decision to land or possibly go-around must be made without hesitation at this point during the approach because the aircraft is close to terrain with the power at lower settings. This transition from instruments flying to a visual landing is ripe for perceptual disorientation due to the abrupt change in attentional demands.

The next chapter presents many of the same aspects of aviation visual perception just shared in this chapter but couched in a human-factors research perspective. Hopefully the examples given in this chapter from a pilot's perspective will help you appreciate the need for the systematic human factors research that is presented there.

References

Beall, A. C. and Loomis, J. M. (1997). Optic flow and visual analysis of the base-to-final turn. *The International Journal of Aviation Psychology*, *7*, 2012–23.

Benson, A. J. (1999). Spatial disorientation—common illusions. In J. Ernsting, A. N. Nicholson, and D. J. Rainford (Eds.), *Aviation Medicine* (3rd edn, pp. 4374–54). Oxford: Butterworth-Heinemann.

Bressey, P. E. (1976). Pilot's view. In R. Hurst (Ed.), *Pilot Error: A Professional Studyof Contributory Factors* (pp. 136–4). London: Crosby Lockwood Staples.

Calvert, E. S. (1950). Visual aids for landing in bad visibility with particular reference to the transition from instrument to visual flight. *Transactions of the Illumination Engineering Society of London*, *6*, 1832–19.

Calvert, E. S. (1954). Visual Judgments in Motion. *Journal of the Institute of Navigation*, *27*, 2332–51.

Flach, J. M. and Warren, R. (1995). Low-altitude flight. In P. Hancock, J. Flach, J. Caird, and K. Vicente (Eds.), *Local Applications of the Ecological approach to Human-Machine Systems* (pp. 651–03). Hillsdale, NJ: Lawrence Erlbaum Associates.

Galanis, G., Jennings, A., and Beckett, P. (2001). Runway width effects in the visual approach to landing. *The International Journal of Aviation Psychology*, *11*, 281–301.

Gibson, J. J. (1950a). *Perception of the Visual World*. Westport, CT: Greenwood Press.

Gibson, J. J. (1979). *The Ecological Approach to Visual Perception*. Boston: Houghton Mifflin Company.

Gibson, J. J., Olum, P., and Rosenblatt, F. (1955). Parallax and perspective during aircraft landings. *American Journal of Psychology*, *68*, 3723–85.

Haber, R. N. (1981). Toward a theory of the perceived spatial layout of scenes. *Computer Vision, Graphics, and Image Processing*, *31*, 2823–21.

Haber, R. N. (1987). Why low-flying fighter planes crash: Perceptual and attentional factors in collisions with the ground. *Human Factors*, *29*, 5195–32.

Haber, R. N. and Haber, L. (2003). Perception and attention during low-altitude high-speed flight. In P. S. Tsang and M. A. Vidulich (Eds.), *Principles and Practice of Aviation Psychology* (pp. 216–218). London: Lawrence Erlbaum Associates.

Hasbrook, A. H. (1971). Anatomy of a landing cue by cue. *Business Commercial Aviation*, *29*, 546–550.

Hodgson, D. A. (1971). Pilot vision during final approach-and-landing in turbojet transport operations. *Aerospace Medicine*, *42*, 2052–08.

Khatwa, R. and Helmreich, R. L. (1998). Analysis of critical factors during approach and landing in accidents and normal flight: Data acquisition and analysis working group final report. *Flight SafetyDigest, Killers in Aviation: FSF task force presents facts about approach-and-landing and controlled-flight-into-terrain accidents*, 17–7.

Langewiesche, W. (1944). *Stick and Rudder: An explanation of the art of flying*. New York: McGraw-Hill.

Leibowitz, H. W. (1988). The human senses in flight. In E. L. Wiener and D. C. Nagel (Eds.), *Human Factors in Aviation* (pp. 831–10). New York: Academic Press.

Lintern, G. (2000). An analysis of slant for guidance of landing approaches. *The International Journal of Aviation Psychology*, *10*, 3633–76.

Lintern, G. and Koonce, J. M. (1991). Display magnification for simulated landing approaches. *The International Journal of Aviation Psychology*, *1*, 597–2.

Lintern, G. and Liu, Y. (1991). Explicit and implicit horizons for simulated landing approaches. *Human Factors*, *33*, 4014–17.

Lyons, T., Gillignham, K., Thomai, C., and Albery, W. (1990). Low-level turning and looking mishaps. *Flying Safety*, 1–5.

Mertens, H. W. (1979). Runway image shape as a cue for judgment of approach angle. Office of Aviation Medicine, Federal Aviation Administration FAA-AM-792–5. Retrieved November 17, 2006 from www.faa.gov/library/reports/medical/oamtechreports/1970s/1979.

Pitts, D. G. (1967). Visual Illusions and Aircraft Accidents. USAF School of Aerospace Medicine Division, Brooks AFB, TX, Report No. SAM-TR-672–8.

Previc, F. H. (2004). Visual orientation mechanisms. In F. H. Previc and W. R. Ercoline (Eds.), *Spatial Disorientation in Aviation, Progress in Astronautics and Aeronautics*, *Vol. 203* (pp. 941–44). Reston, VA: AIAA.

Riordan, R. H. (1974). Monocular visual cues and space perception during the approach to landing. *Aerospace Medicine*, *45*, 7667–71.

Smallman, H. S. and St. John, M. (2005). Naïve realism: Misplaced faith in realistic displays. *Ergonomics in Design*, Summer, 61–3.

Smith, P. A. (1999). Vision in aviation. In J. Ernsting, A. N. Nicholson, and D. J. Rainford (Eds.), *Aviation Medicine* (3rd edn, pp. 4724–88). Oxford: Butterworth-Heinemann.

Tarnowski, E. (2007). From nonprecision to precision like approaches. *Aerosafety World*, 122–1.

Turner, T. P. (2007). Stabilizes approaches: Whether flying the instrument procedure or flying a VFR traffic pattern, a good landing is much more likely after flying a stabilized approach. *Aviation Safety*.

Warren, R. (1988). Visual perception in high-speed low-altitude flight. *Aviation, Space, and Environmental Medicine*, *59*, A1161–24.

Chapter 5
Research on Cues Used for Visual Flying

Of all the skills pilots perform there are two that stand out as particularly demanding and dangerous: landing and low-level flight. In commercial aviation, crashes during the final approach and landing account for more than 33 percent of the total number of accidents and 22 percent of onboard fatalities even though this phase of flight accounts for only 4 percent of the total flight time (Boeing, 2008). Military aviation involves "nap of the earth" flight where the pilot is required to fly as close to the ground as possible between buildings, trees, hills and other terrain features. The high level of danger involved in low-level flight is illustrated by the fact that 55 helicopter fatalities during the Vietnam War were caused by collisions with one particular power line (Marsh, 1985); the only one in all of South Vietnam. Not coincidently, these skills are also the aspects of flying that most heavily rely on visual processing because they involve a relatively close proximity to the ground.

In Chapter 4 we discussed how tasks are performed from a pilot's perspective and introduced some theories about the visual processing involved. The emphasis of Chapter 4 was to demonstrate how pilots operationally maneuver their aircraft based on environmental perception of the visual scene. In this chapter we delve more in depth into the visual cues that can support landing and low-altitude flight (along with some other flight tasks) and review human factors research in this area. Thus, although many of the same terms, concepts, and research are referenced, the focus of the present chapter is from a visual perception research perspective, not the "pilot-speak" perspective that framed Chapter 4.

Visual-motor Control in Approach and Landing

The approach and landing phase of flight can be broken down into three component subtasks: (1) aligning the aircraft with the runway, (2) controlling altitude and airspeed appropriately to contact the end of the runway (the aimpoint), and (3) arresting the descent with a "landing flare". Note in Chapter 4 six aspects of an approach and landing were presented and couched in a pilot's perspective. Because this chapter is concentrated on visual perception research, the research fell into three categories and hence was presented as such. We next consider each of these three tasks in detail.

Aligning the Aircraft with the Runway

Beall and Loomis (1997), introduced in the previous chapter, analyzed the visual information available to the pilot for performing aircraft alignment, commonly called the "base-to-final turn" or sometimes called an "overhead pattern." Figure 4.2 from Chapter 4 depicted and contrasted the base-to-final turn from a visual straight-in approach. When flying a visual straight-in approach, runway alignment is the least difficult aspect for an inexperienced pilot to conquer; whereas, the base-to-final turn maneuver is more perceptually challenging to learn. Initially traveling in a direction that is oblique or perpendicular to the orientation of the runway (the base), the pilot must initiate a turn at the appropriate time and with the necessary turn rate such that at the completion of the turn the direction of travel becomes aligned with the runway (the final). This task becomes particularly difficult for a short final approach. When the base-to-final turn is initiated a short distance from the runway, pilots often overshoot the turn and are forced into making a sudden corrective maneuver. This can lead to an aerodynamic stall resulting in a crash. The base-to-final turn is made even more difficult by the sluggish controls in a typical fixed-wing aircraft: turn rates rarely exceed 10 deg/sec and a 90 deg turn takes upwards of 30 seconds to complete.

What visual information can the pilot use to decide when to initiate the turn and to regulate the turn rate? Calvert (1954) first proposed the idea that the optical *splay angle* (σ) could be used to control this maneuver. As illustrated in Figure 5.1, the splay angle is the angle formed by the centerline of the runway and the vertical at the convergence point on the horizon.

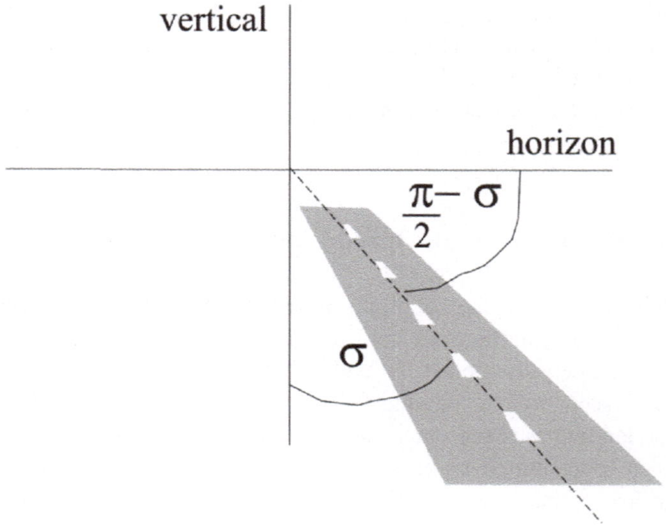

Figure 5.1 Splay angle

If the horizon is visible, splay angle is also equal to $\pi/2$ minus the angle formed by the runway centerline and the horizon. A pilot can use the splay angle to turn into alignment with the runway by turning so that the rate of change of the splay angle ($d\sigma/dt$) is held constant (Beall and Loomis, 1997). Presumably, the initiation of the turn begins when the splay rate exceeds some threshold above this critical value. The main evidence supporting this splay angle hypothesis comes from the work of Beall and Loomis (1997), in which optical variables during real landings were analyzed. Flight data showed that over several landings pilots appeared to hold the splay rate roughly constant during the base to final turn. It would be interesting for future research to examine this task in more detail in a flight simulator, for example by introducing online perturbations in the splay angle by perturbing the position of the simulated runway.

In support of this field research, psychophysical experiments have provided evidence for a neural mechanism that would be sensitive to splay angle, providing the horizon is visible (Gray and Regan, 1996; Regan, Gray, and Hamstra, 1996). This research found that observers could estimate the absolute angle formed by the intersection of two lines and discriminate changes in this angle independently of the orientation each of the two lines. However, it should be noted that these experiments used angles that were presented in the fronto-parallel plane whereas the splay angle lies on the ground plane.

One important implication of this splay rate control hypothesis is that it is not necessary for the pilot to generate a full 3D reconstruction of the world to control the turn. Splay angle can be estimated without any knowledge about the distance to the runway or the aircraft's velocity. Consistent with this analysis, Beall and Loomis (1997) reported that base-to-final turn maneuvers were similar for day and night landings. In the latter case ground texture cannot be used to estimate distance and speed.

From an applied research perspective this finding has some profound implications, namely that the vast majority of research on judgments of distance, speed of self-motion and spatial layout may have limited relevance to understanding the control of visual-motor action. However, the fact that these variables are not necessary for control does not mean that we do not use them when they are available. This issue we will be considered in more detail below.

Controlling Altitude and Glide-slope

As introduced in the previous chapter and emphasized throughout the course of this book, it is a perceptual challenge to accurately perceive height and distance in the aviation visual scene. Figure 4.1 from Chapter 4 depicted a glide-path of 3 degrees, showing specific altitudes above the ground at specific distances from the runway. To do this using only environmental perception is difficult but not impossible. What follows is visual perception research that attempts to untangle the riddle that is aviation visual perception during an approach and landing; as

Schwirzke and Bennett (1991) stated, "what pilots use as cues for landings is still a mystery" (p. 575).

Once aligned with the runway, the pilot next needs to reduce speed and altitude appropriately so that the plane will be in position to contact the ground near the start of the runway (the actual contact is controlled in the landing flare stage discussed next). What visual information could the pilot use to judge whether the current descent rate is sufficient? Field observations and some very clever simulation research conducted by a Boeing engineer Conrad Kraft suggests that the rate of descent is primarily controlled on the basis of perceived altitude (Kraft, 1978). When the Boeing 727 was first introduced into commercial aviation in the late 1960s it was involved in a large number of landing accidents. Kraft's accident analyses revealed that many of these crashes involved landing short of the runway during a night approach over water or other featureless terrain (commonly called a "black-hole" approach). As discussed in detail in Chapter 6, Kraft hypothesized that these crashes resulted from a misperception of altitude due to insufficient visual information.

But wouldn't the altimeter gauge allow the pilot to judge the altitude accurately? The answer is yes; however, pilots rarely consult the altimeter during landing. Kraft's accident analysis concluded that there was no reason to suspect mechanical failure in these crashes. In fact, the grim reality is that 'human error' is the suspected cause of the vast majority (>75 percent) of aviation accidents. This large percentage is not due to negligence on the part of the pilot; workload is very high during this phase of flight involving communicating with ground control, monitoring power settings, and so on. When faced with these multiple demands, pilots typically choose to monitor the movement of the plane using their own senses; after all visual perception is a highly developed, effortless process that serves us well 99.9 percent of the time. This observation highlights the importance of understanding the realities of visual information available for the control of flight even though the same information is also provided indirectly by the aircraft's instruments.

Why do pilots overestimate altitude and descend too quickly during "black-hole" landings? There are multiple sources of visual information that can be used to estimate the rate of change of altitude as illustrated in Figure 5.2. This specific visual illusion is discussed in detail as well in Chapter 6, but more from a pilot's perspective of environmental perception. One source of information comes from the angle formed at the eye by the horizon and an object or edge that is oriented perpendicular to the path of travel as depicted in Figure 5.2A. This angle, called the *depression angle* (δ), can be used to estimate the rate of change of altitude (dY/dt). For small values of δ:

$$\frac{dY}{dt} \approx \left(\frac{d\delta}{dt} - \frac{dZ_g}{dt} * \frac{\delta}{Y} \right) \frac{1}{\delta^2} \qquad [1]$$

where Z_g is the distance between the pilot's eye and the object/edge on the ground (Flach, Warren, Garness, Kelly, and Stanard, 1997). The second source of

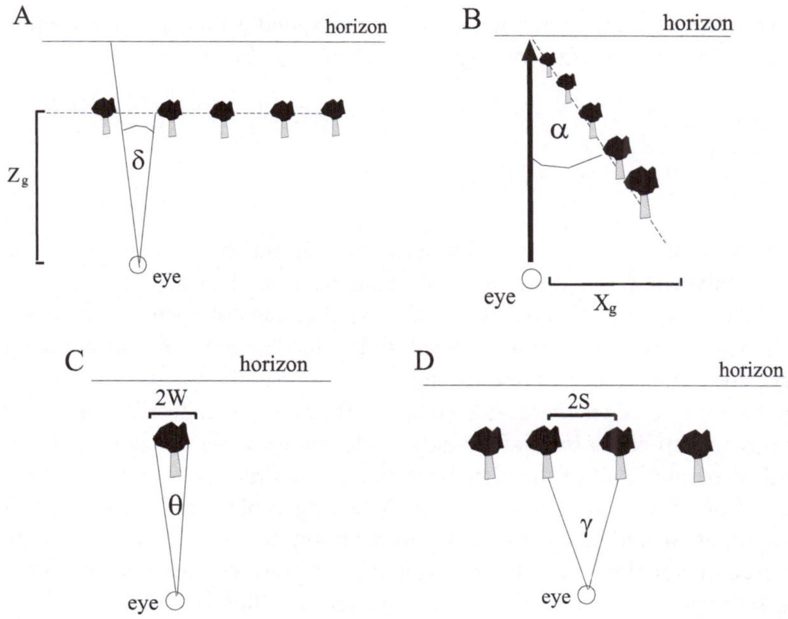

Figure 5.2 Altitude cues during landing

information is based on the angle formed by the motion path and an object or edge that is oriented *parallel to the path of travel* [called the *altitude splay angle* (α) and depicted in Figure 5.2B]. For small values of α:

$$\frac{dY}{dt} \approx \left(\frac{d\alpha}{dt} - \frac{dX_g}{dt} * \frac{\alpha}{Y} \right) \frac{1}{\alpha^2} \qquad [2]$$

where X_g is the lateral separation between the object on the ground and the axis perpendicular to the pilot's eye (Flach et al., 1997). Objects that could provide altitude splay and depression angle information include railways, roads, a grove of trees, a river or the runway itself. The third source of visual information that could be used is the *angular size of a familiar object* on the ground such as a building, tree or the runway itself. For small values of θ:

$$\frac{dY}{dt} \approx \frac{d\theta}{dt} \frac{Y}{W} \qquad [3]$$

where the physical width of the object on the ground is 2W and θ is angular size of the object's retinal image. This information source is illustrated in Figure 5.2C.

Finally, if the physical layout of objects on the ground is known (e.g., when flying over a city) the *texture density* (Figure 5.2D) can also be used as:

$$\frac{dY}{dt} \approx \frac{d\gamma}{dt}\frac{Y}{S} \qquad [4]$$

where 2S is the physical separation between adjacent texture elements (e.g., the spacing between rows of trees) and γ is the angular size of the gap between adjacent texture elements. Equations [1–4] are all scaled in units of eyeheights/sec that is,, they indicate the number of times the altitude will change by the current height of the eye above the ground in one second.

Before we consider these information sources in more detail, it is now clear why the mishap pilots in Kraft's study could not estimate the rate of change of altitude accurately: all of the visual correlates of altitude rely on the presence of visible terrain features during the approach. During a "black-hole" landing only the runway lights would be clearly visible to the pilot; therefore only the information expressed in equation [3] would be available. As presented later in Chapter 6, the black-hole approach is a form of "featureless terrain" illusion. Furthermore, for the initial part of the descent [i.e., at a distance of 32 km (20 miles) from the runway] the angle formed by the lights on a 40 m (131 ft) wide runway would be a mere 0.042 radians and its rate of change would be 0.108 rad/sec for an 100 m/sec (220 mph) approach speed. Because it has been shown that observers cannot reliably estimate time to contact (Gray and Regan, 1998) or approach speed (Hoffmann and Mortimer, 1996) for such low rates of expansion, it is unlikely the rate of change of the angle formed by the runway could be used to reliably estimate the rate of change of altitude; however, this has not been empirically tested.

A closer look at these information sources provides some important insights into perceptual-motor control during flying. A major problem associated with using depression and altitude splay angle cues is that the estimate of the rate of descent based on depression angle will be altered by changes in the speed of forward motion (dZ_g/dt) and the estimate of the rate of descent based on altitude splay angle will be altered by changes in the speed of lateral motion (dX_g/dt). The main problems associated with using familiar size and texture density are that they require that the physical layout of the environment remains constant. For example, if a pilot is using the angular size of the runway to estimate altitude and he/she assumes the width of the runway (i.e., S in equation [3]) is 60 m (196 ft), altitude will be dangerously overestimated when landing at an unfamiliar runway that is only 40 m (131 ft) wide. This particular estimation error has been identified as a cause of several crashes involving novice pilots (Galanis, Jennings, and Beckett, 2001; Mertens and Lewis, 1982). This topic of over-estimating a visual glide-path to landing is again addressed in Chapter 6.

Experimental research on the perception of altitude has primarily used two experimental tasks: (1) altitude maintenance and (2) judgments of the direction of altitude change following an occlusion period. Flach and colleagues (Flach, Hagen, and Larish, 1992; Flach et al., 1997) studied the relative contribution of depression and altitude splay angles in a simulated altitude maintenance task. Participants were required to track a constant altitude in the presence of simulated fore-aft, up-down and right-left wind disturbances. As predicted by the cross-talk between dZ_g/dt and dY/dt in equation [1], root mean square (RMS) error was significantly higher when flying at high speed over a terrain with only depression angle cues (lines perpendicular to the direction of motion) than when flying over a terrain with only altitude splay angle cues (lines parallel to the direction of motion). Somewhat surprisingly they also found that RMS errors were higher for a grid terrain (which contains both cues) than for the parallel line terrain. Flach and colleagues proposed that this was due to the fact that the addition of perpendicular lines in the grid terrain introduces noise into the perception of altitude.

Kleiss and Hubbard (1993) used an occlusion technique to investigate the importance of ground terrain and the presence of 3D objects on altitude judgments. After flying at a constant altitude for 20 seconds, the visual display was blanked for 3 sec (mimicking what would occur if a pilot flew through a bank of clouds). When the display reappeared participants judged whether the perceived altitude had increased, stayed the same or decreased. Randomly positioned trees were used as 3D objects and terrain texture patterns were random noise. This randomization substantially limits the effectiveness of depression and altitude splay angle information. Judgment accuracy improved as tree density increased. Texture density was also positively related to response accuracy; however, performance at high terrain texture densities was not as good as performance at high tree densities. These findings suggest that the information sources expressed in equations [3] and [4] may be important for control of altitude.

Unfortunately, to our knowledge there has not been any research that has systematically manipulated all of the cues to altitude within the same set of experiments. Due to differences in the experimental tasks, display parameters and subject populations used in the studies described above it is difficult to compare the relative contributions of the information sources in equations [1]–[4]. In addition, the experimental tasks used in previous research have only limited relevance to controlling the descent during landing. The tasks used do not address the problem of knowing whether the descent rate is appropriate to land safely. Clearly future research is needed in which observers are required to make simulated landings over terrains for which the optical variables expressed in equations [1]–[4] are systematically and independently varied.

Mertens and Lewis (1982, 1983) researched the black-hole illusion by manipulating runway shape and approach lighting systems during night approaches. They found that higher runway ratios (length divided by width) produced more shallow glide-paths and the addition of an approach lighting system increased a runway's ratio and further induced lower glide-paths. The length/width ratio rather than the length or the width alone was determined to be the controlling variable. Lintern and Walker (1991) examined scene content and runway width in simulator landings, assessing pilot performance from 3.0–0.7 km from the runway. Scene content was qualified (not quantified) as reduced or normal and as a factor it was significant with more shallow glide-paths flown in the reduced content scenes. The scene content and runway width interaction was not significant, however, and may have been confounded by the presence of the horizon in all conditions.

Lintern and Liu (1991) manipulated the horizon and its influence on simulated visual approaches to landing. They found a high implicit horizon induced shallow glide-path performance and that perspective lines weakened the bias of the high horizon and produced more stable performance. Palmisano and Gillam (2005) explored the accuracy of visual touchdown perception in a passive task of a simulated approach to landing on a low-ratio runway. They found that adding an explicit horizon improved precision and reduced judgment bias, and that by adding landing surface orientation cues glide-path performance improved.

Mayer, Mershon, Lim, Chipley, and McAllister (2006) investigated visual factors that influenced pilots when performing off-airport emergency landings. They manipulated visual scenes and quantified terrain objects with levels of 0, 10 trees, or 21 trees/buildings. Mayer, et al. found that judgments made during zero-bank visual straight-in approaches resulted in misperception of landing prior to the emergency landing field, similar to a landing short black-hole approach. Other studies have examined the effect of distance and weighting of certain visual cues (Galanis, et al., 1998; Lintern and Walker, 1991) and mathematical models have attempted to quantify glide-path performance by pilots incorporating aspects of scene detail (Galanis, et al., 1998; Lintern, 2000).

A theoretical paper by Gibb and Gray (2006) outlined a possible theory of aviation visual perception based on terrain orientation as the most salient aspect of a visual approach. Mertens (1981) had recommended research into "extra-runway cues" for visual perception guidance and Perrone (1984) expressed the need to quantify "adequate textural information." Pilots have demonstrated a preference for visual approaches and, consequently, adding more internal cockpit displays would be of little value when the runway is in clear view. Thus, the alternative is to enhance the runway environment to improve perception. It was hypothesized that if random terrain objects could help pilot perception of the runway surface orientation, then possibly a reconfigured approach lighting system could improve perception of the runway surface as well as lower the runway ratio.

Gibb, Schvaneveldt and Gray (2008) investigated the cues used for glide-path control in a flight simulator. Part of the goal of this study was to test the theory proposed by Gibb and Gray (2006). Twenty pilots flew simulated approaches under various combinations of visual cues. Performance was assessed relative to the desired 3-degree glide-path in terms of precision, bias, and stability between 8.3 and .09 km (4.5–0.5 nm). Overall, the results pointed to glide-path over-estimation being influenced by runway ratio, random terrain objects, and approach lighting systems. It was difficult for the authors to declare one particular visual cue as most salient because the pilots perceived the cues differently at different distances from the runway.

In the Gibb et al (2008) study, the terrain orientation theory proposed by Gibb and Gray (2006) was tested and it was hypothesized that higher densities would allow the pilot to judge the runway orientation more accurately. This was not the case, as glide-path performance was degraded (below desired 3 degrees) with added terrain objects. It was concluded that the number of terrain objects used in the study, zero, five, or ten, was insufficient and future research ought to use significantly more random terrain objects to help define the runway's ground plane.

Glide-path performance was improved by reconfiguring runway approach lights systematically on the sides of the landing runway. Approach lighting systems used at airports have the runway lights prior to the beginning of the runway, as shown in Chapter 4, Figure 4.9. The perceptual success for more accurate glide-path performance could be attributed to lowered runway ratio, (width becoming perceptually wider due to the lights) or improved runway surface orientation or enhancing linear perspective.

Gibb et al.'s (2008) study also replicated the work of Mertens and Lewis (1983) in terms of an approach lighting system increasing the perceived runway ratio and inducing glide-path over-estimation. Finally, their study of pilot glide-path performance demonstrated that over the 18 trials, average performance did improve. In other words, after the first two approaches (which averaged 36 m and 20 m below glide-path) the average improved to 5 m above glide-path. This shift in performance demonstrates the possibility of experiential cognition overcoming the visual perception bias.

The Landing Flare

The final stage in landing involves a transition from this controlled descent to contact with the runway in the landing flare. The flare was also presented in the previous chapter. The typical vertical speed during the final stages of the approach [roughly 3–5 m/sec (10–16 ft/sec)] is much too fast for a comfortable and safe landing. The purpose of a flare maneuver is to reduce the vertical speed to an acceptable level just before touchdown. The flare is initiated at an altitude of roughly 3–6m (10–20 ft) by pulling back on the control stick, causing an increase in the angle between the direction of motion and the orientation of the nose of the plane (the 'angle of attack'). Precise timing of this maneuver is critical because a flare initiated too late will not reduce vertical speed sufficiently before contact, and a flare initiated too early can cause the plane to level off or even climb away from the runway.

There are two primary control strategies a pilot could use in during the flare (Mulder, Pleijsant, van der Vaart, and van Wieringen, 2000). First, the flare could be initiated at a constant critical altitude. In practice this could be achieved by initiating the flare when the retinal image of the runway reaches a critical angular size. However, this strategy would not be as robust for variations in vertical speed (e.g., due to a down draft) and could be dangerous in situations where a pilot is landing at an unfamiliar runway that is wider or narrower than expected (see above). A more effective strategy is to initiate the landing flare at a constant value of the time to contact (TTC) with the runway. The visual information that supports judgments of TTC has been studied extensively (reviewed in Regan and Gray, 2000, 2001). The TTC information that could be used for timing the landing flare is illustrated in Figure 5.3. For simplicity, first consider the situation of an aircraft on a straight-in approach to a rectangular object that is oriented perpendicular to the direction of travel (e.g., flying towards a wall). In this scenario, illustrated in Figure 5.3A, the angular subtense of the object's horizontal meridian (θ_h) can be used to estimate TTC as:

$$TTC \approx \frac{\theta_h}{d\theta_h / dt} \qquad [5]$$

provided that approach velocity is constant and θ_h is small (Hoyle, 1957). Note that in this special case the values θ_h and $d\theta_h/dt$ will be equal along the entire vertical extent of the object because the object expands isotropically (i.e., constant shape). As will be discussed below, it has been demonstrated that observers can estimate absolute TTC with a high degree of accuracy in this situation (Gray and Regan, 1998).

Next consider the optical geometry associated with landing a plane that is, a straight line approach to a rectangular object but with a angle of approach that is considerably less than 90 deg. In this situation, illustrated in Figure 5.3B, equation [5] can still be used to estimate TTC; however, an accurate estimate requires

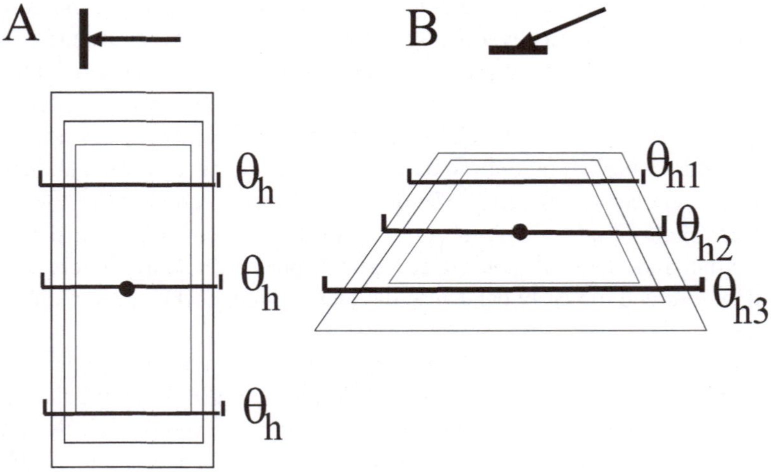

Figure 5.3 Retinal image expansion to estimate time to contact

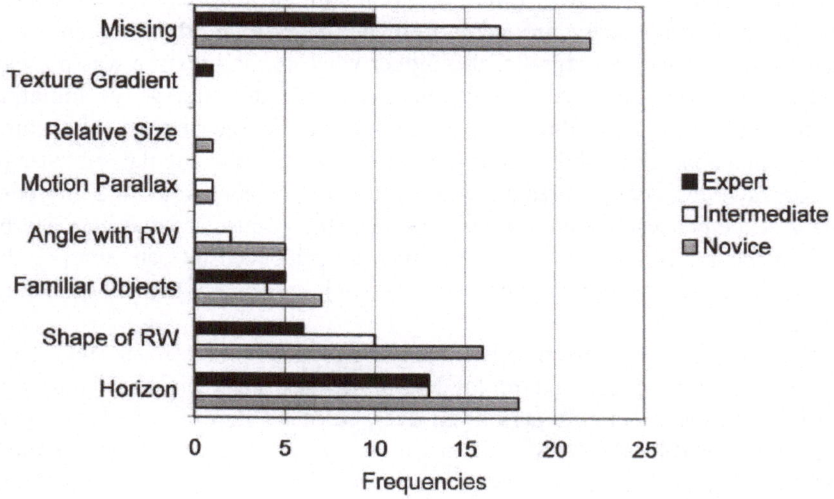

Figure 5.4 Pilot subjective ratings of visual cues used during landing flare

Source: Reprinted with permission from Benbassat, D. and Abramson, C. I. (2002).

that θ_h and $d\theta_h/dt$ be derived from points on the runway's vertical edges that are adjacent to the point of contact (shown as θ_{h2} in Figure 5.3B). Because the runway does not expand isotropically in this situation, TTC estimates based on portions of the runway that are further away than the point of contact (e.g., θ_{h1} in Figure 5.3B) will be overestimates of the actual TTC and TTC estimates based on portions of the runway that are closer than the point of contact (e.g., θ_{h3} in Figure 5.3B) and will be underestimates of the actual TTC. For example, for an approach speed of 60 ms/s and an approach angle of 4 deg, a TTC estimate based on portions of the runway that are 10 m (32 ft) further away than the point of contact will be approximately 290 ms longer than the actual TTC. Over this time period the plane will travel 17 m (55 ft)! More psychophysical research is needed to determine if observers can estimate TTC accurately during non-perpendicular approaches.

Recall from Chapter 4 the discussion on the flare and how pilots do not actually see where they are going to touch down on the runway. Thus, much of this TTC discussion in terms of aviation becomes a "blind TTC" in that once the pilot shifts from an aimpoint to begin the flare the aircraft blocks their view of the runway. Consequently, it becomes even more impressive how pilots have a "feel for the wheels" of their aircraft and know when they are touching down on the runway without seeing the spot on the runway.

Mulder et al., (2000) examined the timing of flare maneuvers in a flight simulator. In this study the movement of the simulated plane was not directly controlled by the observer; instead he/she was only required to press a button to initiate a pre-programmed flare maneuver. The main independent variables were the width of the runway (40 or 60 m) and the presence/absence of texture lines on the surface of the ground and runway. They found that the number of successful landings was significantly higher when ground texture was present. Under these conditions data were consistent with the strategy of initiating the flare at a constant value of TTC. Conversely, in the absence of texture, participants appeared to base the timing of the flare on the angular subtense of the runway. Mulder and colleagues argued that the presence of texture improves performance because it gives more edges near the point of contact that can be used to accurately estimate TTC. However, as evidenced by equation [4], the addition of texture would also serve to improve judgments of the rate of change of altitude.

Benbassat and Abramson (2002) analyzed NTSB records from 1995–1997 and used a pilot questionnaire to determine flare accident rates and their probable causes. Over this period, an average of 89 flare accidents per year was investigated by the NTSB. Not surprisingly, pilots rated the flare as more difficult than several other standard maneuvers including steep turns, takeoff roll, descending and taxiing. Consistent with this perceived difficulty, when asked to estimate the number of flare accidents rate per year, pilots' average estimate was more than double the actual rate. Over 85 percent of pilots reported that they predominately used vision to determine when to initiate the flare with the visual cues used shown in Figure 5.4. Horizon cues and the end of the runway were rated as the most important. These findings suggest that pilots are using a variety of cues to perform this task.

Visual-motor Control in Low Level Flight

Object Detection

During low level flight the pilot's most urgent task is to avoid colliding with objects on the ground. This task involves several different components including knowing the direction one is heading relative to the object and knowing the instant in time that contact will occur. However, the first thing the pilot must do is something we often take for granted, namely he/she must detect that there is an object there in the first place. If an object's retinal image does not differ from the retinal image of its surroundings it is invisible to the pilot and cannot be acted upon. A good example of this is power lines. Along with the accidents in Vietnam described above, it has been estimated that in the U.S. between 1970 and 1979, wire-strike accidents accounted for 208 civilian accidents. The majority of the accidents occurred with clear visibility; the pilot was simply not aware there was an object there to hit! Furthermore, this failure in detection is not restricted to very small objects as a large majority of aviation accidents involve the pilot flying a perfectly functioning plane directly into a ground feature or obstacle: so called "controlled flight into terrain" (CFIT).

What visual properties render an object visible to a pilot? A difference in the luminance contrast between an object and its surroundings can be used for detection; however, this cue will be greatly affected by the veiling glare produced by the sun (Regan, Giaschi, and Fresco, 1993; Regan, 1995). Difference in the textures between an object and its surroundings is another visibility cue that is critical for low-level flight (Regan, 1995). For example, a sloping hill covered with bushes can be distinguished from a grassy valley even though the mean luminance and color of the two areas will be roughly the same.

Finally, when mean luminance, color and texture are similar for different terrain features (e.g., when a grass covered hill is surrounded by grassy terrain), the pilot can use motion parallax to detect the presence of the feature. The retinal image of an object that is further away than the pilot's point of fixation will move in the same direction he/she is moving while the retinal image of an object that is closer than the fixation point will move in the opposite direction. As it turns out, there are large individual differences in sensitivity to motion and texture cues and susceptibility to glare (Regan, 1995). Tests for these abilities developed by Regan and colleagues may be an effective screening tool for evaluating novice pilots (Regan, 1995), as discussed in Chapter 3.

Perception of Direction of Motion and Time to Collision

Once the pilot has detected the presence of an object such as a building or a hill he/she next needs to determine two things for successful collision avoidance: (1) will I collide with the object if I continue to travel in the same direction? and (2) how much time do I have before collision will occur? Consider the case of flying towards a tree. The lateral distance at which the tree will cross the fronto-parallel plane that contains the eyes (the "crossing distance") can be estimated using both monocular and binocular sources of visual information. As shown in Figure 5.5A, if the approach velocity is constant, the crossing distance (X_c) of the tree is given by:

$$X_c \approx \frac{2R(d\alpha / dt)}{d\theta / dt} \qquad [6]$$

where $d\alpha/dt$ is angular lateral speed of the tree's retinal image, $d\theta/dt$ is the rate of change of the tree's angular subtense and R is the tree's width (Bootsma, 1991; Regan and Kaushal, 1994). Equation [6] is available to either eye alone. Psychophysical experiments have shown that humans are sensitive to the information expressed in equation [6] as thresholds for the discrimination of the *relative* direction of an approaching object based on this optical variable range from 0.03 to 0.12 deg (Regan and Kaushal, 1994). To our knowledge the ability of observers to estimate absolute direction of motion in depth using the information expressed in equation [6] has not been investigated.

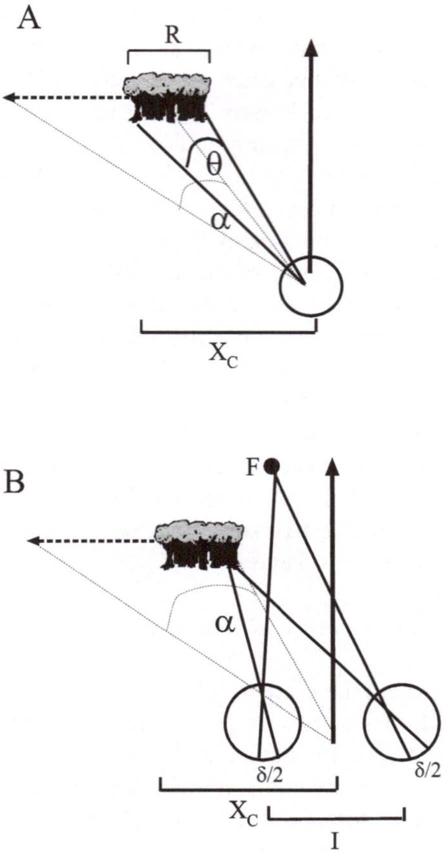

Figure 5.5 Visual correlates of crossing distance

As illustrated in Figure 5.5B, the binocular correlate of crossing distance relies on changing retinal disparity as:

$$X_c \approx \frac{I(d\alpha / dt)}{d\delta / dt} \qquad [7]$$

where dδ/dt is the rate of change of retinal disparity relative to a fixed reference point (F) and I is the interpupillary separation(Regan and Kaushal, 1994; Regan and Gray, 2000). Although it has been shown that humans are sensitive to the information expressed in equation [7] and can use it to make precise discriminations (0.2 deg) of variations in the trajectory of an approaching object (Portfors-Yeomans and Regan, 1996, 1997), it has not yet been demonstrated that equation [7] can be used to make absolute estimates of X_c based on this information source alone. Furthermore, it remains to be tested whether estimates of X_c are more accurate when both equations [6] and [7] are available (as is the case in the real world) than estimates based on either information source alone. One might expect an advantage when both information sources are available given that judgments of absolute TTC are more accurate when binocular and monocular information is combined (Gray and Regan, 1998).

The number of seconds remaining before the tree crosses the fronto-parallel plane containing the eyes [the 'time-to-passage'(TTP)] is also specified by both monocular and binocular sources of visual information. As illustrated in Figure 5.6A, for a tree approaching point P some distance from the pilot's eyes:

$$TTP \approx \frac{\theta}{d\theta / dt} - \frac{\gamma}{d\gamma / dt} \qquad [8]$$

γ is the optical angle at the eye subtended by the current location of the tree and the point at which the tree will cross the fronto-parallel plane containing the eyes (P), and $d\gamma/dt$ is the rate of constriction of this angle (Bootsma and Oudejans, 1993). In the special case where the tree is directly approaching the pilot's eye, γ=0 and

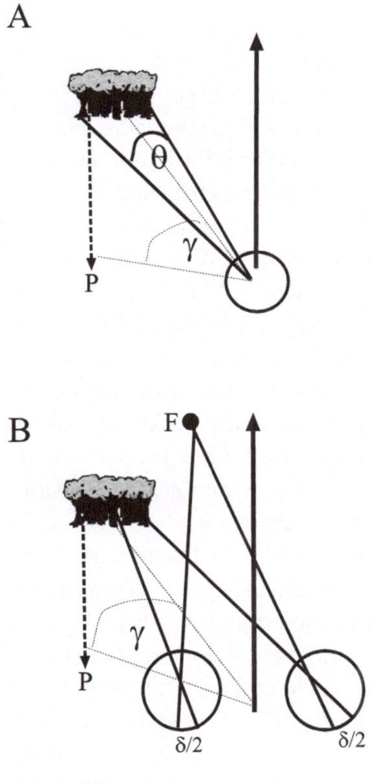

Figure 5.6 Visual correlates of time to passage

equation [8] reduces to the correlate of the time to collision (TTC) commonly called 'tau' after (Lee, 1976):

$$TTC \approx \frac{\theta}{d\theta / dt} \qquad [9]$$

For direct approaches it has been demonstrated that humans can accurately estimate TTC on the basis of equation [9] alone with estimation errors ranging from 2–12 percent of the actual TTC (Gray and Regan, 1998). The problem of judging absolute TTP for an object passing to the side has not been studied in detail. However, Bootsma and Oudejans (1993) have shown that observers can reliably discriminate the *relative* TTP of two approaching objects on the basis of equation [8] alone.

Binocular information about TTP is illustrated in Figure 5.6B. This information source relies on relative disparity information as:

$$TTP \approx \frac{2d\delta / dt}{d^2\delta / dt^2} - \frac{\gamma}{d\gamma / dt} \qquad [10]$$

Estimation of TTP based on binocular information alone has not been previously investigated except in the special case of an object directly approaching the midpoint between the eyes where estimation errors range from 2 to 10 percent of the actual TTC (Gray and Regan, 1998).

At this point a reader familiar with the topic of visual perception may be asking the question: but wouldn't binocular information be ineffective in flying because the objects are too far away? While it is true that the relative retinal disparity (i.e., δ in Figures 5.5A and 5.6A) for a given depth separation between two objects is inversely proportional to the square of the viewing distance, this limitation only applies to judgments of static depth. It is for this reason that we did not mention stereopsis as a cue to object visibility in the discussion above. At a distance of 50 m (165 ft), two objects must be separated by more than 3 m (10 ft) for the depth separation to be detectable. When judging the direction of motion in depth and TTP, or TTC, the relevant retinal image variable is the rate of change of disparity ($d\delta/dt$). Because the magnitude of $d\delta/dt$ is proportional to the approach velocity, the high speeds involved in flying would ensure the value of $d\delta/dt$ is well above perceptual threshold in most situations.

It is evident from the analyses above that precise detection of the rate at which the angular size of an object is increasing ($d\theta/dt$) is important for making judgments about the direction of motion and time to contact. Therefore, it might be expected that more highly skilled pilots would have a greater sensitivity to retinal image expansion as compared to novice pilots. This prediction was tested directly in a unique merging of laboratory and field research conducted by Kruk, Regan and colleagues (Kruk and Regan, 1983; Kruk, Regan, Beverley, and Longridge, 1981). In these studies laboratory measurements of discrimination thresholds for rate of change of size were found to significantly correlate with flying performance in low-level flight and formation flight.

A final point on visual cues in low-level flight concerns the use of texture in flight simulator displays. As discussed above, the presence of texture on object and ground surfaces is critical for the pilot's ability to visually segregate objects from their surrounding. Therefore it is not surprising that considerable effort has been put into adding realistic texture to simulator displays. However, because the addition of texture is "computationally expensive" and can dramatically reduce the display frame rate, shortcuts are often taken that result in visual information that is not consistent with what occurs in the real world. An extreme example of this is *texture mapping*, where an object such as a building or tree is "painted" with a texture pattern that does not change as a function of viewing distance. Even today the expansion of texture elements on objects is rarely simulated. If it is done, it is usually in one or two discrete steps instead of a continuous change. This creates a potential problem because the rate of expansion of the texture elements on the surface of the object provides information about TTC that compliments information provided by the change in the overall angular size of the object (Beverley and Regan, 1983). Indeed, it has been shown that when the texture elements do not expand, TTC is dangerously overestimated (Gray and Regan, 1999). This effect, shown in Figure 5.7, depends on the grain of the texture on the object. The overestimation is larger for objects with large texture elements (e.g., the bricks on the side of building) compared to objects with small texture elements (e.g., the needles on the surface of a pine tree). In the extreme case (texture elements less than roughly 5 min arc) the lack of expansion does not affect judgments of TTC, presumably because the rate of expansion is normally below threshold. Currently there are several more complex problems associated with use of texture displays. For example, the computer graphics technique known as *mipmapping* leads to the undesirable side effects that the luminance contrast of the display is inversely related to texture density (Chaudhry and Geri, 2003) and blurring increases as simulated altitude decreases.

Altitude Maintenance

A minimum requirement for successful low altitude flight is that the pilot be able to keep the aircraft's altitude near a specified value—flying too high can lead to radar detection, for example, whereas flying too low can lead to ground contact. Early research on altitude maintenance in low altitude flight focused primarily on the use of depression and splay angles. Flach et al. (1992) and Flach, Warren, Garness, Kelly, and Stanard (1997) used simple terrain textures composed of lines and grids, and an experimental task that required maintenance of a constant altitude in the presence of simulated fore-aft, up-down and right-left wind disturbances. They found that either depression angle or splay angle could be used for altitude maintenance during simulated low altitude flight, and that the relative cue effectiveness varied across flying conditions. Although depression angle and splay angle have proven important for altitude maintenance (see also Wolpert, Owens, and Warren,1983, and Johnson, Tsang, Bennett, and Phatak, 1989), it should be

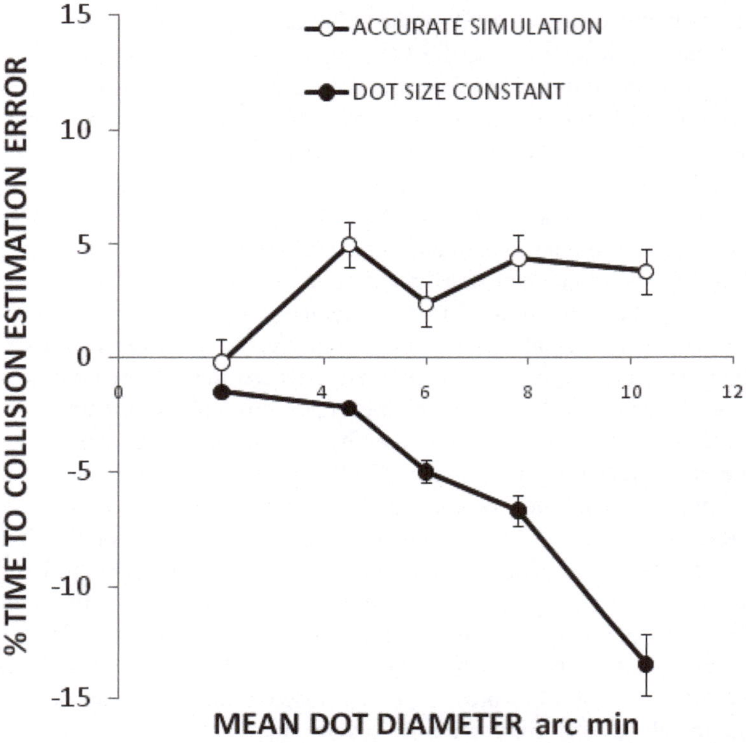

Figure 5.7 Errors in estimating time to contact

Source: Reprinted with permission from Gray, R. and Regan, D. (1999).

noted that the simulated ground textures used in those studies were specifically designed to optimize those cues. A more natural terrain with irregularly spaced 3D objects of varying heights may reduce the effectiveness of those cues; however this has not been empirically tested.

Kleiss and Hubbard (1993) and Kleiss (1995) used an altitude-change detection task to investigate how terrain objects (e.g., trees, buildings, etc.) are used in altitude maintenance. Participants flew over a simulated ground terrain populated with different types of 3D objects positioned randomly and with varied density. Participants first actively controlled their altitude; the display was then blanked and the altitude was changed. The participants' task was to indicate whether their perceived altitude was higher, lower or the same as before the display was blanked. Those authors found that perceived altitude was determined primarily by object density; see Martin and Rinalducci, (1983), for a similar finding. In separate experiments, Kleiss and Hubbard (1993) also found that the absence or presence of detailed texture on the 3D objects and the absence or presence of

2D texture on the ground surface did not significantly affect judgment accuracy. Finally, Winterbottom, Geri, Pierce, and Harris (2001) used an active-control task in a flight simulator to investigate the effects of texture density on altitude-maintenance performance. In that study, terrain textures were obtained from random noise patterns, so that edges, which could be used to detect changes in splay and depression angle, were minimized. After flying over a flat portion of the terrain for ten seconds at an altitude of 50 m (164 ft), participants were required to maintain that altitude as they flew over a sloped section of terrain. The mean altitude deviations ranged from 4 to 10 m (13–32 ft), depending on airspeed, suggesting that participants can use changes in perceived texture density to maintain altitude.

Motion parallax is another cue that may be relevant to altitude control in low altitude flight. Motion parallax refers to the relative movement of objects, which occurs when the observer moves, and is a consequence of the fact that nearer objects move faster across the retina than do farther objects. In the context of movement over a textured surface as in low altitude flight, motion parallax is often referred to as motion perspective (see e.g., Hershenson, 2000). In the case of movement over a textured surface or a scene containing 3D objects, the most salient perceptual cue is the motion gradient formed by differential movement of the texture elements or the 3D objects (Sedgwick, 1986). This motion gradient is a visual cue that could be used for altitude maintenance (Patterson, Akhtar, Geri, Morgan, Pierce, Dyre, and Covas, 2003). Horizontal motion-perspective has been shown to be used for lateral control such as in the control of heading (see e.g., Longuet-Higgins and Prazdny, 1980; Li and Warren, 2000), and there is some evidence that vertical motion-perspective cues may also be used for altitude control (Covas, Patterson, Geri, Akhtar, Pierce, and Dyre, 2005).

Finally, Gray et al. (2008) investigated the use of visual occlusion as a cue to altitude maintenance in low altitude flight. Visual occlusion may be manifested in two independent ways, both of which require the presence of 3D objects. The more common case is the occlusion of objects by other objects (Gibson, 1979; Kaplan, 1969). In the case of a moving observer, the time course of occlusion (and disocclusion) is a cue to both the relative distance among objects and the relative distance from the observer to the various objects. As described above, this form of occlusion is an inherent component of the motion parallax associated with 3D objects. A less often considered case of visual occlusion, which may be relevant for altitude maintenance, is that involving occlusion of the ground plane by 3D objects. As illustrated in Figure 5.8, the amount of ground surface that will be occluded from a pilot's view by 3D objects is inversely related to eyeheight (compare Figures 5.8A and B), and directly related to the width, height (compare Figures 5.8 C and D), and density (compare Figures 5.8A and C) of the 3D objects. The relationship among these variables, as well as the magnitude of occlusion has been modeled by Leung and Malik (1997). In that model, the degree of occlusion is represented by the probability that only the ground surface (i.e., the

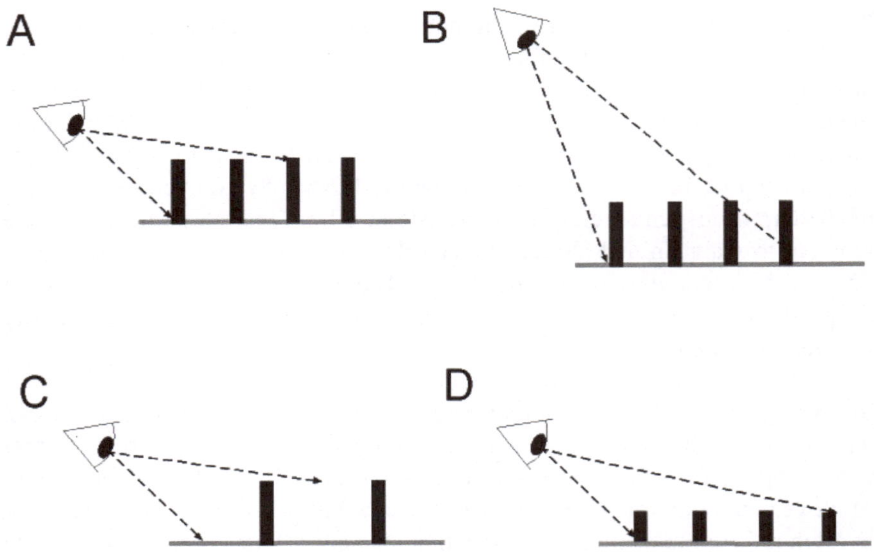

Figure 5.8 Visual occlusion and altitude maintenance

Source: Reprinted with permission from Gray, R., Geri, G.A., Akhtar, S.C. and Covas, C.M. (2008).

space between objects) is not visible within a window through which the terrain is viewed. Thus, a low probability of seeing the ground is associated with a high degree of occlusion, and *vice versa*. This probability can be estimated using the following equation from Leung and Malik:

$$p(ground) \approx e^{-2hdr\tan\sigma} - d\pi r^2 \qquad [11]$$

where h is the object height, d is the object density, r is the object radius, and σ is the angle formed by the line of sight and a line perpendicular to the ground surface and which is proportional to the eyeheight or altitude.

In the study by Gray et al. (2008), participants attempted to maintain a constant altitude during simulated flight over an undulating terrain with trees of various heights, radii, and densities. As would be predicted if participants used occlusion (equation 11), error in attitude was related to the product of tree height and tree density and to the product of tree radius and tree density. So, in other words, when very few trees were placed on the terrain surface (i.e., low density), altitude maintenance was poor if the trees were small and narrow. However, if trees were either tall (large height) or thick (large radius) performance was significantly better. This pattern of performance is what one would expect if pilots are using occlusion as a cue to altitude maintenance because, as illustrated in Figure 5.8, bigger trees occlude more of the ground and provide more information to the pilot.

Due to both computational and display limitations, flight simulator designers must often make trade-offs in display design, which may have important implications for training. For example, in order to achieve a high level of detail on the surface of objects and on the ground terrain while at the same time keeping the frame rate sufficiently high, object density must be kept sufficiently low. Therefore, it is critical to identify the relative importance of the different display parameters (e.g., object detail vs. object density) to flight performance to ensure that these trade-offs result in positive transfer of training between simulated and actual flight. From the results of Gray et al. (2008), there appears to be an important trade-off between both object height and object density, and object radius and object density. If it is necessary to have a simulation with low object density these results suggest that good low-altitude flight performance can be maintained by using taller and/or wider objects.

Control of Heading

Along with longitudinal (altitude) maintenance, a key to safe low-altitude flight is the accurate control of the lateral movement or *heading* of the aircraft. Gibson (1958) was the first to note that heading can be determined by the movement of the stationary objects in the environment (across the retina) that results from one's own movement. This information, called *optic flow*. When one is heading towards a particular object in the environment (e.g., a runway) the image for that object will remain stationary while all other objects will move (or flow) outwards from the location of heading. This point of stationarity is referred to as the *focus of expansion (FOE)*. Gibson (1958) proposed that:

> The center of the flow pattern during forward movement of the animal is the direction of movement" and "to aim locomotion at an object is to keep the center of flow of the optic array as close as possible to the form which the object projects. (p. 187)

Using the FOE to control heading becomes more problematic when observers do not keep their eyes fixated on the aimpoint. In this situation, *optic flow* (as discussed by Gibson) is not equivalent to the *retinal flow* that will be received at the back of the observer's eye. For example, when an observer is traveling on a straight-line path but fixates a feature to the side of locomotor path, the FOE occurs at the point of fixation rather than at the aimpoint (Regan and Beverley, 1982). To recover optic flow from the retinal pattern, processing would be required to subtract the eye movement component. Proposed solutions for the problem of extracting optic flow from retinal flow use of either decomposition algorithms or template methods (Warren and Hannon, 1988; 1990; van den Berg, 1993; Stone and Perrone, 1997; Li and Warren, 2000) or extra-retinal signals that specify the head and eye motion (Royden et al., 1994; Banks et al., 1996; Crowell et al., 1998).

More recently other sources of visual information have been proposed for the control of heading. Li and Warren (2000) reported that the depth cue of motion parallax (see Chapter 2) provided by vertically oriented objects in the environment such as trees is combined with optic flow to judge heading. It has also been proposed that heading can be controlled on the basis of the visual direction of the goal object without the use of optic flow information (Rushton, Harris, Lloyd and Wann, 1998; Wann and Land, 2000).

Patterson et al, (2006) investigated the active control of heading during simulated low-altitude flight. Participants were required to maintain their heading in the presence of simulated crosswinds. The presence of 3D objects (e.g., trees) significantly improved control accuracy and performance was better for higher object densities. These findings further support Li and Warren's (2000) proposal that optic flow is combined with motion parallax when controlling heading.

Finally, as first noted by Loomis and Beall (1998), lateral control in aviation involves more than controlling heading towards a target. Other related maneuvers include negotiating an aperture/passageway, turning into alignment with a straight path (see discussion of base-final above), steering along a curved path, and orbiting. Loomis and Beall argued that many of these behaviors cannot be achieved on the basis of optic flow alone and more research is needed to understand how a pilot's perception of the the 3D spatial layout of the environment is used.

Mid-air Collision Avoidance

The "See and Avoid" concept set forth by the FAA relies on a pilot's ability to vigilantly monitor the out-of-cockpit scene, judge the trajectory and speed of other aircraft, and use this information to avoid collision (regardless of whether they are under IFR or VFR). Between 1991 and 2000 there was an average of 15.6 mid-air collisions per year in civilian aviation, causing an average of 24.8 fatalities per year (Morris, 2005). In over 90 percent of these cases, the NTSB cited a probable cause associated with an error in visual perception such as "failure to see and avoid" or "inadequate visual lookout." In this section, we examine the visual information available to support mid-air collision avoidance in attempt to understand why pilots frequently make errors in this task. It will be demonstrated that the "see and avoid" concept has major physical and psychological limitations.

In order to successfully "see and avoid" a pilot must: (1) detect the presence of another aircraft, (2) judge its trajectory and time to collision, and (3) use these judgments initiate a maneuver that will prevent collision. There are inherent problems associated with each of these steps. Let's consider problem associated with detection first. As illustrated in Figure 5.9, when two aircraft on a collision course are both ascending (panel A) or both descending (panel B) it is possible that

Figure 5.9 Relative bearing of climbing/descending aircraft

Source: Reprinted with permission from Morris, C. C. (2005).

each aircraft can be completely obstructed from view from each other. This would be most likely to occur when the aircraft at the higher altitude is traveling faster. Even if the other aircraft is visible through the windscreen, its angular size (at the distance it needs to be detected for successful collision avoidance) is often too small for reliable detection. It has been estimated that the minimum time required to detect an aircraft, judge the collision course and then successfully maneuver to avoid collision is roughly 12.5 sec (FAA 1983). As illustrated in Figure 5.10, for closure speeds of 200 knots and above, the angular size of the other aircraft will be less than 1 deg (i.e., less than the size of your thumbnail at arm's length) at this critical 12.5 sec time. These small angular sizes are dangerously close to the hypothesized size detection threshold (from NTSB investigations) of 0.2 deg.

Furthermore, at this size, view of the other aircraft can be obstructed from view by wings, posts, visors or even windscreen imperfections. The final problem associated with detecting another aircraft mid-air is that in many situations the image of the other aircraft will be completely stationary in the pilot's visual field. If the relative bearing to another aircraft remains constant and if each aircraft remains at a constant velocity, there will be no relative motion, making the other aircraft appear motionless and less likely to be detected by a pilot. This lack of perceived motion is a serious problem as motion is one of the strongest cues that draw our attention to important objects in our visual periphery.

Even if the pilot can successfully detect the presence of another aircraft in enough time to avoid collision, these are still problems associated with accurately judging where that aircraft is heading and its time to collision. The primary cues available to a pilot for making these judgments would be the monocular cues to TTC and direction of motion-in-depth that are based on changing angular size

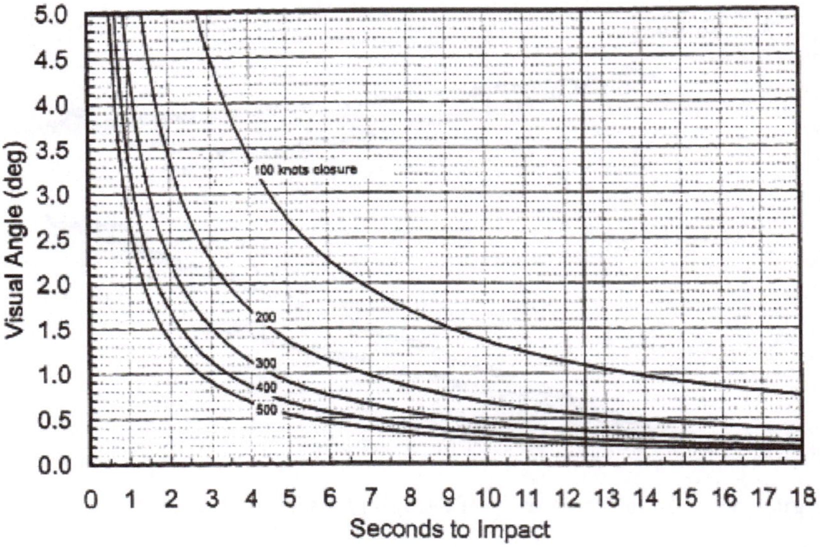

Figure 5.10 Midair collision data for closure rate

Source: Reprinted with permission from Morris, C. C. (2005).

(i.e., equations 6 and 9 above). At the distances and closing speeds involved it is unlikely that the binocular cues to TTC and direction of MID would be reliable in this situation. As illustrated in Figure 5.10, a problem associated with using monocular cues in this situation is that the rate of change of object size (i.e., dθ/dt in equations 6–10 above) is likely to be too small for reliable detection until just before collision. For example, at the closure rate of 300kn in Figure 5.10, the angular size of the other aircraft does not begin changing size appreciably until the TTC falls below roughly 6 sec. Gray and Regan (1998) have shown that when the angular size of an approaching object (and its rate of change) are small, observers tend to overestimate TTC, i.e., they think there is more time until collision than there actually is. This would, of course, be an extremely dangerous misjudgment for a pilot to make.

Campbell and Bagshaw (2002) presented many of these same arguments in terms of how extremely limited pilots' visual systems are in detecting other airborne aircraft. They emphasized scanning techniques because fewer than 10 percent of mid-air collisions occur head-on. However, relative movement (up-down, left-right, for-aft) is needed for our ambient visual system to detect a possible traffic conflict, which doesn't happen with head-on collision paths; the only movement will be a growing of the retinal image at the same location in space. However, the growth in the perceived size of the image is so small, it is very difficult to detect until the last moment. Thus, it is the combination of constant heading, speed, and altitude leading to the lack of relative movement between colliding aircraft that fails to provide visual cues of impending collision. As mentioned previously, closure rates between aircraft are so great that detection is highly improbable even if there is relative movement. Campbell and Bagshaw stated that studies have found it takes approximately 10 seconds for a pilot to detect and avoid oncoming traffic even though the closure rate allows for 36 seconds of "viewing time."

When these perceptual problems are combined with the sluggish response of a typical aircraft, it makes "see and avoid" a very challenging task indeed. And there are also other factors. There can often be confusion about "who is responsible" in this situation between pilots and ATC. Some pilots (particularly less experienced ones) assume ATC is watching over them with radar and will initiate a warning if a collision is impending, while ATC assumes the pilot is vigilantly watching the skies. This was the case between two aircraft that tragically collided over Brazil as detailed in Chapter 7. Pilots in both cockpits assumed that technology within their own aircraft and by air traffic control would keep them from colliding with any another aircraft.

Morris (2005) performed a geometric analysis of 156 mid-air collisions between 1991 and 2000, taking into account the velocities, directions and visual angles associated with each incident. In 88 percent of crashes one of the aircraft was maneuvering and 70 percent both were maneuvering. The high rate of crashes involving these types of scenarios is not surprising given the potential obstructions illustrated in Figure 5.9. Head-tail collisions were the most frequent case. In this study, a probabilistic model of pilot visual scanning was developed. Table 5.1 shows the estimated probabilities of a pilot successfully achieving "see and avoid" for different closure rates. This model compares an ideal (and unrealistic) pilot who is scanning the sky for another aircraft 100 percent of time with more likely scanning rates of 0.66 and 0.33. As can be seen in this table the detection rates are incredibly low.

Table 5.1 Probability optimal observer or real pilot can see and avoid a random converging aircraft

Probability Can See and Avoid Aircraft					
Closure Speed (kn)	**Seconds to Impact when Visual Angle = 0.2°**	**See-and-Avoid Window of Opportunity Duration (s)**	**Optimal Observe Scanning All of Flight Time**	**Theoretical Pilot Scanning 2/3 of Flight Time**	**Theoretical Pilot Scanning 1/3 of Flight Time**
100	681.	55.6	1.000	1.000	0.723
200	34.1	21.6	0.907	0.605	0.302
300	22.7	10.2	0.487	0.324	0.162
400	17.0	4.5	0.276	0.184	0.092
500	13.6	1.1	0.150	0.100	0.050

Simulator research by Andrew (1991) is consistent with this model: motivated pilots (being explicitly monitored for performance by an analyst) detected on average only 56 percent of 64 near mid-air collisions. Clearly, it seems that pilot "see and avoid" is not an effective method for avoiding mid-air collisions. Possible technological advances to aid the pilot in this situation are discussed in Chapter 8.

Prior to leaving this topic of detecting and avoiding airborne hazards it is worth mentioning another hazard to pilots—birds. The detection of an object significantly smaller in size than another aircraft is nearly impossible for a pilot to "sense." Take the example of the miraculous Hudson River water landing. What forced the pilot to use his visual perception capabilities and successfully ditch the aircraft into the river came about due to visual perception limitations of detecting airborne hazards as small as Canadian Geese. Airport managers go to great lengths to ensure birds avoid the runway environment. However, in the case of Flight 1549, the jet engines ingested birds well away from the airport, some 900 m (~3,000 ft) in the air. In terms of pilot expectancy, searching and detecting birds at that altitude is very difficult.

Conclusion

Chapters 4 and 5 have presented the environmental cues available and used by pilots from both a pilot-perspective and a research-perspective. The stage is now set for better understanding of how and why pilots fall prey and succumb to visual and vestibular spatial disorientation. The visual illusions and misperceptions presented in Chapter 6 come about due to pilots' preference for visual flying, the often unreliable environmental cues available, and the overconfidence pilots have in their visual perception capabilities. Chapter 7 then presents a sample of visual misperception mishaps that clearly demonstrate the pilots have, for far too long, been seduced and led astray by their faith in their perception of the environment.

References

Andrew, J. W. (1991). Unalerted air-to-air visual acquisition. Lexington, MA: Lexington Laboratory, Massachusetts Institute of Techology; 26 Nov 1991.

Banks, M. S., Ehrlich, S. M., Backus, B. T., and Crowell, J. A. (1996). Estimating heading during real and simulated eye movements, *Vision Res.*, *36*, 431–443.

Beall, A. C. and Loomis, J. M. (1997). Optic flow and visual analysis of the base-to-final turn. *International Journal of Aviation Psychology*, *7*(3), 201–223.

Benbassat, D. and Abramson, C. I. (2002). Landing flare accident reports and pilot perception analysis. *International Journal of Aviation Psychology*, *12*, 137–152.

Beverley, K. I. and Regan, D. (1983). Texture changes versus size changes as stimuli for motion in depth. *Vision Res*, *23*(12), 1387–1399.

Boeing Corporation (2008). Statistical Summary of Commercial Jet Airplane Accidents. Seattle, WA: Boeing Commercial Airplanes. Retrieved March 13, 2009 from http://www.boeing.com/news/techissues/pdf/statsum.pdf.

Bootsma, R. (1991). Predictive information and the control of action: What you see is what you get. *International Journal of Sports Psychology*, *22*, 271–278.

Bootsma, R. J. and Oudejans, R. R. (1993). Visual information about time-to-collision between two objects. *J Exp Psychol Hum Percept Perform*, *19*(5), 1041–1052.

Calvert, E. S. (1954). Visual judgments in motion. *Journal of the Institute of Navigation*, *7*, 233–251.

Campbell, R. D. and Bagshaw, M. (2003). *Human Performance and Limitations in Aviation*. London: Blackwell Science.

Chaudhry, S. and Geri, G. A. (2003). Display related effects of terrain-texture density and contrast on percieved air speed in simulated texture. In *SID 03 Digest of Technical Papers* (Vol. 24, pp. 276–279).

Covas, C. M., Patterson, R., Geri, G. A., Akhtar, S. C., Pierce, B. J., and Dyre, B. P. (2005). Horizontal motion-parallax is as effective as texture density for altitude control in simulated flight. *SID 05 Digest*, 382–385.

Crowell, J. A., Banks, M. S., Shenoy, K. V., and Andersen, R. A. (1998). Visual self motion perception during head turns, *Nat. Neurosci.*, *1*, 732–737.

Flach, J. M., Hagen, B. A., and Larish, J. F. (1992). Active regulation of altitude as a function of optical texture. *Percept Psychophys*, *51*(6), 557–568.

Flach, J. M., Warren, R., Garness, S. A., Kelly, L., and Stanard, T. (1997). Perception and control of altitude: Splay and depression angles. *Journal of Experimental Psychology-Human Perception and Performance*, *23*(6), 1764–1782.

Galanis, G., Jennings, A., and Beckett, P. (2001). Runway width effects in the visual approach to landing. *International Journal of Aviation Psychology*, *11*(3), 281–301.

Gibb, R. and Gray, R. (2006). The terrain orientation theory for visual guidance of approach and landing. *Proceedings of the Human Factors and Ergonomics Society 50th Annual Meeting*.1661–1665.

Gibb, R., Schvaneveldt, R. S., and Gray, R. (2008). Visual misperception in aviation: Glide-path performance in a black hole environment. *Human Factors*, *50*, 699–711.

Gibson, J. J. (1958). Visually controlled locomotion and visual orientation in animals, *Br. J. Psychol.*, *49*, 182–194.

Gibson, J.J. (1979). *The Ecological Approach to Visual Perception*. Boston: Houghton Mifflin.

Gray, R. and Regan, D. (1996). Accuracy of reproducing angles: Is a right angle special? *Perception*, *25*, 531–542.

Gray, R. and Regan, D. (1998). Accuracy of estimating time to collision using binocular and monocular information. *Vision Research*, *38*, 499–512.

Gray, R. and Regan, D. (1999). Motion in depth: Adequate and inadequate simulation. *Perception and Psychophysics*, *61*, 236–245.

Gray, R., Geri, G. A., Akhtar, S. C., and Covas, C. M. (2008). The role of visual occlusion in altitude maintenance during simulated flight. *Journal of Experimenta Psychology: Human Perception and Performance*, *34*, 475–488.

Hershenson, M. (2000). *Visual Space Perception*. Cambridge, MA: Bradford, pp. 146–149.

Hoffmann, E. R. and Mortimer, R. G. (1996). Scaling of relative velocity between vehicles. *Accid Anal Prev, 28*, 5–421.

Hoyle, F. (1957). *The Black Cloud*. Middlesex, England: Penguin.

Johnson, W. W., Tsang, P. S., Bennett, C. T., and Phatak, A. V. (1989). The visual control of simulated altitude. *Aviation Space and Environmental Medicine*, *60*, 152–156.

Kaplan, G. A. (1969). Kinetic disruption of optical texture: The perception of depth at an edge. *Perception and Psychophysics*, *6*, 193–198.

Kleiss, J. A. (1995). Visual scene properties relevant for simulating low-altitude flight: A multidimensional scaling approach. *Human Factors*, *37*, 711–734.

Kleiss, J. A. and Hubbard, D. C. (1993). Effects of 3 Types of Flight Simulator Visual Scene Detail on Detection of Altitude Change. *Human Factors*, *35*(4), 653–671.

Kraft, C. L. (1978). A psychophysical contribution to air safety: Simulator studies of visual illusions in night visual approaches. In H. Pick, H. W. Leibowitz and J. R. Singer (Eds.), *Psychology from research to practice* (pp. 363–385). New York: Plenum.

Kruk, R. and Regan, D. (1983). Visual test results compared with flying performance in telemetry-tracked aircraft. *Aviat Space Environ Med*, *54*(10), 906–911.

Kruk, R., Regan, D., Beverley, K. I., and Longridge, T. (1981). Correlations between visual test results and flying performance on the advanced simulator for pilot training (ASPT). *Aviat Space Environ Med*, *52*(8), 455–460.

Leung, T. and Malik, J. (1997). Perpendicular texture: Why do we see more flowers in the distance? *IEEE Conference on Computer Vision and Pattern Recognition*. New York: IEEE Press.

Li, L. and Warren, W. H. (2000). Perception of heading during rotation: Sufficiency of dense motion parallax and reference objects. *Vision Research*, *40*, 3873–94.

Lintern, G. (2000). An analysis of slant for guidance of landing approaches. *The International Journal of Aviation Psychology*, *10*, 363–376.

Lintern, G. and Liu, Y. (1991). Explicit and implicit horizons for simulated landing approaches. *Human Factors*, *33*, 401–417.

Lintern G, Walker M. B. (1991). Scene content and runway breadth effects on simulated landing approaches. *The International Journal of Aviation Psychology*, *1*, 117–32.

Longuet-Higgins, H. C. and Prazdny, K. (1980). The interpretation of a moving retinal image. *Proceedings of the Royal Society of London B: Biological Sciences*, *208*, 385–397.

Loomis, J. M. and Beall, A. C. (1998). Visually controlled locomotion: It's depended on optic flow, three-dimensional space perception, and cognition. *Ecological Psychology*, *10*, 271–285.

Marsh, G. (1985). Avoiding the wires. *Defense Helicopter World, June-August*, 22–23.

Martin, E. L. and Rinalducci, E. J. (1983). Low-level flight simulation vertical cues. *AFHRL-TR-83–17, AD133 612*. Williams Air Force Base, AZ, Air Force Human Resources Laboratory, Operations Training Division.

Mayer, C. M., Mershon, D. H. et al. (2006). Visual factors affecting touchdown point judgments during off-airport emergency landings. *International Journal of Aviation Psychology*, *16, 401–418.*

Mertens, H. W. (1981). Perception of runway image shape and approach angle magnitude by pilots in simulated night landing approaches. *Aviation, Space, and Environmental Medicine, 52,* 373–386.

Mertens, H. W. and Lewis, M. F. (1982). Effect of Different Runway Sizes on Pilot Performance During Simulated Night Landing Approaches. *Aviation Space and Environmental Medicine*, *53*(5), 463–471.

Mertens, H. W. and Lewis, M. F. (1983). Effects of approach lighting and variation in visible runway lendth on perception of approach angle in simulated night landings. *Aviation, Space, and Environmental Medicine*, *54*, 500–506.

Morris, C. C. (2005). Midair collisions: Limitations of the see-and-avoid concept in civil aviation. *Aviation Space and Environmental Medicine 76*(4): 357–365.

Mulder, M., Pleijsant, J. M., van der Vaart, H., and van Wieringen, P. (2000). The effects of pictorial detail on the timing of the landing flare: Results of a visual simulation experiment. *International Journal of Aviation Psychology*, *10*(3), 291–315.

Palmisano, S. and Gillam, B. (2005). Visual perception of touchdown point during simulated landing. *Journal of Experimental Psychology: Applied*, *11*, 19–32.

Patterson, R., Akhtar, S. C., Geri, G. A., Morgan, W., Pierce, B. J., Dyre, B. P., and Covas, C. M. (2003). Terrain texture and 3-D object cues in the control of heading in simulated flight. *SID 03 Digest of Technical Papers*, 319–321.

Patterson, R., Geri, G. A., Dyre, B. P., Pierce, B. J., Akhtar, S. C., Covas, C. M., and Morgan, W. (2006). Active heading control in simulated flight based on vertically extended contours. *Perception and Psychophysics*, *68*, 593–600.

Perrone, J. A. (1984). Visual slant misperception and the 'black-hole' landing situation. *Aviation, Space, and Environmental Medicine, 55,* 1020-1025.

Portfors-Yeomans, C. V., and Regan, D. (1996). Cyclopean discrimination thresholds for the direction and speed of motion in depth. *Vision Res*, *36*(20), 3265–3279.

Portfors-Yeomans, C. V., and Regan, D. (1997). Discrimination of the direction and speed of motion in depth of a monocularly visible target from binocular information alone. *J Exp Psychol Hum Percept Perform*, *23*(1), 227–243.

Regan, D. (1995). Spatial orientation in aviation: Visual contributions. *J Vestib Res*, *5*(6), 455–471.

Regan, D. and Beverley, K. I. (1982). How do we avoid confounding the direction we are looking and the direction we are moving? *Science*, *215*, 194–196.

Regan, D. and Gray, R. (2000).Visually-guided collision avoidance and collision achievement. *Trends in Cognitive Neuroscience*, *4*, 99–107.

Regan, D. and Gray, R. (2001). Hitting what one wants to hit and missing what one wants to miss. *Vision Research*, *41*, 3321–3329.

Regan, D., Giaschi, D. E. and Fresco, B. B. (1993) Measurement of glare sensitivity cataract patients using low-contrast letter charts. *Ophthalmic Physiol Opt. 13*, 115–123.

Regan, D. and Kaushal, S. (1994). Monocular discrimination of the direction of motion in depth. *Vision Res*, *34*(2), 163–177.

Regan, D., Gray, R., and Hamstra, S. J. (1996). Evidence for a neural mechanism that encodes angles. *Vision Research*, *36*, 323–330.

Royden, C. S., Crowell, J. A., and Banks, M. S. (1994). Estimating heading during eye movements, *Vision Res.*, *34*, 3197–3214.

Rushton, S. K., Harris, J. M., Lloyd, M. R., and Wann, J. P. (1998). Guidance of locomotion on foot uses perceived target location rather than optic flow, *Curr. Biol.*, *8*, 1191–1194.

Sedgwick, H. A. (1986). Space perception. In K. R. Boff, L. Kaufman, and J. P. Thomas, (Eds.), *Handbook of Perception and Human Performance*, *Vol 1, Sensory Processes and Perception*. John Wiley: New York.

Schwirzke, M. F. J. and Bennett, C. T. (1991). A re-analysis of the causes of Boeing 727 "black hole landing" crashes. The 6th Symposium of the International Journal of Aviation Psychology, 572–576.

Stone, L. S. and Perrone, J. A. (1997). Human heading estimation during visually simulated curvilinear motion, *Vision Res.*, *37*, 573–590.

van den Berg, A. V. (1993). Perception of heading, *Nature*, *365*, 497–498.

Wann, J. and Land, M. (2000). Steering with or without the flow: is the retrieval of heading necessary? *Trends in Cognitive Sciences*, *4*, 319–324.

Warren, W. H. and Hannon, D. J. (1988). Direction of self-motion is perceived from optical flow, *Nature*, *336*, 162–163.

Warren, W. H. and Hannon, D. J. (1990). Eye movements and optical flow, *J. Opt. Soc. Am.*, *7*,160–169.

Winterbottom, M. D., Geri, G. A., Pierce, B. J., and Harris, N. M. (2001). *Low-altitude flight performance as a measure of flight simulator fidelity*. Human Factors and Ergonomics Society 45th Annual Meeting.

Wolpert, L., Owens, D., and Warren, R. (1983). Eye-height-scaled versus ground-texture-unit-scaled metrics for the detection of loss in altitude. *Proceedings of the 2nd Symposium on Aviation Psychology* (pp. 513–531). Columbus: Ohio State University.

Chapter 6
Spatial Disorientation—Cues, Illusions and Misperceptions

> Human beings are not designed for aerial operations, either by day or by night. We can only achieve this by the use of technology and training. Our physiological limitations become very evident under certain flight conditions, and the night environment is perhaps the most significant. (Ostinga, Wolff, Newman, and White, 1999, p. 25)

Actions and behavior in visual space require accurate estimation of the depth and distance of objects. A rich viewing environment is filled with redundant cues providing depth and distance information. This redundancy makes it difficult to determine the dominant visual cues that are used for the control of different types of actions (Warren and Owen, 1982), which in turn makes it difficult to predict the influence of the reduction of available cues. While rich viewing conditions reduce (but do not eliminate) visual illusions and misperceptions, impoverished viewing conditions such as at night and/or in the weather, increase the likelihood that a pilot will experience visual illusions and misperceptions, and pilots often do. Wickens and Hollands (2000), explain the impact of cue reduction in that "there are occasions when the hypotheses we assume do not correspond with reality because the cues are few in number, the assumptions we make about the world are incorrect, or the cues are ambiguous" (p. 143). Previc (2004a) reminded readers that only bats have successfully found a way to fly without the aid of vision.

Both Gillingham (1992) and Previc (2004a) concluded that in assessing mishap statistics one may safely assume that roughly half of all spatial disorientation mishaps have some form of visual misperception as a contributing factor. Following an analysis of controlled flight into terrain (CFIT) data between the years of 1980–1996, Khatwa and Helmreich (1998) reported that the incidence of an approach-and-landing accident at night was three times higher than such an accident during daylight hours. More specifically, they found that disorientation or some sort of visual illusion was cited 21.1 percent of the time as the causal factor across 76 approach-and-landing accidents.

Before continuing with the discussion on spatial disorientation, it must be emphasized that pilots are generally aware that during impoverished visual conditions they are more prone to misperceptions. The paradox of this scenario is that pilots continue to confidently control their aircraft on the basis of visual information and fail to utilize the instruments right under their noses. For example, during a night approach to landing pilots often overconfidently attempt to fly "visual" knowing full and well that they are susceptible to misinterpretation of the runway size, shape and distance. Thus, while visual perception underlies many misperceptions, the cognitive aspect of overconfidence also plays a large role in aviation mishaps; understanding this influence of cognition may help reduce the hazards via education and training. Active training may potentially be the best tool because education alone has not been able to significantly impress upon pilots their susceptibility. By having pilots experience their misperceptions first hand in a training simulator, their perceptual limitations can be demonstrated (Gibb, Schvaneveldt, and Gray, 2008; Gibb, 2007). This chapter will overview spatial disorientation due to visual, vestibular, and non-perceptual factors, and some preventative measures that are recommended to reduce the likelihood of spatial disorientation-related mishaps.

Spatial Orientation and Disorientation Defined

Parmet and Ercoline (2008) clearly state the importance of visual perception:

> Vision is by far the most important sensory modality subserving spatial orientation, especially so in moving vehicles such as aircraft. Without it, flight as we know it would be impossible, whereas this would not be necessarily the case in the absence of the vestibular or other sensory systems that provide orientation information. (p. 148)

Spatial disorientation accidents have fatality rates of 90–91 percent (Krause, 2003; Accident Investigation Board report, 2008), which indicates just how compelling the misperceptions can be. The difficulty in understanding cue misperception is that, due to the complexity and interaction of cues, one cannot fully know what the pilots perceived via their eyes. When we look through our own eyes at the incident or accident, we are prone to hindsight bias and find it difficult to appreciate how the misinterpretation occurred. However, especially when reading the accidents presented in Chapter 7, we urge the reader to fully respect and appreciate that, given the perceptual information the pilots had at the time, they did their best.

We also want to emphasize that visual and vestibular spatial orientation occurs at a very basic neural level, and is largely unconscious. Consequently, as pointed out by Parmet and Ercoline (2008), higher-level neural processing often cannot overcome the spatial disorientation "wheels in motion". Wickens and Hollands (2000) conveyed this same message regarding the difficulty in negating visual illusions of distance and depth perception, "it is important to note that these misperceptions are relatively automatic. It is not easy to use our conscious awareness to de-bias the judgments of relative length" (p. 145). Gray (2006) stated that both pilots and drivers of automobiles rely too heavily on their visual system even in impoverished conditions. Because we have learned that we can rely on visual perception in everyday life it is difficult to ignore environmental perceptions even with awareness of environmental traps.

So what is spatial orientation? Veridical spatial orientation is simply the correct perception of one's location and orientation within his environment. On the most basic level, spatial orientation relates to a directional perspective, an up-down perspective, and where one's feet are relative to the sky and horizon. Also part of spatial orientation is awareness of distance and location from global and local references. In terms of aviation, these concepts are extended to include awareness of where one's head and feet are relative to the sky and ground even if one is flying upside down or experiencing a high G-turn. An additional aviation example is awareness of when one is in straight and level flight compared to a climbing or descending turn. So, how hard can this be? It sounds rather simplistic. If you are not a pilot or have not really paid attention while flying on a commercial airliner it is probably difficult to appreciate. In this chapter and Chapter 7 (which discusses specific spatial disorientation mishaps) we hope to articulate the difficulties of maintaining veridical spatial orientation.

Parmet and Ercoline (2008) described how a pilot attains a "sense of position and motion" from environmental cues via ambient/focal vision as well as vestibular inputs. A pilot's spatial orientation may also be influenced by indirect or synthetic informational sources in terms of a pilot's instrumentation depicting attitude and altitude. Maps can further provide a global perspective of location in one's environment. Mosier, Sethi, McCauley, Khoo, and Oransanu (2007) referred to these different methods of spatial orientation cue assessment as correspondence (probabilistic and intuitive cognition directly from environmental sources) and coherence (deterministic data assimilation). The latter method by which to achieve spatial orientation via aircraft displays is considered a synthetic form of orientation. However, regardless of how the information is acquired, the end-goal is veridical perception.

Recall from the Chapter 2 that spatial orientation is attained by a combination of sensory inputs via the visual system and the vestibular system. The visual system is by far the most dominant system in this process. According to Parment and Ercoline (2008), individuals with ineffective vestibular systems, labyrinthine defectives, can accurately perceive their movement in 3D space and function normally unless visual input is removed. Also, some authors have concluded that the visual system comprises nearly 80 percent of the necessary input required for efficacious spatial orientation (Newman, 2007; Ostinga et al., 1999). Consequently, the purpose of this book is to highlight the capabilities and limitations of a pilot's visual perception system towards successful spatial orientation.

So what is spatial *disorientation*? Below we will share several descriptions and highlight how they build our understanding of spatial disorientation with respect to aviation. Benson (1988) stated:

> Spatial disorientation is a term used to describe a variety of incidents occurring in flight where the pilot fails to sense correctly the position, motion or attitude

> of his aircraft or of himself within the fixed coordinate system provided by the surface of the earth and the gravitational vertical. In addition, errors in perception by the pilot of his position, motion or attitude with respect to his aircraft, or of his own aircraft relative to other aircraft, may also be embraced within a broader definition of spatial disorientation in flight. (p. 277)

This description of spatial disorientation includes both the pilot's perspective of his body's spatial orientation and the pilot's perspective of his aircraft relative to the environment. When flying, pilots must extend their sense of self with to the aircraft as a whole so that they may immerse themselves into the visually guided actions required to move the aircraft and control where it needs to go. Appreciating this aspect of aviation helps explain how pilots can confuse the aircraft's position relative to the horizon and gravitational vertical with their own orientation.

The United States Air Force manual of instrument flying procedures (AFMAN 112–17 Volume I, 2005, p. 355) defined spatial disorientation as:

> SD [spatial disorientation] is an incorrect perception of one's linear and angular position and motion relative to the plane of the earth's surface. Specifically in the flight environment, SD is an erroneous percept of any of the parameters displayed by aircraft control and performance flight instruments. Regardless of a pilot's experience or proficiency, sensory illusion can lead to differences between instrument indications and what the pilot "feels" the aircraft is doing. It should be stressed that disoriented pilots frequently are not aware of their orientation error and upon recognizing a conflict exists, often believe an instrument to be in error. Many crashes occur when pilots fail to recognize that SD is happening or when there is not enough time to recover once a conflict has been properly diagnosed.

We like this definition/description of spatial disorientation because it presents so many aspects of the problem and directly relates to the purpose of this book. First, it clearly states that any and all pilots are susceptible to spatial disorientation … a major take-away of this book. Second, it addresses the idea of how a pilot "feels"—it is irrelevant in hindsight to argue what the pilot should or could or would have perceived. If a pilot "feels" steep relative to the runway or if a pilot "feels" as if he is inverted, that perception will drive the resultant aircraft control inputs (flying is a visually guided behavior). Finally, the above description addresses the lack of awareness that pilots have regarding their current disoriented state. This is key to understanding the seductive nature of visual misperception and disorientation. Flying is a cognitively-demanding task and a pilot's limited attentional resources are pulled in many different directions. Also remember what was mentioned previously; visual and vestibular processing occurs at a very basic and unconscious neural level.

Parmet and Ercoline (2008) explained that spatial disorientation is characterized as an "erroneous sense of one's position and motion relative to the plane of the Earth's surface" (p. 181). The authors then further specified spatial disorientation in an operational pretext of the pilot and the aircraft together; it is "an erroneous sense of any flight parameters displayed by aircraft controls and performance instruments" (p. 181). This elaboration of the spatial disorientation concept includes the pilot's cognition and interaction with the aircraft. The basic control aspects of an aircraft are stick/yoke and throttles to manipulate pitch, bank, and thrust. The performance instruments are related to those controls in terms of the aircraft's speed, altitude, heading, vertical velocity, and general vector. Consequently, if a pilot fails to properly comprehend her control inputs and resultant aircraft performance readings, then that pilot is spatially disorientated.

Types of Spatial Disorientation

There are three types of spatial disorientation commonly discussed in terms of pilots' recognition of their orientational state:

1. Type I: unrecognized spatial disorientation. This occurs when the pilot is completely unaware of a disorientation problem and continues to control the aircraft in such a manner based on misperception of actual orientation. Parmet and Ercoline (2008) described unrecognized spatial disorientation in that, "the pilot is oblivious to the fact that he or she is disoriented, and controls the aircraft completely in accord with and in response to a false orientational percept" (p. 182).
2. Type II: recognized spatial disorientation. The pilot is aware of a problem and may attempt to diagnose it but may not realize the issue is an orientational problem. Parmet and Ercoline stressed that this recognized form of spatial disorientation, "does not mean that the pilot must necessarily realize he or she is disoriented … may only realize that there is a problem controlling the aircraft, not knowing that the source of the problem is SD [spatial disorientation]" (p. 182).
3. Type III: incapacitating spatial disorientation. AFMAN 112–17 (2005) defined Type III as "the pilot knows something is wrong, but the physiological or emotional responses to the disorientation are so great that the pilot is unable to recover the aircraft" (p. 371). Unfortunately, in this case the pilot is completely aware of the disorientation problem but is cognitively unable to select the appropriate control input or physically unable to execute the appropriate control input. Both AFMAN 112–17 and Parmet and Ercoline reference the possibility of vestibular nystagmus occurring and making it difficult to visually perceive aircraft instrument readings or environmental cues.

Given the above descriptions of the "types" of spatial disorientation, it is apparent how difficult it is to accurately classify a pilot's experience. Survey research that asks pilots to reflect upon their own experiences is still unable to determine accurate frequency rates. Since Type I is not recognizable and Type II is recognized but possibly not specifically as spatial disorientation, self-assessment can be very difficult. Even more difficult is the classification after an aircraft accident, with or without the pilot's explanation of events. One clue could be aircraft control inputs given spatial position and aircraft location but that is still conjecture. Possibly crew aircraft conversations of the aircrew experience could shed light on disorientation status. But again, with hindsight and knowledge of the outcome, it is difficult to truly put oneself in the seat of the pilot at the controls and understand his sensations and resulting perceptions.

The above definitions acknowledge many factors that influence spatial disorientation in aviation. Benson (1988) separated them into two categories: input errors and central processing errors. According to Benson input errors occur in four different manners. The first input error is due to external visual cues becoming impoverished. In approximately 60 seconds a pilot can become disoriented with the removal of visual cues during straight and level flight (Benson). The second input error type is from aircraft instrument displays: a coherence error due to faulty instrumentation. A third input error type is when vision is impaired by vibration, G-forces, nystagmus, or blood-alcohol content. The final type of input error for an aviation disorientation episode comes from inadequate or misleading vestibular cues. These cues are misinterpreted due to their improper perception or below conscious-threshold sensations.

Central processing errors contributing to spatial disorientation center on "the heuristic nature of the perceptual process … on the conditional probability of a particular temporo-spatial pattern of sensory incoming information being associated with that event" (Benson, 1988, p. 294). Expectancy, bias, previous experience, cognitive simplifications and levels of arousal can all influence how a pilot perceives his/her spatial orientation.

Although not specifically addressed by Benson (1988), other central processing errors can exist. In Chapter 1 the relationship between a pilot, the aircraft, and the environment was presented in Figure 1.1. The pilot brings "baggage" as well as the capability to perform, and, when combined with the varying environments, different outcomes may result. Gillingham (1992) presented those same factors as contributing to spatial disorientation, but then went on to specifically address two: task-loading and training. Training refers to exposure and preparation for various high-risk environments. This is an area that could greatly be improved still today. Gillingham argued that task loading entailed both psychological and physical components; aircraft performance capabilities, improved canopy/windshields, location and type of flight instrumentation, head-up display (HUD), and increased night flying all contribute to spatial disorientation incidents.

Also feeding into central processing errors, a pilot must often juggle many cognitive tasks during a flight. The limited cognitive resources try to decipher environmental cues and internal instrument displays to form a mental representation of spatial position. The schema or mental model stored in long-term memory is manipulated by working memory, and often, decisions are made for cognitive simplification; that is, heuristics are used. These factors plus real-time time constraints all combine to make a pilot very susceptible to disorientation or misperception. For example, during low-level flying the intent is to maintain a low altitude for navigation or specific mission objectives (enemy radar avoidance or enemy engagement). What can be unique in the low–level environment, however, is distraction from terrain clearance tasks due to preoccupation with mission-related tasks. Thus, a competing cognitive task can decrease conscious perceptual processing of incoming environmental stimuli and lead to a mishap. Even though pilots may be very experienced, given complex cognitive loading, they become susceptible to spatial disorientation.

General Visual Sources of Aviation Misperception

As indicated above, spatial orientation/disorientation is a complex phenomenon and visual inputs dominate misperceptions leading to disorientation. Previc (2004a) articulated four aspects of visual perception that drive misperception issues in aviation. Those four are optic flow, false horizons, degrees of freedom, and night vision devices. Night vision devices are addressed in Chapter 8 and won't be discussed here. Optic flow, as presented in the Chapters 2, 4, and 5, is highly dependent upon speed and altitude as well as terrain features. Regardless of speed, if a pilot is high above the ground, optic flow will not be a strong cue; whereas if at low altitude, high speed is perceived only if there is enough terrain (texture) to provide movement cues. In sum, there are many optic flow variables that can lead different pilots to perceive different interpretations of the visual scene even with a rich viewing environment. If given an impoverished environment, then many more interpretations are possible from restricted visual cues. Consequently, optic flow is a common source for both veridical ego-motion misperception and environmental misperception.

The false horizon, according to Previc (2004a), is another contributor for aviation illusions. In aviation, the location of the horizon is not always apparent given different meteorological conditions. Yet, the location of the horizon is crucial for orientation of the pilot's sense of pitch and bank of the aircraft. Finally, Previc addressed the numerous degrees of freedom for the perception of movement in aviation as a contributor spatial disorientation. On Earth, we are generally limited to two axes of motion (for-aft, left-right), but in aviation, the dimension of altitude is added, which significantly changes visual and vestibular perceptual experiences.

Because they can be misperceived, environmental cues can also be sources of aviation misperception. Environmental cues used to fly a visually guided approach to landing are summarized by the following: (1) the runway, (2) the horizon, (3) aimpoint, (4) field-of-view references, (5) terrain cues, global and local, and (6) pictorial/monocular cues. The majority of the illusions and visual misperceptions stem from pilots' failing to properly perceive two basic properties of their spatial position: (1) height (altitude) above the ground and (2) distance to an obstacle to avoid or a runway to land on. Both height and distance interpretation are challenging because they are interpretations of real 3D characteristics based on a 2D proximal stimulus on the retina.

Aviation-related Visual Illusions

Aviation visual perception allows a pilot to successfully fly (e.g., the 2009 Hudson River ditching), and visual perception may potentially derail a pilot's visually guided behaviors (e.g., the 2002 Tallahassee, FL accident). The Flight Safety Foundation (2000) published a tool-kit as part of their efforts to reduce approach-and-landing accidents. Below is taken from that tool-kit (pp. 103–104), and it provides a great introduction to the issues of visual illusions.

1. Visual illusions result from the absence of visual references or the alteration of visual references, which modify the pilot's perception of his or her position (in terms of height, distance, and/or intercept angle) relative to the runway threshold.
2. Visual illusions are most critical when transitioning from instrument meteorological conditions (IMC) and instrument references to visual meteorological conditions (VMC) and visual references.
3. Visual illusions affect the flight crew's situational awareness, particularly while on base leg and during the turn to final.
4. Visual illusions usually induce crew inputs (corrections) that cause the aircraft to deviate from the vertical flight path or horizontal flight path.
5. Visual illusions can affect the decision process of when and how rapidly to descend from the minimum descent altitude/height (MDA[H]).

What follows in this chapter and what is presented in Chapter 7 directly relates to the above five statements. With regard to #1, later in this chapter the illusions of height and distance are specifically addressed in terms of explaining seven possible reasons for the misperception. Item #2 above is one of the major causes of General Aviation mishaps … pilots unfamiliar with instrument flying finding themselves in an environment that requires instrument skills (the need for indirect perception gained via instrument displays). The statement regarding visual illusions affecting vertical and horizontal aspects of the approach most often results in issues of vertical controls (altitude awareness). Khatwa and Helmriech (1998) found in their study of approach-and-landing accidents that the majority of mishaps on final approach were appropriately aligned (horizontally) with the landing runway; however the pilot's improper vertical control and lack of terrain awareness resulted in the accident. Finally, regarding item #5, the decision and execution at the go or no-go portion of a final approach is a very challenging maneuver when done in impoverished visual conditions. This is a moment of high cognitive task loading and many illusions affect a pilot's perception of landing making the decision less certain. Some accidents presented in the next chapter reflect this challenge.

What follows is a presentation of many aviation-related visual illusions and misperceptions that can lead to spatial disorientation, by which we hope to enlighten the reader with different perspectives of some of the illusions encountered by pilots. Other authors have approached the illusion organization in terms of focal and ambient vision-system illusions. Our presentation is based on the pilot's environmental perspective, interfacing with different aspects of flight and the associated illusions. The major areas addressed are decreased visibility (night and weather), featureless terrain illusion, featureless terrain—black-hole illusion, terrain, runways, terrain surrounding runways, and miscellaneous visual illusions; the vestibular illusions then follow.

Decreased Visibility (Night and Weather)

It has been mentioned that too often pilots confidently fly visual approaches at night or continue upon their navigational route despite limited illumination and visibility. Vision's dominance and the fact that accurate spatial orientation requires visual inputs has been discussed in the previous chapters as well as this one. Chapter 2 also outlined the fact that only the rod receptor system functions in dim lighting, and that system does not support color vision or detailed focal vision. Worse, night removes the unappreciated large portion of the visual information, the aspects of the visual scene that are unconsciously processed by our ambient visual system. This leaves the pilot with an over-reliance on the focal visual system which is physiologically ill-prepared to function in dim lighting. Parmet and Ercoline (2008) eloquently differentiate focal and ambient vision: "focal vision serves to orient the perceived object relative to the individual, whereas ambient vision serves to orient the individual relative to the perceived environment" (p. 150). Thus, the meteorological condition of night removes pilots' ability to orientate themselves with environmental cues because those cues are cloaked in darkness.

Consequently, it should be no surprise that "night" is a major contributor to spatial disorientation because of the impoverished visual cues that lead to inaccurate visual perception. The Flight Safety Foundation Approach and Landing Accident Reduction Tool Kit (2000) simply stated "visual approaches at night typically present a greater risk because of fewer visual references, and because of visual illusions and spatial disorientation" (p. 103). Again, this seems to state the obvious, but given the number of accidents it cannot be stated enough. Throughout this section, chapter, and Chapter 7 is it clearly presented the contribution of impoverished visual conditions (primarily night) on aviation mishaps.

Aviation safety articles have been written that urge pilots to respect the hazards of night flight. One entitled, *Night VFR ... an Oxymoron* (Leland, 2001) and another, *Darkness Increases Risks of Flight*, (Wilson, 1999) pleaded with pilots to take heed of the warnings regarding loss of ambient visual cues. The problem rests in the fact that vision normally seems effortless, rapid and accurate; this inaccurate concept relates back to naïve realism as presented in Chapter 1. In actuality, our vision is simply our best-guess that is heavily influenced by cognitive biases. Consequently, overconfidence combined with impoverished visual environments is a deadly mix.

Wilson (1999) related the adverse impact of the night in two well known aviation accidents, the 1972 crash into the Florida Everglades by Flight 401 (due to selective attention focused on a burned-out nose gear light) and the 1995 American Airlines crash into mountainous terrain near Cali, Columbia (due to navigational mode error). According to Wilson neither mishap would have occurred if the pilots could have seen their terrain, or if global and local terrain cues had been visible for spatial orientation and vertical terrain awareness. At the time of both mishaps visual meteorological conditions existed, however, "the visual impediment was the darkness of night" (p. 1). Ostinga et al. (1999) estimated that the night flying fatality rate was nearly three times the rate for daytime flying conditions. Braithwaite, Douglass, Durnford, and Lucas (1998) reported that for US Army rotary-wing operations, 62 percent of all spatial disorientation mishaps occurred at night.

Visual flight into weather and meteorological conditions by pilots leads to spatial disorientation problems similar to darkness of night. This is very dangerous for pilots not trained or proficient in instrument flying. Consequently, these unfortunate pilots find themselves unexpectedly flying with no horizon or ground reference. Benson (1988) stated that straight and level flight would deteriorate into spatial disorientation within 60 seconds if all visual cues were removed. In Chapter 7 the tragic accident of JFK Jr. is detailed and falls right into this discussion of a visual-only qualified pilot finding himself with no visual references for land, water, shoreline, or horizon. Consequently, spatial disorientation resulted and three fatalities occurred. In JFK Jr.'s situation it is estimated he became disoriented within 30 seconds due to the loss of crucial visual cues (Ostinga et al., 1999). In the United States during 1997, greater than 80 percent of all general aviation weather related fatalities occurred following visual flight into instrument meteorological conditions (Ostinga et al.). More recent data from the Nall Report suggests that for General Aviation flying the probability of a fatality is more than twice as high for night flying than for a mishap during the daytime (Aircraft Owners and Pilots Association, 2008). Unfortunately these findings of the danger of night flight are not new. Buckwalter in 1976 wrote an entire book on night flying and in it he presented data from 1969 that found the incidence of an accident on a "dark night" was eleven times higher than accidents on a moonlight-bright night.

Featureless Terrain Illusion

The Federal Aviation Administration's (FAA) Aeronautical Information Manual defines the featureless terrain illusion:

> An absence of ground features, as when landing over water, darkened areas, and terrain made featureless by snow, can create the illusion that the aircraft is at a higher altitude than it actually is. The pilot who does not recognize this illusion will fly a lower approach. (US DoT, 2001, pp. 81–6)

This definition implies that during the daylight hours, featureless terrain could occur over snow, sand, or calm water. For instance, in Chapter 7, a mishap from 1941 is presented that details a water-landing accident due to the smooth surface of the "water runway." At nighttime, featureless terrain can occur over any type of terrain as long as it is not somehow illuminated. Consequently, *featureless terrain* is a more general term that describes the phenomenon of an environment devoid of visual cues. Featureless terrain with poor lighting can also induce a pilot who is flying low-level at night to lose vertical awareness regarding terrain clearance. The lack of referents prevents the pilot from establishing the ground plane or the geometric slant of the terrain. The lack of ground orientation allows the runway to "float", making it difficult to determine the approaching aircraft's height above the ground, the distance to the runway, and a proper perception of depth (Calvert, 1950). It also becomes very difficult for a pilot to perceive a gradual terrain change if the ground lacks features to help define its slant (Previc, 2004b). Consequently, a pilot cognitively loaded with many other tasks (the "full bucket" concept introduced in Chapter 4) may not recognize the change in the terrain's elevation relative to the aircraft's vector. To employ Parmet and Ercoline's (2008) explanation of spatial disorientation, the pilot no longer has an accurate sense of position relative to the plane of the Earth's surface.

White/brown-out Another form of featureless terrain is brought about by whiteout or brown-out conditions. These are very specific environmental situations brought about by high winds in the presence of snow or sand. Also, helicopters encounter these conditions often during any hover or landing, making their height estimations by vision unreliable. In some cases, the atmospheric conditions may make visual identification nearly impossible even though the terrain may have credible features for perceiving depth and distance.

Featureless Terrain—Black-Hole Illusion

The black-hole illusion is a specific form of the featureless terrain illusion related to an approach to landing. The "black hole" is not the runway but the runway's environment, the featureless terrain surrounding the runway. Wade and Swanston (1991) described how multiple objects in the scene provide information to help interpret the visual angle of a particular object and help support accurate judgments of distance and height information. Featureless terrain prevents a pilot from using objects in the environment to guide a landing. Chapter 1 described the 2002 commercial freight accident at Tallahassee, FL, which concluded that black-hole approach conditions contributed to the pilots landing short of the runway. Also in Chapter 1, Figure 1.3 presented the black-hole phenomenon. As described in the featureless terrain illusion, the lack of visual cues seduces the pilot into "feeling steep" and initiating an unwarranted descent. Often the pilot impacts terrain short of the runway in this illusion. There has been considerable research conducted on this particular visual illusion (see below for a summary) and some aviation researchers believe that it is no longer worthy of discussion. On the contrary, this illusion in particular may be one of the classic illusions that in reality needs further research and updated discussions.

Parmet and Gillingham (2002) as well as Parmet and Ercoline (2008) distinguished featureless terrain from the black-hole illusion in that the former is due to a lack of focal vision detail while the latter results from a lack of ambient vision. Gillingham (1992) made a similar distinction in that featureless terrain results from a lack of focal cues, referents, causing a pilot to overestimate height above the ground due to the absence of recognizable objects and textures to gauge size-distance relationships. The argument for a lack of focal cues in addition to the lack of ambient cues, however, could be made for a black-hole approach.

More specifically, the featureless terrain, black-hole illusion is founded on basic visual perception of environmental cues. Global and local features within a scene provide objects of fixation for focal vision as well as peripheral objects for ambient vision. For instance, a pilot may momentarily attend to a small pond for distance, height, and cue comparison (relative size of environmental cues) using focal vision when it is between the pilot and the runway. That pond provides retinal image comparison relative to the runway of known size. When flying past that same pond, ambient vision unconsciously uses that pond in the periphery to judge motion and overall global movement. Orientation within the viewing scene is gained as the pond, serving as a referent, passes by the pilot's peripheral vision. During night flying, that pond is not seen. Thus both focal and ambient vision can be affected by its absence from the scene.

Parmet and Ercoline (2008) reported that, due to the lack of peripheral cues to "help provide orientation relative to the Earth, the pilot tends to feel that the aircraft is stable and situated appropriately but the runway itself moves about or remains malpositioned (is down-sloping, for example)" (p. 167). Note this description parallels the reference of ambient vision providing Earth-fixed coordinates for stable orientation. They go on to state that the worst-case scenario is when the only lights available are the runway lights and the lights of a city in the distance. This type of black-hole approach may set up a pilot to fly a constant visual null angle, resulting in the pilot arcing below the desired glide-path (a concave approach) and landing short of the runway in Figure 6.1; note its similarity yet subtle difference from Chapter 1's Figure 1.3.

Research on the black-hole illusion deserves an expanded discussion not only because of the large number of aircraft accidents that have included the black-hole

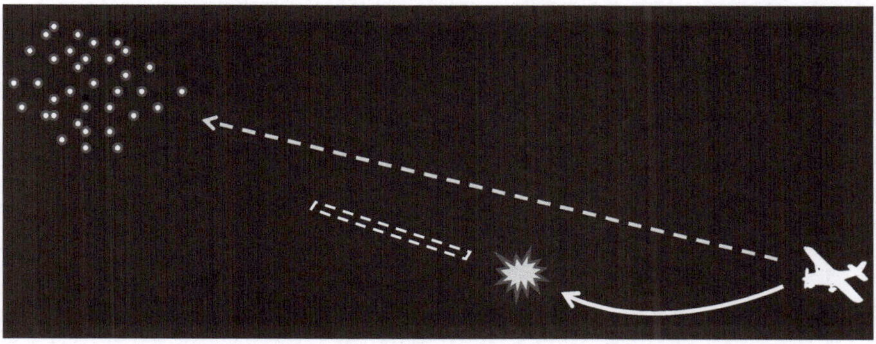

Figure 6.1 Black Hole illusion depiction

Source: Adapted from Parmet and Ercoline (2008).

illusion in the investigative report, but also because there are some inconsistencies in the research that have not been questioned up to this point. For instance, two fairly recent articles demonstrated the problem of handing down explanations. Newman (2007) used a similar figure to Figure 6.1 and described the lack of peripheral cues for inducing the illusion but also mentioned the city-light phenomenon and the visual null theory. Two years earlier, Flight Safety Australia magazine featured an article entitled "Eyeball Error" and presented the black-hole illusion and issues associated with it in terms of featureless terrain as well as city lights beyond the runway. That article stated, " … for complex reasons, not all of which are currently understood, you may fly an approach that is too low" (Newman, 2005, p. 32). The above two examples demonstrate explanations of the black-hole illusion that range from featureless terrain, a lack of peripheral cues, city lights in the distance, a visual null theory, and finally "not currently understood." Readers of these articles are left wondering what the researched-based explanation for the black-hole illusion really is.

Black-hole illusion research: Visual null theory One debated theory is that the black hole arises from pilots attempting to maintain a constant angle of descent to distant city lights, "causing the aircraft to arc far below the intended approach as the aircraft gets closer to the runway" (Parmet and Gillingham, 2002, p. 207). This was shown in Figure 6.1, which is a well-used pictorial of the black-hole approach. The 2007 Australian Safety publication by Newman (2007) discussed the tendency of a pilot to fly the constant visual angle during the black-hole approach, and Kraft's two publications (Kraft and Elworth, 1969 and Kraft, 1978) on this topic have been referenced over and over through the years. Kraft's work (introduced in Chapter 5) initiated when he recognized a common theme across a series of tragic aviation accidents and determined visual misperception was a leading contributor to the mishaps. Kraft's research is often given credit as one of the first to expose the "black-hole effect" to the aviation community and address efforts to educate the pilot force.

Kraft (1978) postulated that if flying horizontally, the visual angle of the city increases as it is approached from a constant, level altitude (Figure 6.2A), and if descending in the vertical (like a helicopter) the visual angle decreases at a constant rate (Figure 6.2B). Thus, if on a 3-degree glide-path approach to the city's airport, the horizontal and vertical components should cancel each other out and

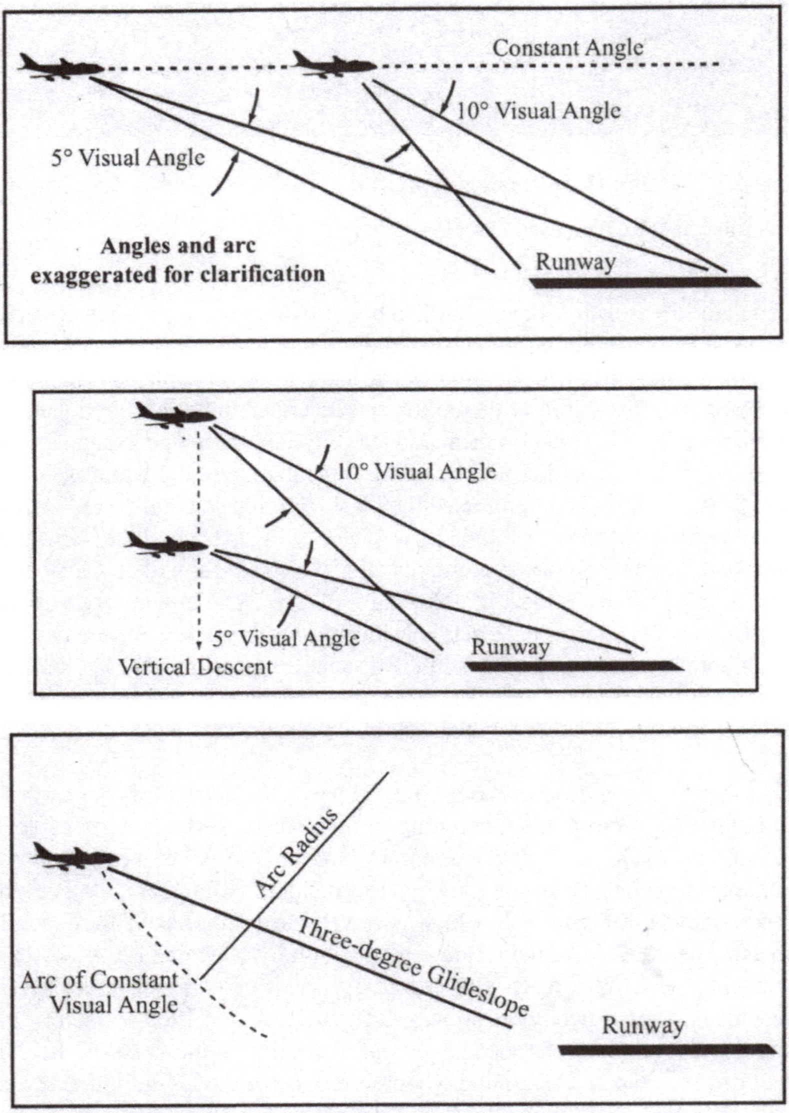

Figure 6.2 Visual null theory of runway perception

Source: From Schiff (1990).

the lights should maintain a constant visual angle (Figure 6.2C). Note that Figure 6.2C exaggerates the concave approach shape due to the constant visual angle theory.

Kraft and Elworth (1969) wrote a short article on night visual approaches that warned of the visual misperception caused by light patterns and sloping terrain. They based their warning on data from a simulation procedure using a simulated 3-degree sloped runway starting 32 km (20 miles) from the runway. The procedure stopped 7.2 m (4.5 miles) short of the runway which was "approximately one mile short of, and 250 feet above the point where relative motion would normally start to favorably influence altitude judgment" (Kraft and Elworth, p. 2). This is an interesting comment with no citation to explain the quantification of values and how a pilot at night could view terrain cues. (One possible explanation may be that the simulator created for Kraft's study may not have allowed subjects to approach the runway any closer than 7.2 km.)

Kraft and Elworth (1969) explained the concave nature of their test-pilots' performance as an attempt to "fly the null"; that is, maintain a constant visual angle of the runway image. The approach-path followed the arc of a very large circle toward the runway, and they stated that this arc maintained a constant visual angle as the pilot descends. In terms of applied aviation safety, the authors also presented valuable information on how to avoid succumbing to the black-hole illusion shared later in this chapter. One last notable point is that the phrase *black hole* was not used in the article.

The 1978 reading was in a textbook, *Psychology: From Research to Practice*, and it presented similar but more detailed information compared to the 1969 article. Kraft (1978) investigated the cause of a series of aviation accidents (presented in Chapter 7). Common to all the accidents was that they occurred at night in visual conditions, the altimeters were working correctly, and the terrain below the approach was dark and featureless (p. 365). Kraft proposed the cause was an over reliance on visual cues, leading the pilots to fly dangerously low approaches and "land" short of the runway—controlled flight into terrain caused by an "error in space perception, that is, a visual illusion" (p. 365). The pilots' confidence in their ability to accurately perceive night environmental cues led to the error, even though cockpit displays would have corrected their misinterpretation.

Kraft (1978) compared three conditions of lights: (1) airport only lights, (2) partial city lights, and (3) total city lights surrounding the airport. Kraft summarized the issue of an illuminated terrain for night approaches (p. 381):

> The perceptual errors occur when the pattern of lights on the ground is irregular and does not outline a terrain variation, and the information about the terrain is not available to the pilot. However, if streets or other regular alignment of lights provide a perspective that will indicate the up-sloping terrain, he will disregard a familiar visual angle, recognizing that he is not approaching flat terrain. If the lights are few enough, or random enough, so that no terrain information is

> available, then the visual angle he customarily uses can be invalid as well as valid. He must determine which applies in this particular approach by depending on some other means of judging altitude and distance.

Kraft determined through the simulator study that pilots flew a concave approach-path arc towards the intended runway. This arc kept the angle of their approach glide-path constant, but was based on visual misperception of the runway, resulting in landing short of the runway. Kraft concluded that an up-sloping runway misled pilots into perceiving their position to be too steep, misled pilots into perceiving their position to be too steep because of less splay, requiring an adjustment to shallow their glide-path, which in turn put the aircraft in a dangerously shallow, unsafe glide-path. Kraft's study also manipulated terrain lighting and demonstrated that with total city lights and a flat surface, performance was improved.

Pendleton (2000) clarified Kraft's work and focused on the basic geometry of the approach angles, relying on the fact that inscribed angles intercepting the same arc of a circle are congruent. Thus, a pilot maintaining the same visual picture of the runway will fly a curvilinear approach to an airfield that initially begins more steeply and then flattens out. The concave-shaped approach, however, will not provide ground clearance the last few miles to the runway.

Pendleton (2000) also presented the black-hole scenario from a pilot's perspective. Pilots' normal interior cross-check of instruments is centered on their attitude indicator (display of ground and horizon); unfortunately this instrument does not provide enough information during a black-hole approach. The difference between a 3 or 4 degree approach is difficult to differentiate on an attitude indicator, and is lethal if uncorrected during a night visual descent.

Finally, according to Pendleton (2000), the black-hole illusion disappears within 3.2—4.8 km (23 miles) from the runway. No reference was cited with this statement, nor was any research presented to justify the assertion, although this claim has been cited in subsequent reports. However, Palmisano and Gillam (2005) manipulated the visual environment within 0.8 km (½ mile) from the runway and found pilots still fell prey to the illusion. The next chapter on aviation accidents also presents investigative reports showing that pilots can be lured into a low approach within close proximity to the runway.

Pendleton (2000) was not the only one to accept the "visual null" theory to explain how a pilot may fly in a black-hole environment. Other authors who accepted and "handed down" (Pitts' 1967 phrase regarding the passing on of unchallenged aviation explanations) this description were Schiff (1990 and 1994) and Wilson (1999). Schiff and Wilson both articulated the inscribed angles theory in very similar fashion. And although they made significant contributions to aviation safety in addressing the dangers of night flying and the illusions that pilots may face, they helped perpetuate the focus on the "visual null" theory. As recently as 2008, Parmet and Ercoline endorsed the visual null theory by stating a "hazardous type of black-hole approach" existed when city lights on up-sloping terrain extend beyond the runway and the rest of the environment is complete darkness (p. 167). However, the visual null theory of the black-hole illusion is not supported in all textbooks. For instance, it was noticeably absent in one fairly recent text (e.g., Previc, 2004a).

Criticisms of the visual null theory The visual null logic does not hold when one examines the typical forward and descending motion components. More specifically, when "on glide-path" the aircraft is traveling at a greater speed horizontally than vertically. For example, flying at 232 km/hr (125 knots) at 3 degrees corresponds to 64 m/sec (211 ft/sec) forward velocity and 3.4 m/sec (11.1 ft/sec) descending vertical velocity—a 19:1 ratio. Consequently, Kraft's visual null explanation is not consistent and fails to address all aspects of the visual approach.

Kraft (1978) further reasoned that, if flying this "visual null," the center of the lights and the pilot's focus resides over the city, and the constant angle puts the pilot on a descending arc towards the lights. Pilots, however, are focused on their runway of intended landing and their aimpoint, rather than on city lights in the distance. Related to the latter point, when a pilot flies towards an aimpoint, the runway and aimpoint expand isotropically. Thus, it should alarm a pilot if his visual angle image fails to grow as he/she approaches it. Similarly, Pendleton's explanation was based on always using the total runway image to calculate the visual angle. However, from a distance, and even as a pilot approaches a runway, the aimpoint is the focus point, not the entire runway's length. The angle being maintained is not the visual angle of the complete runway shape, but rather the glide-path angle or the descent angle to the same spot on the runway relative to the terrain underneath, that is, the angle to aimpoint. This angle should be 3 degrees from the terrain or 177 degrees from the sky above. Also, as demonstrated in Chapters 2 and 5 in the discussion of medial visual angles, the visual angle of the entire runway is too small to make visual judgments.

An additional criticism of Kraft's studies is that he does not provide the length and width ratio of the runway used in his simulations, and that the slope of his runway (3-degree incline) was extreme. Runway slope and perceived ratio can alter a pilot's perception of the runway height and distance during daylight and night conditions (e.g., Previc, 2004a; see below for more discussion of this effect). Kraft's research failed to appreciate the runway's size/shape contributing to the misperception of the pilots. Also, 32 km (20 miles) from a runway to begin a visual maneuver is a bit unrealistic. At 32 km the runway would subtend a visual angle of 0.04 degrees and that would be very difficult for a pilot to use to visually guide his actions. Kraft did clearly demonstrate, however, that given only visual cues of a runway and random lights, a pilot is induced into a dangerously shallow, low approach.

Others have also criticized Kraft's studies and the visual null theory. Schwirzke and Bennett (1991) conducted a re-analysis of Kraft and Elworth's (1969) conclusions regarding black-hole approaches. They challenged Kraft and Elworth's assessment in that the data reported were "not consistent with the curvilinear, low-altitude approaches that would be generated by pilots attempting to maintain a constant visual angle to the runway" (Schwirzke and Bennett, p. 574). The authors' concern was that the visual angles actually flown by the pilots were significantly different for flat versus sloped runway-scene conditions. Schwirzke and Bennett stated that Kraft and Elworth's assertion that pilots attempted to maintain a constant visual angle was inconsistent with their data. Further, accidents may occur outside the up-slope runway/terrain paradigm that was used by Kraft. For example, two black-hole accidents presented in Chapter 7 occurred on down-sloped runways, not up-sloped (St. Thomas Virgin Islands in 1997 and Tallahassee FL in 2002). Down-sloped runways should induce steep approaches, not shallow ones (Parmet and Gillingham, 2002; Previc, 2004a).

Overall, Kraft's (1978) analysis showed that surrounding city lights could improve pilot performance, but that they could also negatively influence glide-path if they were irregular and did not depict the terrain variation. Kraft's safety emphasis of cross-checking the altimeter and use of visual glide-path aids may have been his best contribution to aviation safety; something that still isn't emphasized enough today.

Alternative hypotheses to the black-hole illusion An alternative explanation to the visual null theory, and one more consistent with many of the black-hole approaches or featureless terrain illusion mishaps, is simply the lack of terrain and ambient visual cues. Gray (2006) advocated that our visual system is better at relative discrimination, that is, comparing, than we are at absolute judgments. Thus, when there are no objects in a visual scene to compare against, the runway stands alone and is easily/often misjudged. The black-hole illusion results from focal-only vision attempting to perceive height and distance to the runway from a featureless terrain. The result is the bias of retinal image interpretation due to a lack of environmental context. Parmet and Ercoline (2008), while "handing down" the visual null theory, also described the black-hole illusion in terms of absent ambient cues and a pilot having to make visual judgments using only focal vision for their global orientation. They stated that in this case pilots view the world in an "upright egocentric reference frame" contributing to their misperception of a night approach to landing (p. 151). Following from the lack of terrain and ambient visual cues, the size-depth constancy bias approach attempts to explain the black-hole illusion based on inaccurate size and depth perception of the runway leading to the feeling of being steep, and resulting in a glide-path overestimation. In support of this alternate approach, Roscoe (1980) claimed that most visual illusions were due to "systematic misjudgments of size and distance relationships" (p. 97).

Previc (2004a) also focused on size/depth perceptions, and he may have been the first to fully endorse a theory on the black hole to the exclusion of the "visual null" theory. Previc advocated Perrone's (1984) model of slant misperception based on runway length/width ratio as a leading theory regarding the black-hole illusion. Perrone's theory stipulated that in impoverished visual conditions, pilots focus on the width of the runway in contrast to rich viewing conditions when pilots perceive the entire runway perspective relative to terrain cues. Previc also cited Riordan's (1974) survey of pilots and their preference for the runway size and shape as a visual approach to landing cue and Mertens and Lewis's (1982) conclusion that runway ratio contributes to the bias a pilot may have during their glide-path control.

Perrone's (1984) model has its limits and Previc (2004a) acknowledges them regarding limited distances and explaining all black-hole environmental situations; however, it a useful model for capturing the glide-path overestimation pilots experience during a black-hole approach. For instance, at night, when 3,408 m (10,000 ft, 1.9 miles) from the runway pilots perceive their glide-path to be 5.9 degrees when it actually is 3.0 degrees. Pilots then incorrectly adjust to feel like 3 degrees when in reality they just put themselves on a dangerously low 2.1 degree glide-path (Perrone, p. 1023). Table 6.1 depicts visual angles for the different runway widths and the consistent runway length (2,743 m, or 9,000 ft). The medial visual angle is consistent but the changing runway widths produce different lateral visual angles. These combine for the area of the 2D retinal image produced by the runway. Note in Table 6.1 that the lower runway ratio is, the more accurate Perrone's model is in predicting that the pilot will fly closer to the 3-degree glide-

Table 6.1 Misperception of glide-path model data examples

Runway Length 2,743 m (9000 ft) Runway Width	Distance in km (nm)	Perrone's Equation Calculated GP degrees
22. 9 m (75 ft) wide runway – 120 ratio		
	9.3 (5)	26.8
	7.4 (4)	25.5
	5.6 (3)	23.6
	3.7 (2)	20.6
	1.9 (1)	14.8
	0.9 (0.5)	9.4
	0.5 (0.25)	5.4
45.7 m (150 ft) wide runway – 60 ratio		
	9.3	14.2
	7.4	13.4
	5.6	12.4
	3.7	10.6
	1.9	7.5
	0.9	4.7
	0.5	2.7
91.4 m (300 ft) wide runway – 30 ratio		
	9.3	7.2
	7.4	6.8
	5.6	6.3
	3.7	5.4
	1.9	3.8
	0.9	2.4
	0.5	1.4

path. Thus, perception of glide-slope is more accurate on lower-ratio runways. It is also important to emphasize how small an image a runway projects onto a retina; attempting to base a visually-guided action on such a small image can only lead to bias in perception and action.

The explanation of the black-hole illusion can also be examined in terms of the model of parallel processing for angular size and objective size previously described by McKee and Smallman (1998). This model was mentioned in Chapter 2. Relating this model to visual illusions in the black-hole scenario is fairly straightforward. What results is the absence of the distance information indicated on the right-side of the flow chart which results in the inability to make comparisons (Figure 6.3). Consequently, the only information available for judgments regarding distance and altitude for position as well as target size and shape for glide-path control is a 2D retinal image of the runway. Size, shape, and depth constancy are inaccurate, if not completely absent.

Hence, one can understand how the black-hole illusion cannot simply be explained by the perception of "feeling steep". Too often, authors casually mention

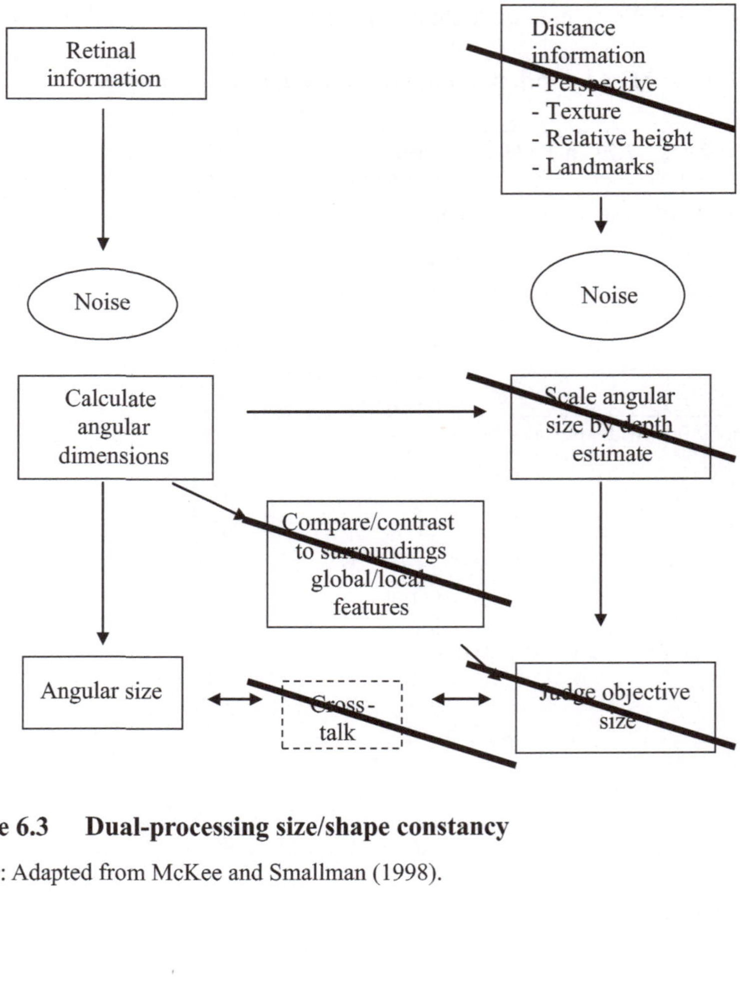

Figure 6.3 Dual-processing size/shape constancy

Source: Adapted from McKee and Smallman (1998).

"constant visual null" without expanding upon the meaning of that and the ensuing concave approach. More current human-factors research on the illusion by Gibb et al. (2008) was presented on this subject in Chapter 5.

Black-hole takeoff illusion (Somatogravic illusion) The black-hole takeoff illusion is not as well known or as well researched, but has unfortunately claimed its share of pilots' lives. If visual cues are available, pilots can correct their momentary spatial disorientation. But if vision is limited, taking off into a dark night with no discernible horizon may induce a pilot to make a deadly control input. This illusion has also been called the dark night takeoff illusion or the pitch-up illusion; more recently it is simply referred to as a general somatogravic illusion (Lessard, Matthews, and Yauch, 2000; Newman, 2007; Parmet and Ercoline, 2008).

A pilot who experiences the black-hole takeoff illusion normally only has vestibular input to determine spatial position. When the aircraft accelerates, otolith organs of the vestibular system accurately sense forward/linear acceleration; however, with no visual information to confirm or deny the sensation, a pilot will "feel" as if he or she is climbing or pitching up. Too much pitch-up in an aircraft, especially close to the ground, is very dangerous due to the potential of a stall (exceeding the aircraft's angle of attack). This illusion of a climb has been so strong that it has induced pilots into pushing over on their yoke to minimize or arrest their climb rate. In reality, the pilot is climbing out at a safe and acceptable angle, but pushing over puts the aircraft in a dangerous position by descending or not climbing rapidly enough relative to the terrain below. This sequence of events is most likely to occur following takeoff over a "textureless" area in conditions of extreme darkness but good visibility (in the meteorological sense), which may induce the pilot to forsake his attitude instrument(s) in favor of looking out (Buley and Spelina, 1970, p. 553). If you cannot understand the power of this illusion, Flight Safety Australia (1999) challenged the reader, "the next time you fly as a passenger in a commercial jet, close your eyes during the takeoff roll and see if you get the sense of pitching up. It is a very powerful sensation, and has led to the loss of many aircraft, both civil and military" (p. 28). Chapter 7 presents some accidents involving the black-hole takeoff illusion.

The black-hole takeoff illusion can be avoided by using a *composite cross-check*, that is, when a pilot uses both inside the cockpit instrumentation and external visual cues to maintain pitch and bank control. Although takeoff is primarily a visual maneuver, pilots who are accustomed to using their instruments frequently check their attitude indicator (for pitch and bank information via an artificial horizon). Thus, experienced pilots use a composite cross-check during takeoff allowing them to maintain a constant pitch angle during the takeoff, somewhere between 7 and 15 degrees of pitch (aircraft dependent). Pilots also have a *vertical velocity indicator* (VVI) that depicts the aircraft's rate of climb or descent. Pilots not accustomed to instruments, because of their visual-only flight qualifications, are less likely to check these and other instruments during a takeoff.

This illusion is similar to the black-hole illusion which could also be avoided if pilots cross-checked their instruments regarding their position from the runway and descent rate. But pilots are overconfident in their visual capabilities and prefer to fly the approach visual and accomplish the takeoff visually. The cognitive aspect of illusions, pilot overconfidence, often is the biggest hurdle to overcome and points not only to education but actual simulator training to de-bias the illusion. The black-hole takeoff/somatogravic illusion is less common than the black-hole illusion experienced during approach due to the fact that during takeoff the nose of the aircraft is high and occludes the outside scene. Thus, normally, the pilot must transition to instruments, mainly the attitude indicator, to maintain a constant pitch angle and ensure the wings don't roll or bank.

This illusion can also occur in level flight or while on a slight descent while configuring the aircraft. For instance, if at level flight the pilot retards the throttles, the sudden deceleration will be processed by the otolith organs of linear deceleration and lead to the perception of a pitch-down. There have been times while configuring the aircraft for landing, in extending the landing gear and/or flaps, the first author momentarily experienced illusory pitch changes and rapid deceleration, leading to some spatial disorientation regarding level flight pitch picture. The corrective action has always been to simply transition to instruments and concentrate on the attitude indicator in terms of the horizon and the aircraft's relative position. The somatogravic illusion is addressed again in the vestibular illusion section of this chapter and is more specifically referred to as the oculogravic illusion.

Terrain

Terrain texture Illusions involving terrain can occur in either rich viewing conditions or impoverished conditions. When the terrain is visible and has features, the size constancy illusion may lead to a misperception of height. For example, when flying over smaller than expected terrain objects (e.g., immature trees rather than old growth forest), the pilot may be induced to fly unusually low. This is shown in Figure 6.4.

Of course the opposite can also occur in terms of flying too high over misperceived terrain. In terms of low-level terrain safety, this direction of misperception is not a real problem, but if trying to stay low to avoid detection, terrain could induce the pilot into flying too high and then being detected. Inaccurate size constancy of terrain objects for landing can also induce a steep approach. A similar problem may occur if terrain changes as a pilot is flying a visual approach to landing. For instance, if accustomed to flying over large trees and then, prior to landing, smaller trees are under the flight path, the pilot may fly an unsafe shallow approach. If terrain perception is acclimated to small trees/shrubs and then, prior to the runway there are much larger trees/shrubs, the pilot may be induced into a steep approach. Parmet and Gillingham (2002) stated that although ambient vision plays a role in some size-constancy illusions, focal vision is dominant in directly perceiving or misperceiving size.

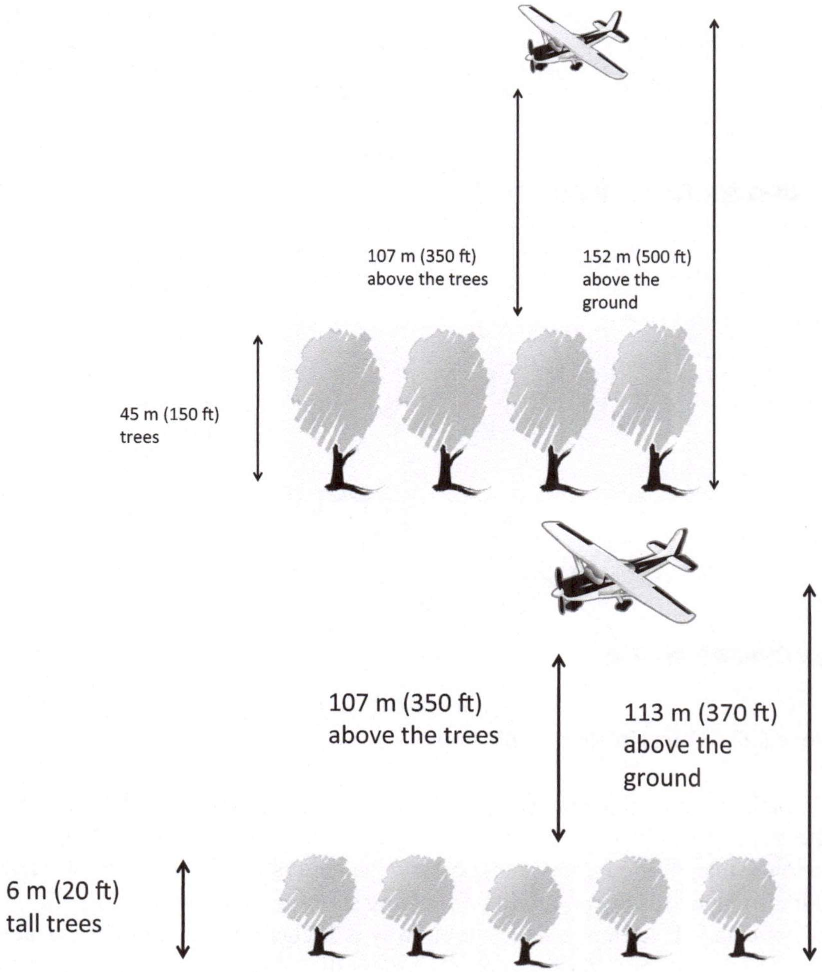

Figure 6.4 Size constancy over terrain

Terrain geography Another series of visual illusions can occur when the geography of the terrain surrounding a runway induces perception of being too steep, being too shallow, climbing out too excessively or failing to climb enough. Again, these disorientating illusions occur during the day; ironically, if devoid of terrain illumination there would be a good chance the pilots would not be prone to these geographical illusions, because if pilots are unable to see the terrain, then they cannot be fooled by the terrain relative to the runway.

For example, Figure 6.5 depicts a runway environment that has an incline prior to the approach end of the runway, with the terrain rising up to the level runway. A pilot approaching this runway would perceive himself to be steep relative to the terrain below. Consequently, as presented in the black-hole illusion, if

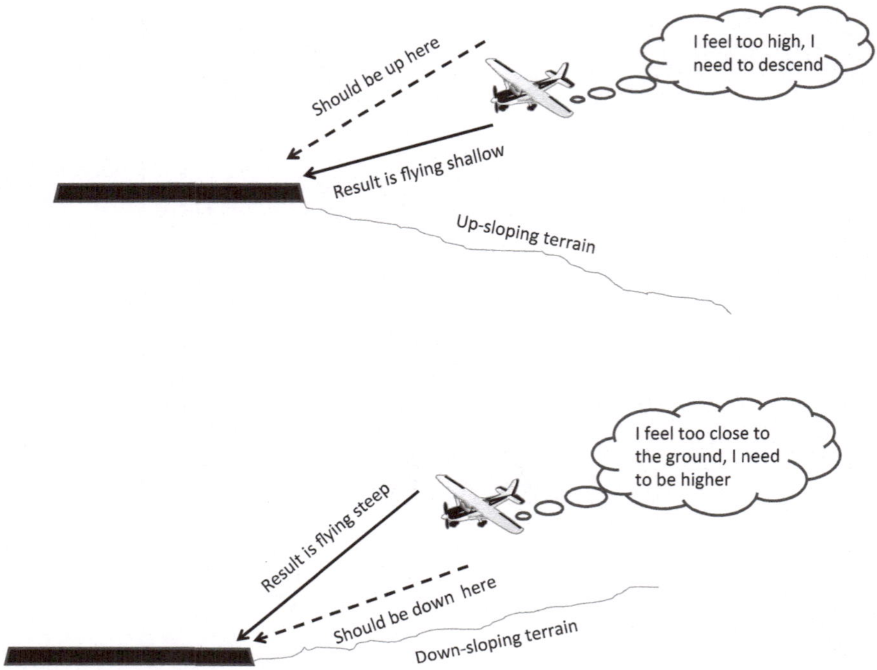

Figure 6.5 Terrain sloping illusions

glide-path is overestimated, the pilot will descend to a more shallow approach angle to perceptually capture the desired glide-path. However, this puts the pilot on a lower than desired glide-path, possibly dangerously low relative to obstacles, and it also increases the possibility of landing short of the runway.

Conversely, if terrain prior to the runway is higher than the airfield elevation (down-sloping terrain), the perceived nearness of the terrain prior to the runway induces the pilots to decrease their descent rate and fly a more steep approach. This is glide-path underestimation. An example of this is on final approach to San Diego's airport in California. The approach-path brings the aircraft right over the downtown area and from there the terrain slopes down toward the runway. Due to the elevated terrain prior to the runway, the pilot may be induced into flying steeply toward the landing runway.

Similarly, perceptions of climbing rate after takeoff can also be influenced by the terrain geography surrounding the runway. The terrain below may possibly create the perception of excessive climb-rate if the terrain falls away rapidly, or if the terrain at the departure end of the runway rises rapidly the perception becomes that the climb is too shallow. In either case, a composite cross-check ought to keep any perception issues in check by confirming appropriate pitch attitude while departing the airfield.

Low-level terrain issues Flying a low-level sortie or low-level navigation, and flying at high speeds lead to special circumstances when optical flow (discussed in Chapters 2, 4, and 5) can help pilots ascertain their future position in space relative to the terrain. Specifically, the point of expansion, the source from which the flow emanates, is probably the most beneficial aspect of optic flow. For instance, if the point of expansion of optic flow rests on a ridgeline ahead of them and no corrective action is taken, then their aircraft will impact that particular spot on the ridgeline.

The challenge of low-level flying using optic flow is that changes in terrain directly influence perception of height and speed. Terrain that becomes sparse provides little information for optic flow. For instance, if flying over a desert or flat snow covered terrain, there is an insufficient number of environmental cues to provide the pilot with height information. This is contrasted with a low-level route that may consists of man-made objects of known size. These objects in the environment can quickly provide a point of reference for the pilot regarding his or her altitude maintenance for object-size comparison. For example, the known size of cars and trucks provides information to the pilot that helps with height and distance perception.

Also, the terrain may change across a flight route and not be a credible source from one navigational leg to the next. For example, the first author is familiar with a low-level route that begins over eastern Arizona and the mountainous terrain of pine trees. However, the cues for height and speed greatly change as the route takes the aircraft to the west over the sparse desert with large rock formations, desert shrubs, and cacti. This was shown in Figure 6.4. If pilots flying this route "calibrate their eyeballs" to the 45 m (150 ft) pine trees in the beginning of the flight, they might misperceive the much smaller desert Palo Verde trees later and become induced to flying closer and closer to the ground based on their original visual calibration. Recall that, close to the ground, the time for an error in terrain clearance and aircraft control input can be just seconds and that decline in altitude can be the difference between life or death.

A cause for concern when flying low-levels near sunrise or sunset is that the low angle of the sun may create shadows which can occlude perception of terrain hazards. For instance, a ridgeline may be masked if it is located between the pilot and a larger terrain feature if the sun's angle is low enough. The unknowing pilot is unable to perceive depth of the terrain ahead and may clip the smaller ridgeline if visual perception is attended to farther-ahead global objects at the expense of foreground dangers.

An additional low-altitude illusion is the "ridge height illusion" (Previc, 2004). It is due to distance parallax, that is, a misinterpretation of motion parallax due to different viewing angles of objects at different distances from a position of altitude. Distance parallax can create the illusory perception that a near object is lower in elevation due to the greater downward angle than a farther object because it is viewed at a smaller angle. In fact, the nearer object is a danger to the aircraft and the farther object is well below the aircraft's altitude.

Haber (1987) detailed a controlled flight into terrain mishap in the Nevada desert that illustrates some of the dangerous realities of high-speed, low-level flight. In this mishap, the pilot accomplished two turns, a 160-degree left turn followed by a 45-degree right turn at 7-Gs, and subsequently impacted the nearer of two ridgelines. Tunnel vision results during high G maneuvers, as the blood flow to the head and eyes is significantly reduced. Such physiological changes while maneuvering at high speed near the ground greatly influence visual perception. Specifically, tunnel vision results in the loss of ambient cues, which in this case made it less likely for the pilot to notice the nearer ridgeline because he was focused on the horizon created by the farther ridgeline.

The Runway: Size and Shape Constancy

Perception based on the visual image of the runway has been shown to be critical for an approach and landing. This is understandable because the runway serves as the primary object of interest for focal vision. The runway's size and shape are most often noted and preferred by pilots as their visual point of reference (Riordan, 1974). The following section will address additional constancy problems in perceiving the runway beyond their likely influence in the black-hole illusion as described above. Keep in mind that the runway may cause visual-spatial disorientation even during day time hours with ample visual cues available, largely due to cognitive issues of prior-experience, training, and expectations from the "norm." At night, however, expectations also come into play regarding the shape/size of the runway via only a runway's outline from runway-edge lighting. This nighttime issue was discussed in the black-hole illusion regarding the perception a pilot may have of a runway's shape and size based on its ratio (length divided by width).

Perceptual constancy can be a confusing term. It does not imply that the retinal image is constant, because when viewing objects while maneuvering in an environment the retinal images change in size and shape. The term *constancy* refers to the fact that despite the change in size and shape of the retinal image, the *perception* of the object tends to remain the same. Recall from Chapter 2 that regardless of your distance to or orientation angle from a door, it is still perceived to be a perfect rectangle, although from some distances and orientations, the retinal image shape is actually a trapezoid.

A specific size-shape-distance error that commonly occurs in aviation is the interpretation of runways. Runway size (length and width) varies from airport to airport, and because the retinal image size of the runway also varies with distance, there are no constant measures a pilot can use to accurately interpret size and distance based on the retinal image. For instance, if a pilot is especially familiar with a runway of a certain size (e.g., 8,000 × 150 ft, length-to-width ratio of 53), his tendency will be to assume that same familiar size when approaching other, unfamiliar runways. In turn, his interpretation of the depth /distance of that unfamiliar runway will be incorrect. For example, if the size of the unfamiliar runway has a lower ratio than the familiar runway (e.g., 5,000 × 300 ft, ratio of

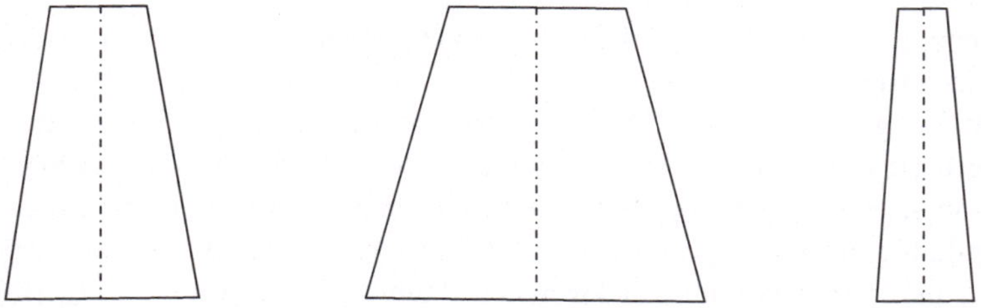

Figure 6.6 Perception of various sized runways

17), then the interpretation of the distance to the unfamiliar runway will be less than reality, because the retinal image size is large and matches a nearer distance to the familiar, small runway. The result is a tendency to flare high and land long or to misperceive their glide-path to be too shallow and initiate an unnecessary climb.

Conversely, if the unfamiliar runway has a higher ratio than the familiar runway (e.g., 10,000 × 100 ft, ratio of 100), the retinal image will be smaller than expected and a greater distance will be assumed (because the familiar runway has a smaller retinal image compared to the familiar runway). The result in this case is a late flare and a hard and short landing, or the misperception of the current glide-path as being too steep, resulting in the pilot initiating an unwarranted descent to shallow their glide-path. Runway sloping can further interact with experience-based runway size/shape expectations and increase the likelihood of misperceptions (see discussion below). Other objects in the visual field (buildings and plants) may help a pilot to assess distance more accurately, but they also vary in size, so they also may only be interpreted correctly if a pilot is familiar with the airport. Nighttime landings increase the likelihood of incorrect runway size-distance perceptions because there are fewer other visible cues for size and depth. The accident described by Ercoline, Weinstein, and Gillingham (1991) demonstrates the negative results of this illusion.

Examples of runways with extreme dimensions were pictured in Chapter 4. For instance, Figure 4.4 depicted the simplistic looking runway at Albuquerque International Airport landing runway 26. This is contrasted with Figures 4.6, 4.8, and 4.10 that have varying shapes and sizes. Other extreme examples are found in terms of a very thin runway in a jungle in Costa Rica (1,100 × 11 m (3,609 × 36 ft) for a ratio of 100); unless a pilot is familiar and proficient at landing on small, narrow runways it could deceive the pilot into feeling steep and fly a low approach. (A runway of this small size would look like a sidewalk to an unfamiliar pilot.) Of course, only if a plane was capable of landing/taking off on small runways would a pilot venture into this particular airfield; however, that same small plane may have departed from a large international airport, thus creating two very different runways for the pilot to perceive. Compared with a very thin runway is runway 04R at Chicago Midway (1,965 × 46 m (6,446 × 150 ft) for a ratio of 43); it is relatively short and wide and could induce a feeling of being shallow and have the pilot incorrectly adjust to a steep approach.

Differentiating between size and shape constancy is tricky when it comes to runway illusions because shape and size are closely related concepts for runway perception. Both glide-angle and runway slope will affect the perception of these characteristics. Depending on the runway slope, the runway will take on a different size and shape. Figure 6.7 depicts two possibilities, a down-sloped runway and an up-sloped runway. The runway slope itself must be fairly steep in order to be directly perceived, but even shallow slopes can influence the perception of the runway size and shape.

For example, imagine the perceptual confusion if a pilot encounters a runway of unfamiliar size/shape combined with terrain illusions. The perceptual directional bias of the induced illusions might counteract each other to eliminate illusion, or combine to make the illusion more compelling and increase the likelihood of a mishap. For example, a long/thin runway may induce a pilot to feel steep and induce a shallow approach. Simultaneously, down-sloping terrain may induce the feeling of being low, leading the pilot to fly a steep glide-path. In such a case, the two inducers at least somewhat cancel out each other. In contrast, consider the scenario at an airfield in Germany (discussed again in Chapter 7) in which the runway size/shape combines with up-sloping terrain to reinforce the illusion

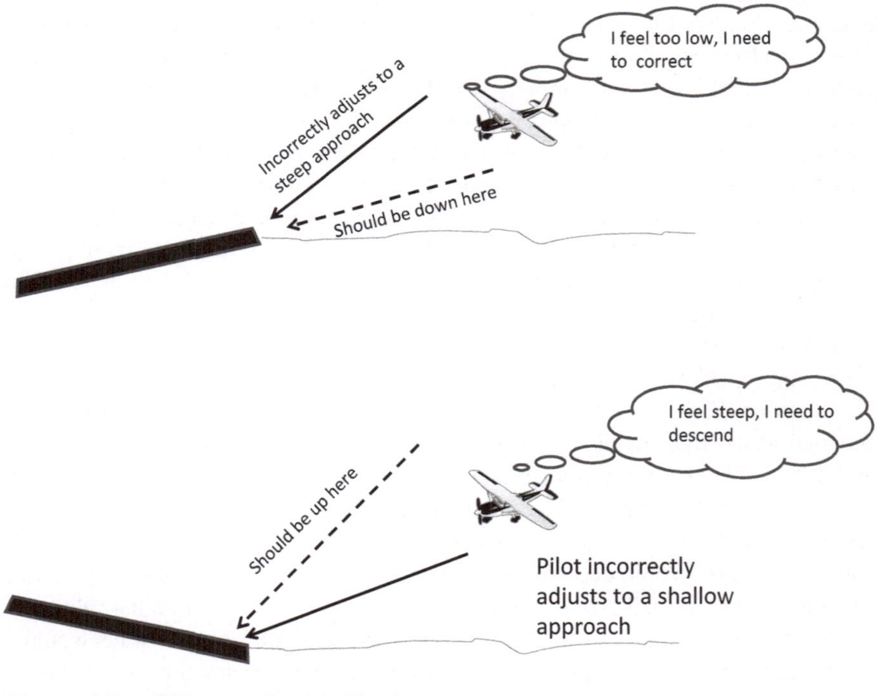

Figure 6.7 Runway sloping illusions

Source: Copyright Boeing (2006).

of being too steep. This redundancy of visual perceptual cues strongly seduces a pilot into a shallow approach. Thus, although the illusions are often isolated for discussion purposes, in reality, pilots may find any combination during an approach to landing. Perceptual problems may be further aggravated if poor weather, pilot fatigue, or emergency actions capture a pilot's limited attentional resources.

A final runway illusion associated with size/shape constancy, that was just briefly introduced, is an illusion that occurs just prior to landing—the flare. The previous visual-perception issues revolve around getting to the runway, the approach. Once at the runway, just prior to landing, the pilot flares the aircraft. This is a tricky maneuver that cuts the descent rate approximately in half as well as cutting 10 knots off of the approach speed to the landing speed of most airplanes. The flare is in some ways a blind maneuver because the nose of the aircraft prevents the pilot from actually viewing the runway and the intended landing spot. Consequently, what is required is a "feel for the wheels" in terms of where and when the aircraft will actually touch the runway. This is accomplished by pilots transitioning their focus to the horizon/end of the runway and using ambient vision to perceive the rate of sink toward the runway.

Size and shape constancy tend to come into play with this flare illusion when there is an extremely wide runway; a pilot is misled into flaring too soon. Similar to the other runway illusions, the flare results from the experience that when the perception of a runway is fairly wide, the aircraft must be either on the runway or just above it. However, for very wide runways, the perception of its great width occurs while still dangerously high relative to the normal height for flare initiation. What may occur is a wing rock and possibly stall just above the runway, which may then result in a "dropped in" landing—a hard landing requiring inspection of the landing gear. Or, worse case, a stall and wing rock with the wing tip catching the runway prior to the wheels. Young student pilots are vulnerable to the flare illusion; the first author witnessed dropped-in landings as an instructor at Williams Air Force Base. Training sorties were often flown to March Air Force Base outside of Los Angeles, CA, where the runway was twice the width of the runway the students were used to at the home station. Despite the pre-flight discussion, invariably some of the student pilots would initiate the flare dangerously high due to the perception of the unusually wide runway.

Conversely, when a pilot learns to control a flare maneuver on a wide runway, a thin runway can trick pilots into thinking that they still have time to descend prior to flaring the aircraft. The pilot fails to even initiate the flare maneuver and unexpectedly the aircraft abruptly hits the runway. The landings are unusually "hard" because neither the descent rate nor the speed was reduced. The first author experienced this while flying into smaller airfields in obscure locations around both the United States and the world. Very narrow runways that looked like sidewalks made it perceptually difficult for a smooth landing due to the feeling that the aircraft was still too high to initiate the flare—only to be proven wrong.

Other Illusions

Effects of fog/haze/rain In addition to simply making it harder to see, fog, haze, and rain can systematically shift perception. Fog or haze often makes runway lights appear more distant. This perception results because distant objects normally appear hazy while closer objects have much more detail. Rainwater can cause image distortion when attempting to see through a wet windshield, leading to improper distance and depth perception. And, as outlined above, if a pilot misperceives depth, then the pilot may make inappropriate control inputs to the aircraft.

False horizon A false horizon occurs when there is a misperception regarding the location of the true horizon due to false horizontal cues. A cloud deck during the day or night, city lights, the beach, mountains, and reflection off water all can disorientate and momentarily confuse a pilot regarding the location of the actual horizon. An example provided by Newman (2005) is a line of lights along a highway on a dark night. If approaching the lights at an angle, a pilot could misperceive his or her attitude to be slightly banked in reference to the false horizon and then put in an aircraft correction. A sloping cloud deck may have the same perceptual effect on a pilot in terms of the horizon.

A final example is given by Previc (2004a), who described the false horizon that occurs when a pilot mistakes a shore line at night for the horizon. This result is because of the water and horizon blending together (discussed below) and the only horizon-appearing line is the shore line. This could become very disorientating as the pilot passes over the shoreline (thought to be the horizon) inducing the perception that the aircraft is climbing.

Blending of the earth and sky This is similar to the false horizon; however, it occurs simply when the horizon is not present or clearly defined. Not that the horizon is false, it is just not obvious to the pilot. Pilots flying over the ocean on a dark night have trouble differentiating where the ocean stops and the sky starts. This certainly must have been the case in the 1999 JFK Jr. accident in which haze occluded the horizon. Another example may occur when the sun, clouds, and shadows reflect off a smooth surface of water making it difficult to confirm exactly where the horizon is location, the splitting between earth and sky.

Dip illusion The dip illusion results from absent ambient cues and occurs during formation flying at night when one aircraft is trailing another. During this illusion scenario, as an aircraft drifts back to "take spacing," the pilot intentionally falls back and below the lead aircraft. In attempting to keep the lead aircraft in the same spot on their windscreen as he drifts back, the pilot may need to decelerate, and in decelerating, the nose of the aircraft must come up. At this point in the maneuver, the small change in pitch attitude and lead's location in the windscreen can result in the trailing aircraft descending excessively below lead. This is dangerous if the formation is flying at low altitudes in mountainous terrain.

Other aspects of this scenario relate to the difficulty in maintaining a constant distance from a point light source (e.g., the lead aircraft) on a dark night by visual reference only. From experience, the first author has found that the trailing aircraft falls back significantly, and then upon realization of how far back he has fallen, the pilot pushes up the power only to find himself now with far too much closure towards the lead aircraft as the point light source quickly grows in his windscreen. This scenario results from a lack of ambient visual cues and is similar to rejoining visually during air refueling.

Leaning on the sun illusion This illusion results from a pilot's expectation that the sun is always high above in the sky. While flying through a cloud, the illuminated portion of the cloud from the sun is perceived to be "up." Previc (2004a) called this an "altered luminance gradient" (p. 294). It is only during midday flying that the luminance of the sun is directly overhead and in the morning and late afternoon the sun is at lower angles. Pilots may fail to account for sun angle while in a cloud and become disoriented regarding true vertical.

Autokinetic illusion If a pilot (or anyone) fixates on a point light source, after a few seconds the light appears to move. This is a result of continuous, small, involuntary repositioning movements of the eyes. According to Air Force Manual (112–17, Volume I, 2005), a stationary light stared at for between 6 to 12 seconds in the dark will appear to move. With no other visual cues available other than the single light source, the normal eye jitter movements get interpreted as movement of the object being viewed (Newman, 2007). The purpose of the involuntary jitter movement is to prevent adaptation to a stationary/unchanging retinal image. Otherwise, during a fixation, if the eyes were perfectly still, after less than a second, the visual scene would completely fade. The jitter is not normally noticeable because the entire visual scene moves just a tiny amount and the visual system discounts the movement. However, in the dark with no other frame of reference, the point of light seems to move due to the lack of a frame of reference. For a pilot, the result of this physiological characteristic is misperception of either their own movement or movement of a light source—either of which can become disorientating.

Vection This is the sensation of self-motion induced by relative movement of viewed objects (Air Force Manual, 2005). Unlike illusions that occur in the absence of ambient vision, this illusion is induced by ambient stimuli: large peripheral motion signals. These peripheral, large-field motion cues are similar to optic flow patterns, and thus, sometimes lead to the false perception of self-motion. An example occurs during formation flying, where it is sometimes a challenge to determine if movement is caused by one's own aircraft or by the lead aircraft. Simulators take advantage of vection to increase the realism of the simulator experience.

Moth illusion The Moth illusion may come into play while a pilot attempts to maintain formation position by focusing on another aircraft's lights. Often the result is that the pilot is "drawn-in" towards the aircraft and ends up dangerously close. Similar to a black-hole illusion, the lack of ambient cues induce the pilot into focal vision only and the pilot fails to accurately perceive depth and distance to the target. One technique to avoid this illusion while flying at night is for the pilot to continuously keep his or her eyes moving between different visual references for formation position, that is, not just on the one light source but other references as well.

The "moth effect" is a term that has also been used to describe the automobile accidents that occur when a car traveling at night on a dark road impacts the rear end of another car parked, with its lights on, on the shoulder or off the road (Uttal, 2006). Uttal provided three possible explanations for this phenomenon: (1) drivers are attracted to the lights, (2) drivers become "hypnotized" by the lights in an 'attentional tunneling" fashion, and (3) drivers tend to steer in the direction they are looking (p. 282).

Chatziastros, Readinger, and Bulthoff (2003) examined the moth illusion and found that drivers tended to steer towards the direction they attended their gaze but that was minimized if more ambient information was provided. In other words, ambient visual cues provided optic flow and global cue information for visual feedback. Uttal (2006) concluded the same. Since driving, like flying, is a visually controlled behavior, if ambient cues are removed the visually guided action may be prone to error.

Focal traps A focal trap can be any salient visual cue that captures a pilot's focus. For example, a dirty windscreen may trap a pilot's attention on the windscreen rather than scanning for traffic conflicts further out in front of the aircraft. Consequently, he may not be able to detect hazards such as other air traffic or birds directly in front of the aircraft.

Empty-field myopia Related to focal-traps, empty field myopia results from a lack of objects in an environment upon which to fixate or focus upon. When this occurs, the eyes naturally gravitate toward their resting spot, or their dark focus, which is approximately 12 m out in front of our eyes. Consequently, the eyes are not accommodated (focused) for detecting (sensing) any objects in middle-to-far distances.

Break-off phenomenon According to Newman (2007) this is a rare illusion and results in pilots feeling detached from the environment and their aircraft, for example feeling as if they were sitting on the wing watching themselves fly. Benson (1988) states that this sensory disturbance is normally experienced in single-seat piloted aircraft when flying at a high altitude over featureless terrain for lengths of time. The dissociative sensation would normally disappear once attention was shifted and/or identifiable global features became present in the pilot's environment. This phenomenon may be more likely to occur during low workload phases of flight.

Reasons for Misperception of Altitude and Distance

Aviation visual perception research has provided an assortment of hypotheses regarding the underlying causes of the misperception of altitude and distance. Gibb (2007) presented seven reasons (Table 6.2) that together or alone may contribute to a pilot feeling steep, and shed some light on why pilots misperceive altitude and distance possibly resulting in controlled flight into terrain. All seven reasons apply to approach and landing scenarios, while several of them are also applicable to low-level navigation. The first four of these reasons have already been explained in detail above in this chapter as well as in Chapter 2. The final three will be detailed below.

Table 6.2 Seven reasons for misperception of altitude and distance

1. Constancy of apparent size-shape-depth
2. Conflict between familiar and relative size
3. Overestimation of visual angles in the medial extent
4. Deficient terrain orientation cues
5. Misperception of slant, optical versus geographic slant
6. Perception of approach lighting systems
7. Tendency toward equidistance

Source: Gibb (2007).

Reason 5 Misperception of slant, optical versus geographic slant. Geographic slant, that is, slope of the terrain (Previc, 2004b), involves both optical slant and perception of height or angular position (Mertens, 1978a). Because distance, depth, and orientation cues are absent in impoverished conditions or featureless terrain, optical slant becomes the sole remaining cue to actual slant. Optical slant, however, is based on line-of-sight relative to the surface (Mertens, 1978a) and if the surface is not available optical slant is also not an adequate cue. Even if optical slant is available, Mertens (1978b) found that subjects overestimated the optical slant value. Consequently, if cultural or runway approach lighting illuminates the terrain enough to allow for optical slant perception, it may only further contribute to height and distance misperception.

Reason 6 Perception of Approach Lighting Systems (ALSs). Approach lighting systems were developed to extend the runway environment towards the pilot during the transition from instrument to visual conditions in low visibility. Acquiring the runway image and flying a night visual approach to landing using the approach lighting system in good visibility may perceptually increase the apparent runway ratio and cause the runway to appear narrower than it really is (Mertens and Lewis, 1983, Perrone, 1984). This misperception of the runway's size/shape may in turn promote glide-path overestimation.

Reason 7 Tendency toward equidistance. According to Gogel (1965), the equidistance tendency occurs when objects appearing near each other in the visual scene are perceived to be at the same distance in the absence of other visual cues. Thus, this tendency may occur in terrain with minimal visual cues or impoverished viewing environments. For slanted-in-depth objects, the equidistance tendency favors the foreshortened, frontal plane—resulting in a perception of slant over-estimation. For a low-level scenario, if minimal objects are present (e.g., on a barren terrain environment), those few objects may be perceived as being at the same distance, thus confusing the pilot's perception. This explanation is related to the lack of distance cues and familiar/relative size cues (Reasons #2 and #4).

Other authors have also created summaries of causes of spatial disorientation and aviation illusions. The next two tables share the summarized spatial disorientation illusions and characteristics as presented by Previc (2004) and the Flight Safety Foundation, Approach and Landing Accident Reduction, FSF ALAR (2000). Each summary uses a different organizational principle and together they serve as a reference as well as comparison of disorientation perspectives. Previc groups his illusions into the three categories of "Caused by distorted ambient vision", "Caused by absent ambient vision", and "Display related." FSF ALAR focuses on depth/altitude misperceptions and the resultant pilot actions and results.

Vestibular Illusions

Often in aviation, when spatial disorientation is mentioned, pilots first think of those illusions associated with vestibular spatial disorientation. Hopefully the sections above clearly indicate how visual perceptions/misperceptions can lead to spatial disorientation. In fact, if adequate ambient and focal visual cues are available, often these vestibular illusions can be minimized if not altogether avoided. However, we also acknowledge the strong connection between the visual and the vestibular senses, which makes it imperative that our discussion also includes vestibular illusions.

Recall from Chapter 2 that equilibrium, (leading to an absence of a movement sensation), occurs when the fluid in the canals, the perilymph, does not move at a rate above threshold, and this threshold was quantified as 2 degrees/second. Above this rate the fluid motion will cause the cilia in the chambers to bend enough to release neurotransmitter chemicals and start the sensation process, ultimately leading to the perception of motion. Because the fluid motion can fall below

Table 6.3 Visual illusions and their characteristics

Illusion	**Characteristics**
Caused by distorted ambient vision	
False Horizons	-Polar lights, nighttime roadway -Receding shoreline -Declined horizon at high altitude -Ground-sky confusion caused by Lights or terrain features
False Surface Planes	-Sloping cloud deck; Rising Terrain -Foreground ridges; Crater illusion
Inversion/luminance	-Low sun angle over water -Misperception of moon positions; -Lean on the sun
Vertical/optical-flow	-Hovering over water, snow -Rotating lights; Airspeed/altitude illusions
Misjudgment of terrain features	-Misperception of terrain feature heights -Terrain-density illusions
Caused by absent ambient vision	
Day IMC	-Featureless terrain, white/brown-out -Haze/fog; vection IMC formation flight -Dip illusion
Nighttime landings	-Approach/runway light illusions -Black hole approach -Runway size/shape, slope illusions -Surrounding terrain illusions
Illusory motion of fixed targets	-Oculogyral illusion; Oclugravic illusion -Elevator and autokinetic illusion
Display related	
Refractive	-Windscreen magnification -Spectacle distortion; color impairments Caused by sunglasses/visors
Collimated flight displays	-Accommodative micropsia -Mandelbaum effect; Cognitive capture
Night-vision devices	-Reduced visual acuity, contrast, depth Perception, -False brightness; Shadowing illusions

Source: From Previc (2004a).

Table 6.4 Factors that cause visual illusions and pilot actions

Factor	Perception	Action	Result
Narrow or long rwy Rwy or terrain prior upsloping	too high, feel ste ep	push over or descent	land short CFIT
Wide/short rwy Rwy or terrain prior downsloping	too low, feel shallow	pull up or climb	steep final, or land long
Bright rwy lights	too close or maybe steep	push over or descend	land short CFIT
Low-intensity lighting	farther away, shallow	pull up or climb	steep final or land long
Light rain, fog, haze Mist, smoke	too high, feel steep	push over or descend	land short CFIT
Enter fog	abrupt pitch up	push over or descend	land short CFIT
Flying in haze	farther away	pull up or climb	land long

Source: From FSF ALAR Toolkit (2000).

threshold even when the body is in motion, disorientation/vestibular illusions can occur. We have organized the vestibular illusions in terms of somatogyral illusions (disorientation of rotational movement) and somatogravic illusions (disorientation of translational movement or orientation with respect to gravity).

Somatogyral Illusions

1. Graveyard Spiral This somatogyral illusion results from habituation to a constant angular motion, especially in the dark or other visually-limited conditions. For example, when a pilot initiates a turn to the left, he or she initially will perceive that motion because the fluid in the semi-circular canal will cause the cilia to bend opposite the turn direction. However, if a constant left turn rate is maintained, the pilot becomes habituated (sensory adaptation occurs), and the pilot no longer feels as if he or she is in a left turn due to a state of equilibrium of the cilia (the fluid motion has matched the body motion). Eventually the pilot rolls-out of the turn; however, a motion after effect occurs because the fluid momentarily continues to move in the direction of the turn, which leads to the sensation of motion in the opposite direction. Thus, the roll-out is perceived to be a turn in the opposite direction from the original turn entry. Consequently, the pilot may initiate another left turn and cannot perceptually get out of it; hence the pilot is in a continuous spiral towards his or her "grave."

2. *Graveyard Spin and Gillingham Illusion* These somatogyral illusions are similar to the spiral but occur during an actual spin (extreme version of the spiral) or roll, respectively both of which can be disorientating maneuvers even with visual cues available. The same adaptation after effect occurs, but in different rotational axes. The Gillingham illusion was only recently named and appeared as a contributing factor in a 2008 F-16 spatial disorientation mishap report in Chapter 7 (Accident Investigation Board report, 2008). Part of any spin or roll recovery is visual confirmation of the direction prior to the recovery control input. For instance, the normal spin recovery involves (1) throttles to idle, (2) allowing all controls to momentarily go to their neutral positions (stick and rudders), (3) confirm the direction of the spin and apply the opposite rudder to counter the spin, once the spin return the rudder to the neutral position, and (4) recover from the dive. Hence, it is critical to visually confirm spin direction because if the pro-spin rudder is applied it will further increase the spin, making it more difficult to recover the aircraft.

3. *Oculogyral Illusion* Both of the above illusions highlight the interplay between the visual and vestibular systems. According to Newman (2007), when recovering from a spin/spiral, nystagmus may occur, which involves involuntary oscillatory eye movements. This oculogyral illusion occurs when the visual field appears to move (due to the involuntary eye movements by the pilot). This perceived movement further reinforces the illusion of rotation. This is a crucial point to consider because the vestibular illusion is falsely confirmed via visual information. One last point to note, however, is that if there are ample global and local environmental cues available, it will be apparent that the eyes are moving because the objects in the environment will "shake." Thus, those same environmental cues may possibly be used to counter the perception of the false rotation. Either way, the complexity and integration of the vestibular and visual systems is demonstrated by these illusions.

4. *Coriolis Illusion* The coriolis illusion results from a quick head movement in different axes from which the aircraft is moving and creates a severe tumbling sensation by the pilot. According to Krause (2003) this is considered very dangerous because, even if recognized by the pilot, it may not be recoverable. Newman (2007) explained this illusion as a cross-coupling of the semi-circular canals; one canal may signal a deceleration (from head movement opposite the turn direction) and another canal may signal acceleration (from head movement up or down). According to AFMAN 112–17 (2005) their description of the coriolis illusion is that during high turn-rates, abrupt head movement may cause pilots to perceive motions in which they are not actually engaged (e.g., tumbling). For example, if a pilot is in a prolonged turn in one plane, the fluid in those canals eventually reaches equilibrium. If the pilot then moves his head in a second plane of motion (up/down) the sensation can be one of a third plane of movement.

The first author recalls spinning in an archaic but effective spatial disorientation training device in the 1980s. That spatial disorientation trainer required the participant to switch radio frequencies after some time of spinning. The purpose of the radio frequency change was to have the pilot move his head in a different plane of motion and experience the coriolis illusion. The objective of the spatial disorientation training was to experience the illusion and learn to then transition to and fully concentrate on the instruments (specifically the attitude indicator). The first author has actually experienced this illusion a few times during an instrument approach in the weather (with no visual cues available). One cannot overstate the tumbling head-over-heels sensation that is extremely difficult to ignore while attempting to focus/concentrate on an instrument approach, knowing that the aircraft is in close proximity to the ground. To overcome this feeling, focus was maintained on the instruments, especially the attitude indicator. Eventually, the visual dominance of the attitude indicator displaying zero bank and a slight level descent overcame the sensation of tumbling during an approach and landing.

5. The Leans One of the most commonly reported forms of spatial disorientation is the leans, and it occurs because of confusion regarding the location of true vertical relative the aircraft. The leans may occur when an inadvertent bank, below threshold level, is entered into by an unsuspecting pilot or if a bank is maintained for a long time (allowing for equilibrium to be reached in the semicircular canals). Newman (2007) simply described the leans as, "manifested by a false sensation of roll … and is so named because it may cause pilots to lean to one side in order to cancel out the false sensation" (p. 9).

For example, the following scenario is common for a pilot experiencing the leans. It begins with a sub-threshold bank unnoticed by the pilot, to the left for instance. This bank is maintained long enough so that what little vestibular disruption there was regains equilibrium and the pilot believes he or she is straight and level. According to AFMAN112–17 (2005) this illusion is often experienced during formation flying in and out of the weather, or at night or when a pilot is focusing on the lead aircraft and not referencing his or her own attitude indicator. It also may occur when a pilot is simply distracted by other cockpit duties. What happens next is that the pilot looks at either the attitude indicator or outside at the horizon and sees a slight bank. To correct the aircraft the pilot then rolls to the right to bring it back to straight and level. However, because he or she had achieved vestibular equilibrium in the left bank, now rolling wings level to the right provides the false sensation that the aircraft is overbanked to the right. To "fix" this false sensation, this orientational disconnect, the pilot leans to the left.

Somatogravic Illusions

1. *Oculogravic Illusion* This illusion occurs during a rapid acceleration or deceleration of the aircraft, and without visual input, may be perceived to be a change in pitch, in addition to speed. It was briefly mentioned in the discussion of the black-hole takeoff illusion. The acceleration during takeoff is interpreted

as a pitch-up of the aircraft, a change in attitude. This misperception may cause the pilot to input a control to lower the pitch of the aircraft. Another example of this type of illusion may occur when the pilot reduces the throttles or extends a speed-brake. The perception may be of a lowering of the nose, or a pitch-down of the aircraft. Even a passenger on a commercial aircraft may experience this illusion when the pilot retards the throttles prior to pushing over on the yoke to start the descent. You feel, hear, and perceive a nose down sensation due to the deceleration.

2. *Elevator Illusion* This illusion occurs from a reduction in descent that is misperceived to be a climb (due to a translational motion after effect similar to the rotational after effect that occurs in the graveyard spiral). The pilot may counter this feeling by pushing the nose of the aircraft down. The opposite can also occur: an abrupt vertical deceleration can be interpreted to be a descent. This illusion may also inadvertently be experienced by a pilot encountering up or down drafts while flying over the mountains or near strong storms.

3. *Inversion Illusion* An abrupt change from a steep climb to a level-off may stimulate otolith organs to signal a change in gravity and linear acceleration, creating the illusion of tumbling backwards. Too counter this sensation the pilot will further push the nose of the aircraft down, which may actually intensify the sensation.

Other Illusions

1. The *G-excess Illusion* results from a complicated interaction of vestibular inputs during a high-G turn (Newman, 2007). What occurs is that there is a perception of under-banking, and the pilot is induced into further their degree of bank angle. The result is potentially deadly if near the ground due to the ensuing descent rate that develops during a high-G turn. Newman detailed an example scenario:

> A pilot who enters a turn at a level of G greater than the normal 1 G turn, and then looks back into the turn, may experience a phenomenon where they feel that the initial angle of bank is reducing … During a +2 G turn, a pilot may experience an apparent underbank of at least 10 to 20 degrees. In order to maintain the desired bank angle, the pilot may apply more bank, with the unintended consequence being a significant overbank phenomenon. (p. 10)

2. The *Giant Hand Illusion* has been described by pilots as the perception that their control input is being countered (as if a giant hand is on the airplane), not allowing the airplane to accept the input. It is classified as a somatosensory illusion (AFMAN 112–17, 2005). This illusion has been described by Krause (2003) as the result of cognitive dissonance, the mind and body fighting one another in an extremely disorientating situation. It has also been described as a, "subconscious reflex behavior, generated by vestibular or somatosensory inputs that interfere with the pilot's conscious control of the aircraft" (AFMAN112–17, p. 370). The

illusion may occur in either the pitch or roll axis. According to the AFMAN 112–17, research has been unable to replicate the illusion on the ground and it is most commonly noted during night air refueling operations.

Non-perceptual Causes of Spatial Disorientation

The US Air Force's Instrument Flight Procedures, Air Force Manual 112–17, presents several non-perceptual causes or contributors to spatial disorientation, which are similar to Gillingham's presentation of the effects of task loading and training. Awareness of these factors may possibly help pilots be conscious of when they may be more at risk to succumb to spatial disorientation. The primary factors that may make a pilot more prone to spatial disorientation are personal factors, workload, inexperience, proficiency, instrument-flying experience, and phase of flight.

Personal Factors (Emotional and Physical)

Recall from Chapter 1, Figure 1.1, the pilot brings to each flight his or her own personal baggage, factors that may impair them from flying their best on any given day. Human-factors training for pilots often stresses that pilot scannot expect to always have their "A-game" or be at their best every day. If pilots can appreciate this fact, then they become more aware of how personal/private events in their life can creep into their flying performance.

Physical factors on the pilot such as G-forces, extreme temperatures climates (heat or cold), or hypoxia also may influence a pilot's ability to ward off spatial disorientation. Fatigue and illness may leave a pilot even more susceptible to spatial disorientation. All of these personal factors can greatly affect a pilot's cognitive processing and his ability to handle attentional demands in challenging conditions.

Workload

Limited mental resources and cognitive task-saturation was alluded to in personal factors above. The information processing model presented in Chapter 2 directly relates to the ability of a pilot to juggle all the competing demands a pilot must perform. It begins with filtering the bombardment of external stimuli (from the aircraft and the environment) to perceive those truly reliable and credible cues, match them to a mental model or create a new model for accurate situational awareness, and then decide upon a course of action and execute that decision.

Inexperience

A pilot who is not as comfortable in a particular aircraft or flying in general will have a more difficult time accomplishing the tasks of flying and interfacing with the aircraft, leaving less time for orientation awareness.

Proficiency

A phrase often used in flying is "I am current but not very proficient." This relates to pilots and their assortment of monthly, bi-monthly, quarterly, and semi-annual flying maneuver requirements that are established as a minimum to ensure safe aircraft operation. Just because a pilot is "current" by a regulatory standard, however, does not necessarily mean that the pilot is truly proficient at that particular flying maneuver. Total flying time does not equate to proficiency … sometimes a younger pilot with less total flying time is more proficient than a more senior pilot. According to US Air Force 112–17, pilots coming off a break in flying are at a less proficient status and are more susceptible to spatial disorientation.

Instrument Experience

It is difficult to differentiate between proficiency and instrument flying experience because, even though any pilot is susceptible to spatial disorientation, it is certainly understandable that a newly qualified instrument pilot simply lacks the experience of flying in challenging conditions and is therefore more prone to misperception. As stated above, even an experienced non-proficient pilot can suffer a spatial disorientation episode; however, the odds are that a newly minted instrument pilot is more susceptible, given weather conditions and night environments, than a more experienced pilot. Taken to the extreme, a visual-only general aviation pilot encountering weather or night conditions is severely at risk because of her total inexperience, education, and training in using instruments.

Phases of Flight

Depending upon whether a pilot is taking-off, flying low level, formation flying, performing an air-to-ground mission, or accomplishing an approach to landing, various environmental conditions during different phases of flight may make a pilot more prone to spatial disorientation. The different phases of flight or mission types require significantly different cognitive requirements of the pilot in terms of workload, mental model formulation, and a higher demand for updated and accurate situational awareness. These changing levels of cognition and attentional demands can create different levels of distraction and channelized attention which may result in spatial disorientation.

Preventing Illusions

As was mentioned at the beginning of the chapter, it is very difficult to consciously overcome the unconscious processing occurring in the vestibular and visual systems. Spatial disorientation can be so powerful and insidious that an entire crew can fall prey to illusions and spatial disorientation or a crew member can be perfectly oriented but fail to realize in a timely manner that another crew member is disoriented (Lyons, Ercoline, O'Toole and Grayson, 2006). Possibly the reason vestibular and especially visual disorientation has not been given a higher stage in pilot and aircraft development is that it is believed that it can be easily avoided. One is tempted to simply state, "stay on instruments" or "don't go visual." The reality is that pilots love to fly visually, by the "seat of their pants" and to trust their instincts, and that they are overconfident in their abilities. Pilots will tolerate hours at cruise flying FL310 (flight level 31,000 ft) with the autopilot on just for the opportunity of 5 minutes of hand-flying during the visual approach and landing. The fun of flying is actually to visually control the aircraft with your own inputs, not to monitor and interface with the cockpit's computer. Pilots also enjoy flying lower because that increases the perception of speed (optic flow). Many pilots also fall into the trap of thinking they can salvage a bad approach because they have the experience and they don't need the plane's computers. Keep these thoughts in mind as you read some of the prevention topics below as well as the accidents presented in Chapter 7.

The US Air Force Manual 112–17 (2005) listed three main ways a pilot can prevent a loss of spatial orientation: training, proper flight planning, and knowledge of procedures. Neubauer (2000) also listed training as his first approach to mitigating spatial disorientation, and Mathews, Previc, and Bunting (2002) found that those pilots who experienced in-flight training were better at recognizing and identifying different forms of vestibular and visual disorientation than those who simply received classroom training. Training commonly occurs through classroom discussions, annual requirements of the Instrument Refresher Course, and even safety journals as well as flight simulators and actual aircraft sorties. To clarify, passive learning, whether it is called education or training, consists of simply watching and listening; whereas active learning is in a simulator or an aircraft and involves experiencing the illusions. Traditionally in aviation passive learning has been used, however given the consistent contribution of spatial disorientation in mishaps it is past the time to shift the learning paradigm to an active based curriculum.

Unfortunately aviation organizations seem to be going to the other way in their training of pilots. Gibb and Olson (2008) pointed out that, at least in the US Air Force, the requirement for continued education on spatial disorientation was extended from every 3 years to every 5 years. Further, although pilots are required to have annual training during an Instrument Refresher Course that covers topics such as spatial disorientation and visual illusions, it is a classroom-only setting and part of an 8-hour course on instrument-related issues, so spatial disorientation receives nowhere near the necessary emphasis. However, the first author has experienced the difficult balance of trying to add more and more training requirements into a syllabus of instruction without the addition of money to increase the available flying time allotment. In-flight training is recommended but with it also must come the money to support its inclusion into the formal syllabus of instruction.

More importantly, the reality is that classroom training is insufficient alone. Experiential training of actual scenarios provides hard evidence to pilots that they can fall prey to the many illusions regardless of their experience level. Gibb (2007) and Gibb et al. (2008) have recommended simulator training to specifically address a variety of spatial disorientation and visual illusion scenarios. This actual training (as opposed to classroom discussion) would greatly enhance the respect pilots of all experience levels show towards their visual and vestibular limitations. As discussed so far in this book and to follow in Chapter 7, no pilot is immune from spatial disorientation and an annual simulator experience demonstrating this fact serves as a clear reminder for pilots to not be so confident in their abilities. This would be valuable for commercial, military, and general aviation pilots.

A shared flying experience by one pilot demonstrates the power of experiential-learning. The account below of flying into a very challenging airfield depicts how a safe learning environment can prove to be the best learning tool.

> The most prominent visual illusion that I ever experienced occurred flying into a day-only VFR [visual flight rules, no weather/clouds to obscure visibility] airfield in Alaska. The runway was situated along a valley and was extremely up-sloped. The up-slope of the runway was 8 percent and I tried to visually make it look like a normal 3-degree glide-path. At approximately 2 miles from the runway my instructor asked me how my setup looked. I replied it looked all right. He then asked me to check my touchdown zone elevation [the runway's elevation] and it was only then that I realized I was level with the runway's approach end. In other words I was 600 ft below my desired altitude due to the illusion of the sloping runway; I had over-corrected my visual glide-path and was seduced into a dangerously shallow approach angle. The instructor's technique of allowing me to safely experience this illusion instilled in me the necessity to calculate the normal glide-path and associated altitudes and distances from the runway. Or in other words, it taught me not to solely rely on vision for a visual approach. (C. Hays, personal communication, 2009)

Granted this is an extreme example of a sloped runway and in this case military C-130 pilots have the opportunity and mission requirements to become familiar with these types of landing environments. The main take-away of this story is that the pilot learned in the airplane with the proper supervision by his instructor pilot rather than while sitting in a classroom.

In addition to training, Neubauer (2000) specifically listed two additional areas that have been traditionally used in military aviation to mitigate spatial disorientation accidents: policy directives and engineering solutions. He noted that, with just over a quarter of all US Air Force accidents having spatial disorientation as a contributing factor, the time to act is *now* (note he was pushing for this nearly a decade ago). He further noted that increased cockpit workload resulting from add-on interface technology as well as mission diversification has countered the improved human-factors integration of newer aircraft. One could argue that the new aircraft and improved human factors should eliminate spatial disorientation as a contributing factor but clearly this is not the case. In response to the historically high percentage of spatial disorientation occurence in the fighter/attack aircraft, Neubauer advocated the automatic ground-collision-avoidance system as one engineered solution. This system involves an automatic fly-up command prior to terrain impact if the pilot is unaware or unconscious due to a G-induced loss of consciousness. He concluded with, "serious consideration must be given to new engineering technologies as prevention tools for SD [spatial disorientation] mishaps, especially in the future, with more expensive aircraft" (p. 33).

In a similar vein, Mathews et al. (2002) concluded their research with the following well-stated comments:

> As aircraft become more agile and greater sensory demands are placed on pilots (HMD [helmet mounted devices], greater use of night vision devices, etc.) the incidence and severity of SD [spatial disorientation] are likely to increase unless more effective countermeasures are introduced. Better training would undoubtedly help, but is unlikely to be enough in and of itself. More effective orientational symbology, whether presented visually (e.g., on HMDs) or non-visually (e.g., tactile vest and 3-D audio), may help pilots maintain spatial orientation and aid in recognizing and recovering from unusual attitudes should pilots become disoriented. Ground collision avoidance systems also have a role to play in reducing the incidence of controlled flight into terrain.

Thus far the discussion has been at a more macro-level of spatial disorientation prevention. Next is a more specific set of rules for pilots to follow. One obvious aspect is to properly plan the flight—a top-down pre-processing technique. Map study is a skill that all aviators should develop and utilize prior to a flight so that expected terrain and topographical changes are noted at relevant points on the flight. Preble (1983, as cited in Previc, 2004a) presented four additional suggestions for pilots flying low-level to avoid falling prey to visual illusions (p. 298):

1. Be alert for impending terrain changes.
2. Increase their terrain-clearance altitude as their workload increases [bucket gets full].
3. Make proper use of the altitude warning and terrain-avoidance systems.
4. Never lose sight of the horizon for more than a brief instant while turning at low level and then only to quickly cross check their instruments.

As was alluded to previously in this chapter, Kraft's work proposed excellent suggestions for pilots to avoid misperception during an approach and landing.

Specially, Kraft and Elworth (1969, p. 4) listed the following as hazards that should serve as warnings to any pilot:

1. A long straight-in approach;
2. Unfamiliar runway ratio (length/width relationship);
3. Runway elevation below surrounding terrain;
4. Navigational facility not colocated on the runway/airfield.

(Note that #3 in both above played a role in numerous mishaps described in Chapter 7 and both #3 and #4 were involved in a 1997 mishap in Guam.)

Newman (2007) specifically focused on the overall air-worthiness of the pilot. Often pilots get caught-up with the maintenance status of their aircraft and the airplane's air worthiness, but neglect their own mental and physical readiness to fly. Newman also detailed flight planning as essential to ward off spatial disorientation. The more familiar pilots are with their route and destination, the more prepared they will be to deal with decision-making options and alternative courses of action; pre-flight planning should include aspects such as terrain, airfield specifics, lighting, forecast weather, sunrise/sunset times and approach options. Finally, Newman suggested continuous education and training for all pilots.

Newman (2007) also addressed prevention of in-flight episodes of spatial disorientation, with the most obvious solution being to transition to instrument flying and/or transfer aircraft control to another pilot if in a crew aircraft. Also, immediately finding visual flight conditions can quickly remedy disorientation if a pilot inadvertently flew into a cloud or weather condition. However, odds are if a visual-only pilot is in a cloud it will be very difficult for him to get out of the weather. Newman concluded his overview of spatial disorientation with the following advice:

> The truth of the matter is that disorientation can affect any pilot, any time, anywhere, in any aircraft, on any flight, depending on the prevailing circumstances. Experience of disorientation does not mean it won't every happen again. It does, however, allow the disorientation phenomenon to be recognized more readily in the future. Awareness and preparedness are key elements in preventing the SD [spatial disorientation] accident. (p. 25)

We support the notion of education, awareness, and experiential training for the reduction of spatial disorientation mishaps. The US Air Force continues to work spatial disorientation training into real-world scenarios and actual aircraft training. For instance, in 2006, during helicopter training flights, different scenarios were presented to the pilots and crew during night flying (G. Hover, personal communication, 2009). Especially enlightening to the crew was the "hover demonstration" that resulted in different disorientating experiences for each crew member dependent upon their position in the helicopter; the loss of visual reference during the hover brought on the illusions.

Over the years momentum has been made toward improving spatial disorientation training devices. There are currently devices made both in the United States and in other countries that address experiential learning scenarios, some better than others. However, the difficulty is convincing the appropriate decision-makers to allocate funding for the spatial disorientation training devices. Although a device of this type would not be inexpensive, relative to the cost of lives and resources lost due to misperception causing mishaps it is a worthy and necessary investment. Recent research efforts to improve pilots' perception during critical phases of flight has taken place in regards to helicopter landing pads in water environments. In Chapter 7 a 2006 fatal mishap is presented that highlights the difficulty even an experienced helicopter pilot has in flying visually to a landing pad in degraded conditions. More recently, in February of 2009, a similar non-fatal mishap occurred when a helicopter impacted the water while attempting a night visual approach to a helicopter landing pad (Daly, 2009a). According to Daly (2009b) helideck lighting has provided an area for aviation perceptual improvements. Based on a United Kingdom Civil Aviation Authority (UK CAA) survey researchers discovered that helicopter pilots could not differentiate the white and yellow lights of the helideck from the oil rig itself, the touchdown spot in the middle of the landing pad had the look of a "black hole," and flood lights intended to assist pilots during dark night landings were actually too bright. Daly (2009b) cited a UK CAA researcher explaining their lighting configuration manipulations, "the science of visual perception is not fully understood so [to] some extent we have to do this by empirical processes" (p. 1). Trials found that the floodlights were ineffective and that a circle of green lights and a large hollow "H" in the center provided the most accurate perception. The green circle of lights especially were effective in providing various images of ellipses to the pilot dependent of their approach angle to the platform. Also, global positioning system (GPS) approaches are being tested to bring helicopter pilots to a point in space, similar to a decision-height, that allow the pilot to assess if sufficient visual cues are available for a safe landing. Hence, applied research is helping in the prevention of visual spatial disorientation.

In the end, the best solution for preventing spatial disorientation from having tragic results is to stay on instruments and trust aircraft displays and instrumentation for indirect perception of the environment. At the very least, back up visual maneuvers with altitude and distance equipment for redundant cues and ensure eyeballs are calibrated properly. It is suggested that the General Aviation pilots' regulations be changed to limit non-instrument pilots from flying at night, at least when it is a "dark night" with minimal horizon cues and minimal moon/cultural lighting available. This restriction might induce more visual-only pilots to pursue their instrument ratings and also pay more respect to the dangers of impoverished flying environments.

Conclusion

The illusions and misperceptions presented above exist because our visual and vestibular systems' physiological capabilities were not designed for human flight. Unfortunately, since the early 1900s, we have learned about many of these perceptual limitations because of accidents and incidents. Granted, technology has improved our visual perception abilities in some areas (e.g., night-vision goggles); however, that same technology has additionally introduced new human limitations into the disorientation discussion. As obscure as some of the presented illusions may appear when reading simple descriptions, Chapter 7 presents incidents and accidents that resulted from the described sequence of events. Accurate cue perception allows for the appropriate aircraft control input, but as this chapter demonstrated, veridical visual perception is not always possible in aviation.

References

Accident Investigation Board (2008). F-16C, T/N 870–347, 93d Fighter Squadron, 482d Fighter Wing, Homestead Air Reserve Base, FL, United States Air Force.

Aircraft Owners and Pilots Association (2008). 2008 Nall Report. Aircraft owners and pilots association air safety foundation. Retrieved May 13, 2009 from http://www.aopa.org/asf/publications/08nall.pdf.

Benson, A. J. (1988). Spatial disorientation—general aspects. In J. Ernsting and P. King (Eds.), *Aviation Medicine* (pp. 2772–96). London: Butterworths.

Braithwaite, M. G., Douglass, P. K., Durnford, S. J., and Lucas, G. (1998). The hazard of spatial disorientation during helicopter flight using night vision devices. *Aviation, Space, and Environmental Medicine*, *69*, 10381–044.

Buckwalter, L. (1976). *The Pilot's Night Flying Handbook*. Garden City, NY: Doubleday & Company.

Buley, L. E. and Spelina, J. (1970, May). Physiological and psychological factors in "the dark night takeoff accident." *Clinical Aviation and Aerospace Medicine*, 5535–56.

Calvert, E. S. (1950). Visual aids for landing in bad visibility with particular reference to the transition from instrument to visual flight. *Transactions of the Illumination Engineering Society of London*, *6*, 1832–19.

Chatziastros, A., Readinger, W., and Bulthoff, H. H. (2003). Environmental variables in the "moth effect." Abstract, 10th Vision in Vehicles conference, Granada, Spain. Retrieved 26 November 2008 from www.kyb.tuebingen.mpg.de/publication.html?publ=2366.

Cocquyt, P. (1953). Sensory illusions. *Shell Aviation News*, *178*, 192–4.

Crassini, B., Best, C. J., and Day, R. H. (2003). Misperceiving extents in the medial plane. In J. Andre, D. A. Owens, and L O. Harvey, Jr. (Eds.), *Visual*

Perception: The Influence of H. W. Leibowitz (pp. 97–123). Washington, DC: American Psychological Association.

Daly, K. (2009a). Losses: A tale of three accidents that cause alarm for North Sea aviation community. Retrieved November 17, 2009 from http://www.flightglobal.com/articles/2009/09/21/332473/losses-a-tale-of-three-accidents-that-cause-alarm-for-north-sea-aviation-community.html

Daly, K. (2009b). Lighting upgrades enhance North Sea helicopter safety. Retrieved November 17, 2009 from http://www.flightglobal.com/articles/2009/09/22/332475/lighting-upgrades-enhance-north-sea-helicopter-safety.html

Ercoline, W. R., Weinstein, L. F., and Gillingham, K. K. (1991). An aircraft landing accident caused by visually induced spatial disorientation. *The 6th Biannual International Symposium of Aviation Psychology*, 6196–23.

Federal Aviation Administration (n.d.). Spatial disorientation: Seeing is not believing. Retrieved August 3, 2006 from www.faa.gov/pilots/safety/pilotsafetybrochures.

Flight Safety Foundation (Aug–Nov, 2000). Flight Safety Foundation approach and landing accident briefing note 5.3—Visual Illusions.

Gibb, R. W. (2007). Visual spatial disorientation: Re-visiting the black hole illusion. *Aviation, Space, and Environmental Medicine*, *8*, 8018–08.

Gibb, R. W. and Olson, W. (2008). Classification of Air Force aviation accidents: Mishap trends and prevention. *The International Journal of Aviation Psychology*, *18*, 3053–25.

Gibb, R. W., Schvaneveldt, R., and Gray, R. (2008). Visual misperception in aviation: Glide path performance in a black hole environment. *Human Factors*, *50*, 6997–11.

Gillingham, K. K. (1992). The spatial disorientation problem in the United States Air Force. *Journal of Vestibular Research*, *2*, 2973–06.

Gogel, W. C. (1965). Equidistance tendency and its consequences. *Psychological Bulletin*, *64*, 1531–63.

Gray, R. (2006). Vision in flying, driving, and sport. In M. R. M. Jenkin and L. R. Harris (Eds.), *Seeing Spatial Form* (pp. 1211–51). Oxford: Oxford University Press.

Haber, R. N. (1987). Why low-flying fighter planes crash: Perceptual and attentional factors in collisions with the ground. *Human Factors*, *29*, 5195–32.

Kern, T. (2002). *Controlling Pilot Error: Approach and Landing*. New York: McGraw-Hill.

Khatwa, R. and Helmreich, R. L. (1998, November–December). Analysis of critical factors during approach and landing in accidents and normal flight: Data acquisition and analysis working group final report. *Flight SafetyDigest, Killers in Aviation: FSF task force presents facts about approach-and-landing and controlled-flight-into-terrain accidents*, 17–7.

Kraft, C. L. (1978). A psychophysical contribution to air safety: Simulator studies of visual illusions in night visual approaches. In H. L. Pick, H. W. Leibowitz,

J. E. Singer, A. Steinschneider, and H. W. Stevenson (Eds.), *Psychology: From Research to Practice* (pp. 3633–85). New York: Plenum Press.

Kraft, C. L. and Elworth, C. L. (1969). Night visual approaches. *Boeing Airliner*, 2–4.

Krause, S. S. (2003). *Aircraft Safety: Accident Investigations, Analyses and Applications*. New York: McGraw- Hill.

Leland, D. (2001). Night VFR … an oxymoron. Retrieved 4 Dec 08 from https://www.aero-news.net/EmailArticle.cfm?ContentBlockID=ba42840f-a6194–e3a-97099–e7732dc90da.

Lessard, C. S. (2000). Spatial disorientation: Dealing with aeronautical illusions. *IEEE Engineering Medicine and Biology*, *19*, 252–7.

Lessard, C. S., Mathews, R., and Yauch, D. (2000). Effects of rotation on somatogravic illusions. *IEEE Engineering in Medicine and Biology*, 596–5.

Lyons, T. J., Ercoline, W., O'Toole, K., and Grayson, K. (2006). Aircraft and related factors in crashes involving spatial disorientation: 15 years of US Air Force data. *Aviation, Space, and Environmental Medicine*, *77*, 7207–23.

Matthews, R. S. J., Previc, F., and Bunting, A. (2002). USAF spatial disorientation survey. Paper presented at the Research and Technology Organization and Human Factors and Medicine Symposium on Spatial Disorientation in Military Vehicles: Causes, Consequences and Cures. La Coruna, Spain. Retrieved March 17, 2006, from www.spatiald.wpafb.af.mil/hfm/mp-0860–7.pdf.

McKee, S. P and Smallman, H. S. (1998). Size and speed constancy. In V. Walsh and J. Kulikowski (Eds.), *Perceptual Constancy: Why Things Look as They Do* (pp. 3734–08). New York: Cambridge University Press.

Mertens, H. W. (1978a). Comparison of the visual perception of a runway model in pilots and nonpilots during simulated night landing approaches. *Aviation, Space, and Environmental Medicine*, *49*, 10431–055.

Mertens, H. W. (1978b). Perceived orientation of a runway model in nonpilots during simulated night approaches to landing. *Aviation, space, and Environmental Medicine*, *49*, 4574–60.

Mertens, H. W. and Lewis, M. F. (1982). Effect of different runway sizes on pilot performance during simulated night landing approaches. *Aviation, Space, and Environmental Medicine*, *53*, 4634–71.

Mertens, H. W. and Lewis, M. F. (1983). Effects of approach lighting and variation in visible runway length on perception of approach angle in simulated night landings. *Aviation, Space, and Environmental Medicine*, *54*, 5005–06.

Mosier, K. L., Sethi, N., McCauley, S., Khoo, L., and Oransanu, J. M. (2007). What you don't know can hurt you: Factors impacting diagnosis in the automated cockpit. *Human Factors*, *49*, 3003–10.

Neubauer, J. C. (2000). Classifying spatial disorientation mishaps using different definitions. *IEEE Engineering in Medicine and Biology*, 283–4.

Newman, D. G. (2005). Eyeball error. *Flight Safety Australia*, 273–2.

Newman, D. G. (2007). An overview of spatial disorientation as a factor in aviation accidents and incidents. Australian Government, Australian Transport Safety Bureau, Aviation research and analysis report—B2007/0063.

Ostinga, J., Wolff, M., Newman, D., and White, S. (Nov–Dec, 1999). What killed JFK Jr? Lessons for VFR pilots. *Flight Safety Australia*, 222–9.

Palmisano, S. and Gillam, B. (2005). Visual perception of touchdown point during simulated landing. *Journal of Experimental Psychology: Applied*, *11*, 193–2.

Parmet, A. J. and Ercoline, W. R. (2008). Spatial orientation in flight. In J. R. Davis, R. Johnson, J. Stepanek, and J. A. Fogarty (Eds.), *Fundamentals of Aerospace Medicine*, (pp. 1422–05). Philadelphia: Lippincott, Williams, and Wilkins.

Parmet, A. J. and Gillingham, K. K. (2002). Spatial orientation. In R. L. Dehart and J. R. Davis (Eds.), *Fundamentals of Aerospace Medicine* (pp. 1842–44). Philadelphia: Lippincott, Williams, and Wilkins.

Pendleton, L. D. (2000). The black hole approach: Don't get sucked in! Retrieved February, 15, 2006 from www.avweb.com/news/airman/1824021–.html.

Perrone, J. A. (1984). Visual slant misperception and the 'black-hole' landing situation. *Aviation, Space, and Environmental Medicine*, *55*, 10201–025.

Pitts, D. G. (1967). Visual Illusions and Aircraft Accidents. USAF School of Aerospace Medicine Division, Brooks AFB, TX, Report No. SAM-TR-672–8.

Previc, F. H. (2004a). Visual illusions in flight. In F. H. Previc and W. R. Ercoline (Eds.), *Spatial Disorientation in Aviation, Progress in Astronautics and Aeronautics*, *Vol. 203* (pp. 2833–22). Reston, VA: AIAA.

Previc, F. H. (2004b). Visual orientation mechanisms. In F. H. Previc and W. R. Ercoline (Eds.), *Spatial Disorientation in Aviation, Progress in Astronautics and Aeronautics*, *Vol. 203* (pp. 941–44). Reston, VA: AIAA.

Riordan, R. H. (1974). Monocular visual cues and space perception during the approach to landing. *Aerospace Medicine*, *45*, 7667–71.

Roscoe, S. N. (1980). *Aviation Psychology*. Ames, IA: The Iowa State Press.

Schiff, B. (1990). Visual illusions can spoil your whole day. *Flight Safety Foundation Accident Prevention*, *47*, 1–4.

Schiff, B. (1994, January–March). Black hole approach. *Boeing Airliner*, 162–0.

Schwirzke, M. F. J. and Bennett, C. T. (1991). A re-analysis of the causes of Boeing 727 "black hole landing" crashes. *The 6th Symposium of the International Journal of Aviation Psychology*, 5725–76.

United States Air Force (2005, January). *Instrument Flight Procedures, Air Force Manual 11-217, Volume 1.*

United States Department of Transportation (2001). *Federal Aviation Regulations and Aeronautical Information Manual, 2002 Edition*. Newcastle, WA: Aviation Supplies and Academics.

Uttal, W. R. (2006). *Human Factors in the Courtroom: Mythology versus Science*. Laywers and Judges Publishing Company.

Wade, N. J. and Swanston, M. (1991). *Visual Perception: An Introduction*. London: Routledge.

Warren, R. and Owen, D. H. (1982). Functional optical invariants: A new methodology for aviation research. *Aviation, Space, and Environmental Medicine*, *53*, 9779–83.

Wickens, C. D. and Hollands, J. G. (2000). *Engineering Psychology and Human Performance*. Upper Saddle River, NJ: Prentice Hall.

Wilson, D. R. (1999). Darkness increases risks of flight. *Flight Safety Foundation Human Factors and Aviation Medicine*, *46*, 1–8.

Chapter 7
Aviation Mishaps: Misperception of Visual Cues

> Accidents do not occur because people gamble and lose, they occur because people do not believe that the accident about to occur is at all possible. (Wagenaar and Groeneweg, 1987, p. 596)

In this chapter we will review research and accident reports related to aviation visual spatial disorientation; the research reports quantify the role of visual spatial disorientation and the mishaps presented qualify the contributing role of visual spatial disorientation. As previously stated, aviation accidents are multi-causal, but if visual spatial disorientation contributions can be reduced or eliminated, then progress can be made towards decreasing the total number of mishaps by removing one link in the chain-of-events. Also, by enhancing the awareness of pilots to the danger of visual spatial disorientation, we can possibly alert pilots to *the accident that is about to occur* and give them a chance to live.

Thus far, the topics presented have included basic aviation techniques, the physiology of the visual and vestibular systems as well as visual perception theories, followed by specific aviation spatial disorientation types and illusions. Now the focus turns to the role that those illusions and false perceptions contributed to mishaps. By stepping through nearly 70 years of mishaps it is apparent that this problem is not going to go away. All aviators are susceptible, even the most competent and experienced pilots, and more needs to be done to reduce the number of accidents due to spatial disorientation.

Different types of mishaps can be categorized by a description of the pilot(s) and/or aircraft status at the time of the accident. A Controlled Flight into Terrain (commonly called by its acronym, CFIT) mishap is defined as an airworthy aircraft unintentionally impacting either the terrain or a man-made obstacle. CFIT mishaps are the leading type of worldwide commercial jet accident according to a Boeing study in 2005, with 57 fatal accidents (Figure 7.1) resulting in 3,735 fatalities (Figure 7.2) within the years 1987 to 2005. Controlled flight into terrain can occur during the takeoff phase, the en route phase, or (more commonly) the approach and landing phase of flight. In their assessment of mishaps from Boeing statistics in 1959–2004, Veronneau and Evans (2004) highlighted several types of mishaps as being strongly correlated with spatial disorientation. These included controlled flight into terrain, loss of control and under/overshoot landing. Note that both Figure 7.1 and Figure 7.2 show that controlled flight into terrain and loss of control are two of the top three

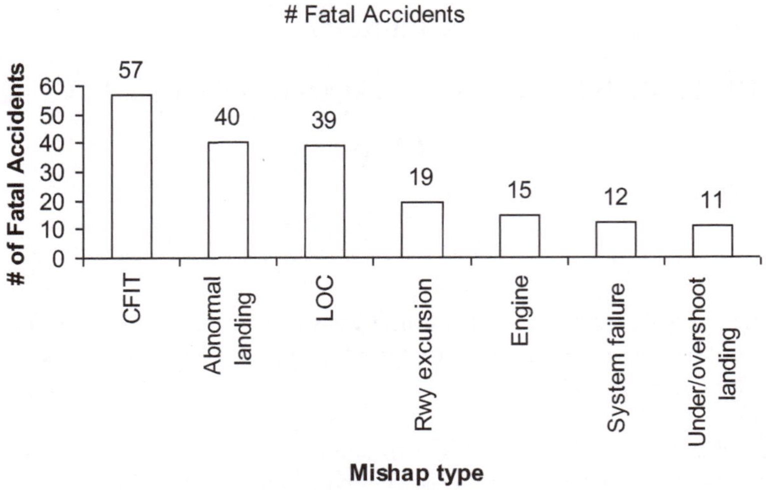

Figure 7.1 Fatal accidents by mishap type

Source: Copyright Boeing (2006).

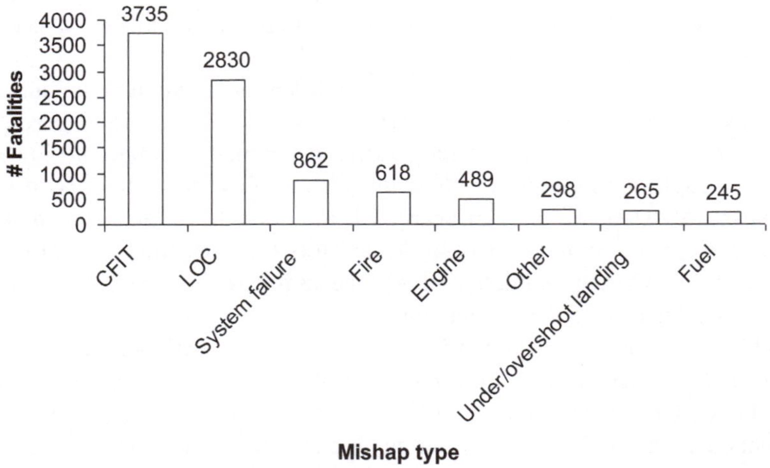

Figure 7.2 Fatalities by mishap type

Source: Copyright Boeing (2006).

mishap-cause types, and that also listed is "under/overshoot landing." Thus, spatial disorientation seems to be a major cause of fatality-related mishaps. A discussion on the underreporting of spatial disorientation will be presented later in the chapter.

An updated Boeing study in 2008 examined the worldwide commercial airline jet fleet between the years 1998 and 2007. It was reported that for the total number of fatalities and fatal accidents, loss-of-control-categorized mishaps moved into the #1 spot with 1,984 onboard fatalities from 22 fatal accidents. CFIT mishaps were the second deadliest with 1,137 fatalities from 18 fatal accidents. Loss of control and CFIT are the mishap categories used by the Boeing research; however, it is difficult to determine what caused the pilots to lose control and/or fly an airworthy aircraft into the terrain. Often, when examining accident reports it appears that some form of spatial disorientation or pilot misperception contributed to the mishap sequence. For example, Figure 7.3 examines the human aspect of aviation mishaps by presenting data between 1996 and 2005 that depict the flight crews' contribution to the majority of accidents within the worldwide commercial jet fleet.

Although accident reports indicate that accidents most commonly occur during the approach and landing phase of flight, some accidents do occur in both takeoff and en route phases. Military aviation accidents, unlike commercial aviation, have a relatively larger percentage of en route mishaps due to flight at low altitude. This difference is a function of operational mission requirements and the increased risk of exposure to terrain hazards.

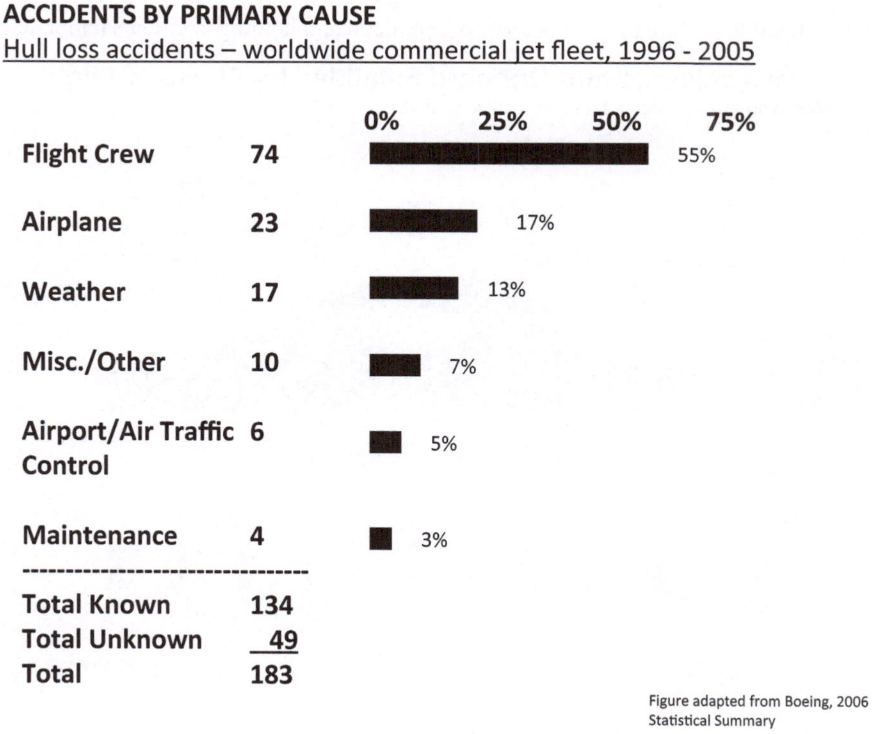

Figure 7.3 Accidents by primary cause

Source: Copyright Boeing (2006).

During takeoff, low-level flight, and an approach and landing, there is little margin for error because of the close proximity to the terrain. An important problem with visual perception in aviation is that sometimes, even when actively looking outside the aircraft at the terrain, a pilot can misperceive altitude. Consequently, the pilot finds him/herself in the worst case scenario—low to the ground with no margin for error.

According to a Boeing (2008) report on worldwide commercial jet fleet accidents that occurred between 1998 and 2007 (Figure 7.4), the takeoff and initial climb-out phase of flight accounts for only 2 percent of total flight time. However, statistics demonstrate that 19 percent of the accidents and 29 percent of fatalities occur within this phase of flight. The en route, or cruise phase, of flying accounts for 57 percent of the total flight time, but only 9 percent of all accidents and 19 percent of all fatalities. Figure 7.4 also illustrates that the approach and landing phase comprises only 4 percent of the total flight time while accounting for 36 percent of all accidents.

The Flight Safety Foundation in 1998 published an extensive review of worldwide aviation approach and landing accidents and controlled flight into terrain mishaps. In the journal, Khatwa and Helmreich (1998) reported a study of 287 worldwide fatal approach and landing accidents between 1980 and 1996. They stated that more than 75 percent of approach and landing accidents happened

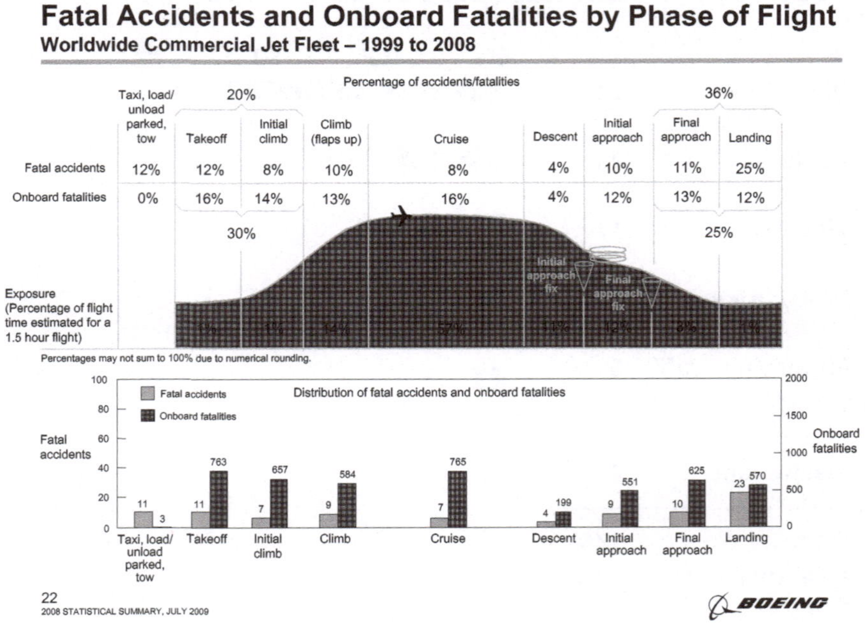

Figure 7.4 Accidents by phase of flight (Statistical Summary of Commercial Jet Airplane Accidents—Worldwide Operations 1959–2008)

Source: Copyright Boeing.

when a precision approach aid was either not available or not used (e.g., no glide-path indicator used), and that the rate of approach and landing accidents at night was approximately three times the rate during daylight. The most common primary factor contributing to the accidents was "omitting a required action" or an "inappropriate action." The second most common contributing factor was a "lack of positional awareness in the air."

Khatwa and Helmreich (1998) also examined 76 approach and landing accidents and incidents between the years 1984 and 1997, primarily from North America and Europe. Sixty percent of the mishaps occurred during poor visibility. In 74 percent of the cases, the Captain was flying and the primary causal factor 74 percent of the time was poor decision-making, judgment, and airmanship. Fifty-one percent of the mishaps were due to a lack of vertical awareness by the pilots; that is, a lack of spatial orientation within the flying environment. Rounding out the list of contributing factors was the presence of disorientation and illusions as causal in 21 percent of the accidents. Taking into account that 60 percent of the reported mishaps involved poor visibility, 74 percent of the time poor airmanship/judgment was displayed, and 51 percent lacked vertical awareness, visual perception played a role in many more than the 21 percent formally classified by the accident investigative teams.

One additional result from Khatwa and Helmreich (1998) is shared to emphasize the visual environment within which pilots operate, and how it may add to an already task-saturated situation. They reported that of the 279 fatal approach and landing accidents worldwide between 1980 and 1996, 103 mishaps had weather (other than poor visibility or runway excursions) and 89 mishaps had poor visibility cited as circumstantial. Thus, out of 279 fatal accidents, 192 (69 percent) involved environmental factors influencing visual perception and spatial orientation.

Research on Spatial Disorientation

A medical metaphor best depicts the usefulness of relying on accident investigation results to improve aviation safety: the mishap report is similar to an autopsy, in that anything gained is too late to help that particular pilot (International Civil Aviation Organization, 2002). Improvements made based upon a fatal mishap will "fix the last accident" but odds are slim that the same latent and active errors will align again for an exact replica of that previous mishap. The effort must be made but that effort is reactive compared to a proactive approach. An annual physical of a healthy person is a proactive approach to ensure good health and ward-off any potential negative health issues prior to them becoming life threatening Incident reports and flight assessments are proactive measures of pilots' operating practices and procedures. Aviation visual perception survey research is one way to collect information on pilots' experiences prior to an accident occurring and can be used to implement proactive changes.

What follows is a presentation of research statistics and survey results describing the past and current state of aviation safety for visual perceptual issues. Included for the reader's perspective are the results of all previously documented spatial disorientation mishaps, visual as well as vestibular-related. Pay note to the dates of the research as well as the group of pilots assessed. Repetition should be noticed as the findings all demonstrate that the aviation community has not sufficiently addressed the problem of spatial disorientation. In fact Mathews, Previc, and Bunting (2002) stated that little has changed in three decades.

Vinacke—1947

Vinacke (1947) interviewed 67 pilots in US Naval Aviation Squadrons and asked them to share their experiences of spatial disorientation. His findings of 77 illusory experiences were categorized into five general types: visual, non-visual, conflicting sensory cues, dissociative or recognitional, and emotional. The visual illusions were further broken down into seven types of visual spatial disorientation: confusion of lights, splitting of lights, *autokenesis* (single light appears to move), depth perception illusions/problems, relative motion, perspective illusions (ill-defined horizon), and some visual hallucinations. Depth perception problems were specifically discussed by pilots in terms of night flying during a "black night" (p. 314). Specifically cited was the difficulty in judging height above the ground during a landing on a dark night or when flying over smooth water. The result for all these conditions was an over-estimation of height above the terrain. Non-visual illusions presented by Vinacke (1947) included failure to perceive rotation or the after-effects of rotation, false sensations, after-effects of rotation, correct perception with a wrong reference point, and hallucinations. Vinacke's early efforts to categorize illusions positively contributed to researching visual and vestibular disorientation.

Vinacke (1947) also addressed the issue of pilot experience in that even senior pilots, despite their wealth of flying hours, were still prone to illusions. Often these illusions are encountered so infrequently, the opportunities to adapt to the encounter and learn the recovery are insufficient to truly become proficient. This is where training programs need to be better developed to ensure both novice and expert pilots can learn the appropriate response. Vinacke expanded upon this theme:

> No matter how good an adaption a pilot may have to his flying environment, he can still suffer from illusions (or other sorts of "vertigo") if conditions are such as to disrupt his conscious behavior. Thus, there is always the possibility that sudden entry into instrument weather, or entry into clouds, or a momentary lapse of attention, or a gesture of bravado, et cetera, will temporarily "throw him off" enough for confusion to result. It should be remembered in this connection that a pilot very seldom encounters the unusual conditions which contribute to illusions as well as other forms of "vertigo," and even then the conditions do

> not last long, as a rule: hence, his opportunities to develop adaptive patterns of behavior are very limited. (p. 323)

Barnum and Bonner—1971

Barnum and Bonner (1971) summarized the status of US Air Force mishaps by describing the typical pilot involved in a spatial disorientation mishap: "he will be around 30 years of age, have 10 years in the cockpit, and have 1,500 hours of first pilot/instructor pilot time. He will be a fighter pilot and will have flown approximately 25 times in the three months prior to his accident" (p. 898). The authors based their statement on a study of mishaps from 1958–1968, in which spatial disorientation was cited as causal in 6 percent of the total number of accidents and in which 11 percent of the fatalities occurred. It is noted that this report simply examined spatial disorientation; no attempt was made to differentiate the type of spatial disorientation experienced by the pilots. One last point regarding Barnum and Bonner's study is that they clearly articulated that succumbing to spatial disorientation is not just a hazard for inexperienced pilots—any pilot can experience the debilitating effects of disorientation.

Hodgson—1971

Hodgson (1971) reported on safety issues during visual approach to landings. He drew attention to the fact that worldwide there were 35 approach and landing accidents from 1958–1967, and of those, 27 had visual perception as a contributing factor. Hodgson further explained that 14 approach and landing accidents occurred at night and 12 of those were on Non-Precision or Visual Approaches. The high number of mishaps during the non-precision and visual approaches is very similar to what Khatwa and Helmreich found in their 1998 study.

Lyons and Freeman—1988

Lyons and Freeman (1988) reported on spatial disorientation in US Air Force mishaps in 1988. They found that, of 53 Class A mishaps, 8 mishaps involved spatial disorientation. Of these, 7 occurred in poor visibility, 2 involved visual illusions and 3 involved a mixture of visual and vestibular problems. In addition to the 8 mishaps involving spatial disorientation, there were 6 controlled flights into terrain accidents during low-altitude flight with 2 attributed to the misperception of terrain features and 2 attributed to channelized attention and/or distraction during a turn maneuver.

Holland and Freeman—1995

Holland and Freeman (1995) examined US Air Force mishaps from 1980–1989 and found that 270 of 356 operationally-related mishaps had loss of situational awareness and/or spatial disorientation as contributing factors. These accidents cost a total of 437 fatalities and $2.05B in resources. Because the authors used situational awareness as part of their classification process, it may be difficult to tease apart situational awareness and spatial disorientation in terms of causality. The aircraft that had the most accidents in this timeframe was the F-16 with a total of 59, for a mishap rate of 2.86 per 100,000 flying hours. The phase of flight during which F-16s most often crashed were during air-combat maneuvers (44 percent) and low-level flying (19 percent).

Holland and Freeman (1995) explained that originally the F-16 was intended to be used as a daytime air-to-air fighter aircraft but its mission expanded to include air-to-ground as well as significant increases in night flying. A dangerous phase of flying any air-to-ground aircraft occurs during the high-G turn and egress maneuver just before or after ordinance delivery. Pilots' curiosity to "score" their own bomb drops may cause them to become spatially disoriented while looking over their shoulder or transitioning back to flying after checking out the result of their dropping a bomb. The first female fighter pilot fatality (Bull, 1997) during a training mission may have had this sequence of events as a contributing factor. She was flying an A-10, an air-to-ground fighter-attack aircraft, during a night mission wearing night-vision goggles, became disoriented, and flew into the ground.

Before continuing, a point needs to be discussed regarding situational awareness, simply called SA. The concept of situational awareness is related to the concept of spatial disorientation and is defined by Jones and Endsley (1996) as, "perception of the elements in the environment within a volume of time and space, the comprehension of their meaning, and the projection of their status in the near future" (p. 507). In pilot-speak, situational awareness is what just happened, what is happening, and what is going to happen; that is, past-present-future in terms of the spatial and temporal environment. "Situational awareness" was a loosely used pilot term that crept into academia and research circles and has since been clouded in description, identification, and assessment problems (e.g., Dekker (2005) went so far as to call situational awareness a folk model). However, "situational awareness" has become such a broad, general term, that it often loses its usefulness when describing causal influences in a mishap. For example, the first author found that, when classifying 124 US Air Force Class A mishaps, nearly every mishap had aspects of situational awareness involved. Thus, the concept of situational awareness does not help to discriminate between the sequences of events leading up to an accident mishap, because the commonly-adopted definition lacks precision. Veronneau and Evans (2004) also brought this point to light by calling the term an "in-vogue categorization" in the 1980s and early 1990s in assessing mishaps (p. 211). They acknowledge situational awareness as a psychological phenomenon; however, they caution the overuse of the term as it may mask the appropriate *cause* of the mishap, a loss of spatial orientation. Verroneau and Evans recommend the deeper interpretation of data when the term situational awareness is used. This discussion is presented later in the chapter in terms of the underreporting of spatial disorientation.

Holland and Freeman (1995) attempted to explore mishaps involving both the loss of situational awareness as well as spatial disorientation. Consequently, they examined any mishap that contained channelized attention, distraction, and/or task saturation in the investigation's report in their quest for loss of situational awareness accidents. This approach helped the authors more precisely define situational awareness and seek contributing factors to accidents within their broad classification. Table 7.1 was taken from the authors and summarizes their work regarding human factors contributions to F-16 mishaps. The highest probability of mishap occurrence occurred due to channelized attention followed by visual restriction, misperception of speed/closure rate, visual illusion, misinterpreted instrument reading, and over confidence. We highlighted those items due to their relevance to this book's topic. As has been alluded to previously, common pilot cognitive limitations are overconfidence as well as excessive motivation to succeed. These two limitations are key factors that interact with visual impairments to increase risk for a pilot.

Table 7.1 Human factors contributions to F-16 mishaps

Contribution	Probability of Occurrence
Channelized Attention	61%
Haze/darkness/visual restriction	30–40%
Misperception of speed/closure rate	
Visual Illusion	20–30%
Task saturation/distraction	
Excessive motivation to succeed	
Overconfidence	10–14%
Misinterpreted instrument reading	

Source: Adapted from Holland and Freeman (1994).

Cheung, Money, Wright, and Bateman—1995

An investigation of Canadian aviation mishaps from 1982–1992 examined 14 Class A mishaps and found that in ten cases visual illusions and visual limitations directly contributed to the accidents (Cheung, et al., 1995). Of those ten, eight were caused by a lack of visual cues while flying over a featureless terrain such as smooth water, frozen or snow-covered lake, and this lack of cues made it difficult to judge the altitude of the aircraft. One of these mishaps, a 1991 C-130 black-hole illusion accident, is detailed later in the chapter.

Braithwaite, Durnford, Crowley, Rosado, and Albano—1998

In a study of US Army rotary-wing (helicopter) operations from May 1987 to September 1995 of all Class A, B, and C mishaps (970 total), 30 percent of them were reported to have spatial disorientation as a contributing factor (Braithwaite, et al., 1998). These authors reported that 62 percent of all spatial disorientation mishaps occurred at night. The leading factors contributing to the spatial disorientation mishaps were broken down into three areas: misjudgment, sensory difficulties, and problems with a night vision device (NVD). Specific misjudgments included misjudgment of obstacle clearance (84 percent), misjudgment of altitude (48 percent), and misjudgment of descent rate (32 percent). Sensory difficulties were attributed to insufficient visual cues (23 percent) and visual illusions (3 percent). The night vision device problems were described as visual limitations (29 percent) and symbology issues (21 percent). The leans (described later in the chapter) were often described as minor spatial disorientation experiences but also were associated with 44 percent of the "worst ever" experience.

Comparing the objective data with the survey of pilot reports brought out an interesting finding. Although visual illusions were a factor in only 3 percent of the accidents, pilots felt they played a role in 83 percent of their worst-ever experiences of spatial disorientation. Overall, the 1993 survey of 299 pilots revealed that 78 percent reported having experienced spatial disorientation during their careers. Despite some lack of clarity in interpreting the results, what is crystal clear is that for US Army rotary-wing operations, 30 percent of the accidents had spatial disorientation as a significant factor.

Sipes and Lessard—2000

Sipes and Lessard (2000) surveyed 141 students at the United States Air Force Advanced Instrument Course, an aviation graduate-level course for experienced pilots. The survey covered 38 different types of spatial disorientation, both vestibular and visual. The pilots, who each averaged 2,886 flying hours (standard deviation 1,271) simply responded to the survey instrument with the number of times they had experienced a particular visual or vestibular illusion. Incidentally, included in the survey was the phrase "loss of situational awareness."

The leading visual spatial disorientations noted by the pilots were in order from most common to least: black-hole illusion 79 percent, blending of sky and earth 63 percent, autokenesis 54 percent, sloping clouds or terrain 52 percent, brownout 45 percent, and whiteout 30 percent. The leading vestibular disorientational experiences were the leans 94 percent, Coriolis illusion 52 percent, misperceived self-moving on formation 43 percent, giant hand 35 percent. Table 7.2 depicts a portion of their findings and displays many of the illusions presented in Chapter

6 to highlight that, as odd and random as some of the illusions' descriptions may have sounded to the reader, they actually do occur to pilots. Note the illusions near the bottom of the table, such as "lean on the sun" and "moth effect." Also, one has to wonder about the pilots who reported graveyard spins/spirals … obviously, they did not reach the full "graveyard" portion of those illusions which demonstrates the limits of self-reported data.

Table 7.2 Spatial disorientation frequency data

Name of Disorientation	# of Pilots	% of Pilots
Leans	132	94
Black hole illusion	111	79
Blending of sky and earth	89	63
Autokinesis	76	54
Coriolis illusion	73	52
Sloping clouds or terrain	73	52
Brownout	64	45
Misperceived self-moving on formation	61	43
Giant hand	49	35
Whiteout	42	30
Flicker vertigo	39	28
False pitch up from acceleration	38	27
G-excess illusion	32	23
Dip illusion	31	22
False pitch down with deceleration	31	22
Elevator illusion	30	21
Graveyard spiral	19	14
Inversion illusion	16	11
Nystagmus	15	11
Lean on the sun	14	10
Break-off phenomenon	8	6
Moth effect	7	5
Graveyard spin	1	1

Source: Adapted from Sipes and Lessard (2000)

Neubauer—2000

A report of all US Air Force accidents from fiscal years 1994–1998 in which spatial disorientation was cited found that of 148 total Class A mishaps, 18 (12 percent) were attributed to spatial disorientation as the primary cause (Neubauer, 2000). Class A mishaps are defined as resulting in a cost over $1 million, the destruction of an aircraft, or the death of a pilot. When a more liberal interpretation of spatial disorientation contributions were included in the analysis, the percentage of mishaps attributed to spatial disorientation rose to 27 percent. It is interesting that, given previous discussions on how difficult it is to categorize and/or classify spatial disorientation mishaps, this article published in 2000 presented two sets of results using narrow and broad definitions of these types of accidents. Again, this demonstrates the difficulty in identifying mishaps in terms of a definitive cause. Neubauer stated, "depending on the training received and which definitions were used, medical officers could have different views on what is or isn't SD [spatial disorientation]" (p. 30).

Spatial disorientation occurred in all phases of flight: takeoff roll (5 percent), maneuver (37 percent), cruise (27 percent), low-level (8 percent), and final approach (14 percent). The types of US Air Force aircraft involved in the mishaps attributed to spatial disorientation were led by fighter/attack (82 percent), with trainer (7 percent) and helicopter (5 percent) far behind in spatial disorientation occurrence. Similar to the findings of Holland and Freeman (1995), Neubauer (2000) concluded that the high number of F-16 mishaps attributed to spatial disorientation accounting for 61 percent of all fighter/attack mishaps and 50 percent of all spatial disorientation accidents. More useful information than percentages is to present the mishap rate, that is, accidents per 100,000 flying hours. This mishap rate for the F-16 was 1.05 for the broader spatial disorientation definition and a 0.42 rate for the narrow definition. The A-10 aircraft, specifically an air-to-ground mission, had a similar spatial disorientation mishap rate of 1.15 for the broad definition. In contrast with those two fighter aircraft that maneuver in close proximity to the ground, the F-15 aircraft with its air-to-air mission only had a mishap rate of 0.5 per 100,000 flying hours. Thus, it is apparent that the F-16 and A-10 communities in particular need to address the problem of spatial disorientation.

Mathews, Previc, and Bunting—2002

Mathews et al. (2002) documented the reports of 2,582 US Air Force pilots on their experiences with spatial disorientation. The authors pointed out that due to the high percentage of fatalities in spatial disorientation mishaps it is difficult to accurately classify these types of accidents. Also, due to the changing nature of aviation, technology, and operational definitions of spatial disorientation it is difficult to compare and contrast research findings. A survey based upon Sipes and Lessard's (2000) study was developed to collect data on pilots' spatial disorientation experiences. The difference between the studies, however, was that Mathews et al. asked for disorientational experiences only in the current airframe. They used the following definition for their survey:

> An incorrect perception of linear/angular position, or of motion, relative to the Earth's surface or another aircraft, sufficient to affect performance, situational awareness or workload—however slight that effect may be. (p. 7.3)

The surveys were collected between August 1999 and January 2000. The average number of flying hours for pilot participant sample was 1,815 and average time in their current aircraft was 842 hours. Only 0.3 percent of the pilots reported not having received spatial disorientation training and education. Those that did report having had education on the topic received it via lecture, ground demonstration, and in-flight demonstration.

Over all types of US Air Force aircraft the most common kinds of spatial disorientation are presented in Table 7.3. The results are similar to Sipes and Lessard (2000); however, the table reports one-time exposure to a particular illusion as well as aircraft type. Fighter aircraft often had the highest frequency of reported disorientations; however as noted in Table 7.3, there are a few instances for which multi-engine aircraft or rotary-wing aircraft lead the prevalence rate in some illusory categories.

It is difficult to tease apart visual and non-visual disorientation because often the lack of visual cues (ambient or peripheral cues) further promotes the incorrect reliance on vestibular senses, which in turn leads to vestibular spatial disorientation. Of the non-visual illusions, the leans, Coriolis and G-excess were most commonly reported. Of the visual illusions, loss of the horizon, sloping horizon, and the night-approach black hole were the most often encountered. It is interesting that featureless terrain was not addressed specifically in this survey instrument. Of all the pilots surveyed, only 22 percent used night-vision devices, but of those, 72 percent reported spatial-disorientation issues. It is anticipated that future spatial-disorientation research will find significantly increased use of night-vision devices by pilots, leading to the possibility that the number of reported disorientational experiences will increase. In terms of aircraft specific illusions, Table 7.4 presents the summarized totals of spatial disorientation type by aircraft

Table 7.3 Most frequently experienced illusions

Disorientational Description	% Reporting 1 Incident
Leans	76
Loss of horizon	69
Sloping horizon	66 (ME highest)
Coriolis	61
Night approach – black hole	58 (ME highest)
Misleading altitude cues	50 (RW highest)
False sense of pitching up	44
Giant hand	38 (RW hightest)
Misjudgment of position – night	38
Elevator illusion	37
Autokinesis	37 (ME highest)
False sense of pitching down	37
G-excess	36
Loss or horizon – sand, snow, dust	33 (RW highest)
Graveyard spiral	32 (RW highest)
False sense of yaw	32
Inappropriate use of sun	24

ME – multi-engine; RW – rotary wing; fighter aircraft had the highest frequency reported unless annotated above

Source: Adapted from Mathews et al (2002).

Table 7.4 Summarized spatial disorientation by aircraft type

Aircraft Type	Most Common	2nd Most Common	3rd Most Common
Fighter:	Leans	Blending of earth and sky	Misjudged formation
Multi-engine	Black hole	Sloping horizon	Leans
Trainer	Leans	Blending of earth and sky	Coriolis
Rotary wing	Undetected drift	Misleading altitude cues	Brown/white-out

Source: Adapted from Mathews et al (2002).

type. Note the aircraft and mission specific differences in the pilot experiences involved with various types of spatial disorientation.

Holmes et al.—2002

A survey of United Kingdom military pilots also led to the conclusion that spatial disorientation is a significant hazard in aviation (Holmes, et al., 2002). Their research was very similar to Sipes and Lessard (2000) and Mathews et al. (2002) in terms of methodology and results. Out of 606 pilots, the most reported types of visual spatial disorientation were loss of horizon due to atmospheric conditions 82 percent, misleading altitude cues 79 percent, sloping horizon 75 percent, distraction 66 percent, and night approach to landing 60 percent. Similar to the previous survey findings, the top two non-visual illusions cited by pilots were the leans 92 percent and Coriolis 66 percent. Statistical analysis revealed different effects for factors from the survey.

Holmes et al. (2002) concluded that the high rate of spatial disorientation was due to poor crew coordination and distraction. To improve upon these in-flight hazards aviation safety needs to improve spatial disorientation awareness and provide airframe-specific training because different aircraft and missions appear to lead to different types of disorientation.

Holmes et al. (2002) concluded that the greater number of those who experienced spatial disorientation were pilots who had received in-flight training. This initially counterintuitive finding can be interpreted as those who were appropriately and realistically trained were better at recognizing their experience (more aware) and could then properly classify it. This speaks well to the successful efforts for in-flight, realistic training. This same finding also suggests that spatial disorientation is often underreported due to some more naïve pilots not accurately recognizing that spatial disorientation had occurred. The significant differences between types of training on the level of reported spatial disorientation (in-flight versus ground/class room type), further suggests that not all training is equal with respect to recognizing spatial disorientation. Because time since last training did not affect the amount of reported spatial disorientation, it seems that the type rather than the frequency of training is most crucial.

Chimonas, Diamantopoulos, Markou, and Stathogiannis—2002

The Hellenic Air Force surveyed their pilots during their annual flight physical in 2002 and found that, of 407 pilots, one-third did not report spatial disorientation. However, of the remaining two-thirds, nearly 70 percent had experienced from 2 to 10 different types of visual spatial disorientation. Blending of earth and sky was experienced by 38 percent of the pilots and sloping horizon by 23 percent. It is difficult to believe that in many years of flying one-third of those pilots had no spatial disorientation experiences to report; perhaps there was some hesitation to openly report instances.

McGrath, Rupert, and Guedry—2002

McGrath et al. (2002) reported that the US Navy in fiscal year 2001 had 19 Class A mishaps. While only 26 percent involved spatial disorientation as a major contributing factor, they accounted for half of all fatalities. In 2004, the US Navy dedicated an entire issue of their aviation safety magazine, *Approach*, to the topic of spatial disorientation. One article reported that out of a total of 120 US Naval Aviation mishaps in fiscal years 1997–2002, 22 involved spatial disorientation and these took 23 lives and cost $475M (Webster, 2004).

Veillette—2004

In a report on business jet operations worldwide from 1991–2002, Veillette (2004) noted that a high percentage of accidents involved controlled flight into terrain. Within that timeframe there were 251 business jet accidents with 67 fatal accidents. Of the more than 1,138 people on board the accident aircraft, 28 percent received fatal injuries. Twenty-seven (40 percent) of the 67 fatal accidents were CFIT and of those 82 percent occurred in mountainous terrain. Of the CFIT mishaps, 13 were during non-precision approaches, while 4 each occurred for precision and visual approaches.

There were 104 accidents during approach and landing, and spatial disorientation and/or visual illusions were cited as causal factors in 16 accidents. Of those 16, six were CFIT mishaps involving a sloping runway in mountainous terrain. Veillette (2004) cited Telluride, Colorado as having a runway with a 1.9 percent positive slope from the midpoint, which can create the perception that they are too steep on final approach. As a result of the perception of being too steep, a pilot may unnecessarily descend and hit terrain prior to the runway, (as presented in the mishap in 2002 discussed in Chapter 1 and other mishaps presented later in this chapter). Other business jet CFIT mishaps were attributed to black-hole runway environments, whiteout conditions, and rain-covered windshields that also induced a shallow approach.

US Air Force and Lyons, Ercoline, O'Toole, and Grayson—2006

The US Air Force Safety Center reported that spatial disorientation was a significant factor in 20 percent of all major mishaps between 1990–1997, and of those, 19 percent were associated with fatalities (AFMAN 11–217, 2005). In a more recent analysis of US Air Force accidents, 1990–2004, Lyons, Ercoline, O'Toole, and Grayson (2006) reported 11 percent were attributed to spatial disorientation and it was a factor in 23 percent of all night-accidents, once more connecting vestibular and visual spatial disorientation. The role of night is a significant factor as the number of night spatial-disorientation mishaps is nearly the same as during the day (27 versus 28); however, substantially fewer night flights occur, so the prevalence rate is really much higher at night. More serious is the very high rate of fatalities for spatial disorientation mishaps (57 percent overall) and that is even higher for night accidents (81 percent).

The F-16 was again highlighted in the study, accounting for 39 percent of all spatial disorientation mishaps; however, this is not surprising given that the F-16 has been involved in the majority of sorties flown by US Air Force aircraft (Lyons et al., 2006). This fact should not excuse the number of mishaps involving this aircraft. This was the third study (Holland and Freeman, 1995; Neubauer, 2000) to highlight the F-16, yet it is not clear what progress is being made to improve safety for those pilots. Spatial disorientation mishaps should *not* be the cost of doing business.

Lyons et al. (2006) also highlighted the difficulties in accurate disorientation mishap reporting by describing many accident investigations that involved spatial disorientation but failed to formally cite spatial disorientation in the one-line narrative. This demonstrates two drawbacks of the current safety system. One, accidents get "labeled" and "stove-piped" into certain mishap categories, (sometimes incorrectly), and two, information is then reported in macro-safety studies which often serve to propagate the inaccurate classification.

Gibb and Olson—2008

Gibb and Olson (2008) examined 124 US Air Force Class A mishaps comprising 5 mishap types: 31 controlled flight into terrain, 17 approach and landing accidents, 15 loss of control, 19 spatial disorientation, and 42 mid-air collisions. The authors analyzed the mishaps from the previous US Air Force accident investigation system in terms of a new Department of Defense Human Factors Analysis and Classification System (DoD HFACS, 2005). The DoD HFACS classifies a mishap into the following four levels: acts, preconditions to unsafe acts, supervisory influences, and organizational influences. Within the acts category, perception errors were present in 56 of the 124 mishaps as part of the causal chain. Perception errors, as defined by DoD HFACS, are (1) when an individual acts or fails to act based on an illusion or (2) misperception or disorientation state such that this act or failure to act creates an unsafe situation (2005, p. 1–3).

In terms of the latent errors of preconditions to unsafe acts, the authors found that the perceptual precondition of "Misperception of Operational Conditions" was present in all five mishap types, and is described as when an individual misperceives or misjudges altitude, separation, speed, closure rate, and aircraft location within the operational conditions and leads to an unsafe situation (p. 1.13). Although this is a broad classification, it does capture many aspects of spatial disorientation presented in previous research as well as the contributing factors discussed in the mishap investigations discussed next.

US Navy—2008

The US Naval Aviation safety center (2009) provided updated data on their mishap statistics regarding spatial disorientation within the fiscal years 2000 to 2007. Within that period of time there were 18 Class A spatial disorientation defined mishaps, of which 10 where helicopters, 7 were jet aircraft, and 1 was a propeller plane (EC-2). The time of day in which the mishaps occurred was primarily at night with 8/10 helicopter mishaps, 4/7 jet mishaps, and the EC-2 crash all happening after dark. The incapacitation and tragic consequences of spatial disorientation is demonstrated by the fact that although just 10 percent (17/177) of all Class A mishaps from 2000–2007 were categorized as spatial disorientation they accounted for 40 percent (65/162) of the total number of Class A fatalities. Also, keep in mind that spatial disorientation is often underreported and these statistics represent a strict categorization of mishap "type".

As mentioned in Chapter 1, data provided by the US Naval Aviation Safety Center (2009) entitled "Aeromedical 'Why' Causal Factors" found that between the years 1990 and 2008 the number one human factor involved with nearly 80 Class A mishaps was spatial disorientation. Fatigue was found causal in roughly 50 mishaps, G-induced loss of consciousness and visual illusions were both found causal in approximately 24 Class A mishaps.

US Air Force—2008

We acquired mishap data from the US Air Force Safety Center that covered the years 1993 to 2009. Data mining searches were accomplished using search words such as "spatial disorientation," "SD," "illusion," and "depth perception." Those searches resulted in 32 Class A mishap reports. It is noteworthy that only two mishaps resulted in dates prior to 2001, one in 1996 and one in 1999. The authors are confident that this reflects a failure of accident investigation formal reports citing these terms rather than an absence of these types of contributing factors being present in accidents. Again, this lends support to the theory of underreporting of visual and vestibular contributions to aviation mishaps. The more recent mishap investigation reports cite visual misperception as causal in 14 of the 32 since 2005.

Of these 32 mishaps, 12 mishaps resulted in fatalities for a total number of 30 fatalities. Out of the 32 mishaps, 16 occurred during the day, 1 at dawn, 1 at dusk, and 14 at night. Of the 14 night mishaps, 7 involved pilots wearing night-vision goggles. The types of aircraft involved in these mishaps were as follows: fighter aircraft—16; helicopters—6; crew aircraft—4; trainers—4; and unmanned aerial vehicles (UAVs)—2. This search seems to have failed to highlight adequately the actual number of UAV mishaps, which is largely due to it being a new system both from the aircraft, operator, system interface, and employment perspective and this search. More information regarding UAV mishaps is presented in Chapter 8.

Research Summary

Within aviation safety circles, spatial disorientation is often taken to refer only to vestibular spatial disorientation. Both Previc (2004) and Gillingham (1992), however, estimated that in roughly 50 percent of all spatial disorientation mishaps *visual* disorientation played a role in the chain-of-events leading to the accident. Veronneau and Evans (2004, p. 204) summarized the lack of accurate mishap-reporting by stating "one can safely assume that underreporting of SD [Spatial Disorientation] occurs in all categories of accident investigations, as the investigating body often will not identify an SD factor as significantly contributing to the mishap." Verronneau and Evans further articulated the accident investigation and mishap categorization of the interactive nature that exists between in-flight loss of control, spatial disorientation, and visual illusions. They advocated that inappropriate classification of accidents adds to the confusion regarding just what "type" of accident occurred.

An example of mis-classification driving accident data assessment is Shappell and Wiegmann's (2001) report on 195 Civil Aviation controlled flight into terrain data from 1993–1994 using their Human Factors and Accident Classification System. They were surprised that, of those 195 mishaps, only 17 percent were the result of "Perceptual Errors" and only 6 percent were due to "Spatial Disorientation." Also, they stated that approximately half the accidents happened during visual meteorological conditions or daylight. Although Shappell and Wiegmann's classification system has greatly contributed to enhancing aviation safety issues, in this case their model's limitations constrained their conclusions. The authors went as far as questioning the utility of terrain-awareness technologies (e.g., ground-proximity warnings systems) because they concluded that *mishap type* was not worthy of resources due to the low numbers of mishaps assigned to that causal factor.

The above example highlights how underreporting a phenomenon may inadvertently reduce resources and lead to additional loss of lives. Any contributing factor, regardless of its perceived significance, is worthy of research efforts and the best available resources. In fact, the measurable existence of VMC and daylight mishaps gives credence to the strength of visual illusions. Combine those visual illusions with night or other reductions in visibility and the power of the illusions substantially increases. It is crucial that the aviation community keeps this in mind, as well as the costs and loss of life that often result when spatial disorientation occurs.

The following section of this chapter contains a review of accidents in which visual spatial disorientation played a contributing role in the sequence of events leading up to the accident. As you read keep in the mind the connection between visual and vestibular systems for veridical orientation. Also, because this book is intended to become a reference source for researchers and pilots, many mishaps are presented. They are presented to emphasize that spatial disorientation is a major and continuing challenge for aviation. We hope that the research summary above, as well as the mishap descriptions given below, will motivate the aviation community to continue the research and use that research to develop enhanced training and awareness programs.

Past Mishaps Attributed to Visual Spatial Disorientation

While reading the following accounts we suggest that the reader look for differences between the "older generation" mishaps and the more current (last 10 years) of mishaps. Is it correct to say that the human perceptual limitations that caused accidents many years ago are the same perceptual limitations that also caused recent accidents? Despite technological advances, we believe that, indeed, the underlying perceptual challenges have been consistent and remain worthy of consideration.

San Juan—1941 (night, featureless terrain—water, misperception of height and distance)

One of the earliest aviation accident reports that cited visual spatial disorientation as a contributing factor was an accident that took place October 3, 1942 (Civil Aeronautics Board [CAB] report, 1942). A Pan American Airways "flying boat" Sikorsky S-42B aircraft was making a twilight water landing at San Juan Harbor, Puerto Rico. The passenger-carrying flight was scheduled to depart Miami, Fl, with stops en route to San Juan in Cuba and the Dominican Republic. They were scheduled to arrive in San Juan prior to sunset; however, a late departure from Miami and other en route delays had the pilots unsuccessfully trying to play "catch-up" on their schedule during the long day.

The pilot was 45 years old and a highly experienced aviator with 11,284 flying hours, of which 1,500 flying hours were in Sikorsky aircraft. He had 690 flying hours at night and had previously made 6 night landings in San Juan Harbor. In contrast, the copilot was only 24 years old and had 583 flying hours.

A night water-landing is a unique maneuver and the landing launch crew prepares the "water runway" depending upon the wind direction. On this particular evening, San Juan Harbor had a water-landing preparation with seven reference landing-lights running east to west and parallel with the wind. A green light was on the downwind end (the approach end) and a red light was on the upwind end (the end of the landing area), with white lights in between, all together extending 610 m (2,000 ft). This lighting configuration allows for the pilot to land right of the lights; thus the pilot in the left seat keeps the lights in view during the course of the landing. The landing-launch team positioned itself at the approach end (the green light), and shone a searchlight parallel to the row of white lights to provide an area for the pilot to aim for, between the white lights and the searchlight beam. (This description of the water landing-zone certainly illustrates the need for the historical color-vision assessment methodologies, many of which are still in use today—see Chapter 3.)

In this particular situation, the pilot chose to land opposite from what the landing-launch team had configured, from west to east; thus the Sikorsky was approaching the red light. The pilot chose to land in an easterly direction because this was how he entered the Harbor area and he determined the wind was negligible due to failing to detect any drift in nearby smoke rising from fires. Consequently, the searchlight beam was not used as it would have been shining in the face of the landing pilot.

The aircraft landed in an unusually nose-low attitude, violently throwing the passengers forward and sideways in their seats. Thus, the accident board's conclusion determined that the aircraft struck the water with the nose of the aircraft lower than usual and with some drift to the right. The aircraft was destroyed and two fatalities occurred.

The subsequent investigation found that the aircraft had no mechanical or structural problems and the probable cause was that the pilot failed to "exercise requisite caution and skill in landing" (p. 13). The CAB report stated a probable cause was also "the smooth surface of the water which rendered difficult the captain's depth perception as well as the exact determination of any lateral movement of the aircraft, constituted a substantial contributing factor" (p. 13). The formal report described the visual misperception:

> The existence of a glassy surface is frequently conducive to misjudgment of height above the water as well as, to a lesser extent, misjudgment regarding the attitude of the aircraft. Another factor tending to lessen depth perception was the presence of a bright moon nearly directly overhead. This had the effect of illuminating the smooth surface of the harbor with sufficient light to decrease the effectiveness of the aircraft's landing lights. (p. 12)

In overestimating the height above the water during landing, the pilot may have initiated or continued to fly an overly-aggressive descent-rate toward the water-landing surface (accounts for the nose-low attitude at impact). The cause of this misperception was the lack of ambient visual cues that would have provided depth information to the pilot. It is interesting that, even in 1941, the occurrence of visual-spatial disorientation by the pilot and the ecological cues available were clearly stated in the formal accident report. Another factor worth mentioning was the impressive amount of flying time the pilot had and he still succumbed to the visual illusion.

This mishap illustrates a common theme that occurs throughout many accidents, and that is the combination of night, misperception of height and distance, and a very experienced pilot at the controls. Granted, not all of the following mishaps occur with these same conditions present; however, time and time again environmental conditions set up the unexpecting pilot for failure. Another common theme with the mishaps is pilot fatigue and/or pilots attempting to "expedite" the landing, or hurry and land. The 2002 mishap described in Chapter 1 had this as well in that the crew accepted the more conveniently aligned runway even though there was not a precision approach available, rather than the more time consuming flight-path to land on the other runway.

Finally, the successful ditching of the commercial aircraft in the Hudson River in comparison with this mishap is certainly even more impressive considering the pilot did not have any engine power available with which to make corrections. In contrast, the 1941 mishap aircraft was designed to land in the water and had full power of the engines at the touch of the pilots; however, the result was still complete destruction and some fatalities.

LaGuardia—1952 (day weather conditions, featureless terrain—water)

On January 14th, 1952, Northeast Airlines Flight 801 crashed in Flushing Bay, New York, 1.1 km (3,600 ft) short of runway 22, LaGuardia Field, at approximately 9:03 AM (CAB report, 1952). The flight had departed Boston earlier that morning. The aircraft, a Convair, received major damage; fortunately, however, only 5 of the 33 passengers were seriously injured. The weather that morning was 518 m (1,700 ft) ceiling with 2.4 km (1.5 mile) visibility. The pilot, who was monitoring the first officer flying the approach, had nearly 14,000 flying hours, 2,400 in Convairs, and had flown approaches into La Guardia five or six times a month for the last ten years. The first officer, flying the instrument approach, had approximately 5,100 flying hours. In the final moments of the flight, the last 300–400 ft in altitude, the first officer was transitioning from an instrument cross-check (looking inside only) to visual references (looking outside) for the landing. The pilot stated he saw the runway lights ahead of them; however, the first officer flying glanced up and failed to see them. Given the weather conditions, they should not have descended below 450 ft unless positive identification of the runway environment was made. (The issue of the pilot flying and the pilot not flying (monitoring) and their coordination at the crucial transition between instrument and visual procedures was and still is today an area of concern.)

The aircraft had no engine or structural malfunction. The Civil Aeronautics Board accident analysis suggested the possibility that the pilot transitioned visually sooner than he should have, and may have fallen prey to a sensory illusion. The last 5.6 km (3.5 miles) of the approach was over water, as shown in Figure 7.5. Note that three of the possible runway approach directions are surrounded by water.

The formal accident report stated: "the surface of the water was glassy, limiting its use as a medium of depth perception" (p. 11). The combined effects of the lighted runway, obscured horizon, and flying over water removed ambient visual cues and may have led to inaccurate perception of altitude and distance to the runway, inducing the pilot to initiate a premature descent. A similar mishap occurred a year later on April 20th 1953, in San Francisco Bay, CA (CAB report, 1953). The aircraft involved, a Western Airlines Douglas DC-6B, was flying over the bay between San Francisco and Oakland. Unfortunately, in this mishap only two of the ten occupants survived and the aircraft was destroyed and it sank.

Figure 7.5 LaGuardia International Airport, New York

Source: With permission from Michael Mantoudis. http://www.airliners.net/photo/0903770/

Missouri—1955 (night, misperception of height and distance)

On the night of March 20th, 1955, American Airlines Flight 711, a Convair aircraft, crashed in an open field near Springfield, MO (CAB report, 1955). Twenty-two out of 35 survived the accident. The formal accident investigation concluded that:

> The probable cause of this accident was a descent to the ground while approaching the airport caused by the crew's inattention to their flight instruments and a

> possible sensory illusion giving them an erroneous impression of the attitude of the aircraft. (p. 8)

Flight 711 was accomplishing a circling maneuver prior to their ground impact. A circling maneuver is a visual maneuver within the runway environment to position the aircraft with the active landing runway at the airfield. The ceiling (bottom of the clouds) on the night of the mishap was approximately 152 m (500 ft) above the ground; thus, the pilot flew an instrument approach to penetrate the weather (get under the clouds) then initiated the circling maneuver to align themselves with the landing runway. It was during the circle, "flying over flat, dark, sparsely lighted terrain in somewhat restricted visibility," that disorientation may have occurred due to a lack of ambient cues to assist with aircraft attitude information and perceived height above the ground (p. 7). Although the pilot flying had 9,670 flying hours and the first officer had 1,922 flying hours, their experience level could not overcome the lack of ambient visual cues needed for reliable visual perception.

Transitioning to instruments is the best and most common method to ensure safe flight while maneuvering in impoverished visual environments. Unfortunately, however, a circling maneuver is a purely visual maneuver and transitioning to instruments to complete the maneuver is not an option. The pilot could have possibly initiated a missed approach and transitioned to instruments, but that in hindsight is not worth addressing. Obviously the pilot felt that sufficient veridical perception was possible.

Minneapolis—1958 (night, black-hole takeoff illusion)

In August, 1958, a Northwest Airlines flight, a Douglas DC-6B, crashed shortly after takeoff from Minneapolis, MN (CAB report, 1959). Fire destroyed the aircraft; however, there were no fatalities. This accident occurred at night and the pilot was attempting a visual takeoff. According to the safety investigation report, the takeoff was normal (rotated the aircraft at 115 knots), gear was retracted when airborne, at 135 knots the flaps were retracted, and then at 155 knots, the pilot called for a reduction in power at about the same time their vision was obscured outside by landing lights reflected against the fog. The pilot turned off the landing lights at about the same time the copilot saw a fence in front of them and shouted to pull up the aircraft. They failed to react in time, hitting the fence and coming to a stop approximately 1,372 m (4,500 ft) from the departure end of the runway.

Witnesses stated that the airplane initially flew an expected climb rate until approximately 30 m (100 ft) above the ground when the plane then gradually descended until ground impact. On the night of the accident the sky was clear with 4.8 km (3 miles) visibility and low fog. In some areas around the airport the fog reduced visibility to 1.6 km (1 mile). The takeoff was toward an open unlighted area. The safety investigation report stated that the removal of ambient visual cues by darkness and reduced visibility, combined with the acceleration and climb-out after takeoff, contributed to a sensory illusion. The report stated:

> The forward acceleration of the aircraft after takeoff causes a sensation of nose-up tilt because the pilot cannot distinguish between the direction of gravity and the resultant of gravity and aircraft acceleration … If it is also very dark and the direction of takeoff is away from a built-up lighted area, there is nothing to be seen which can give a horizon reference and the pilot is now very likely to get this false impression of the attitude of the aircraft in pitch. (p. 6)

With no visual references to correct his inaccurate perception, the pilot pushed forward on the aircraft's yoke, pitching down towards the ground. In other words, the pilot misperceived the attitude of the aircraft and over-reacted to the false sensation of pitching up. Both pilots, despite their experience (the pilot had 12,376 flying hours and the copilot had 9,089 flying hours) were unaware of their descent rate. Although the report presented the sensory illusion discussion at length, the official probable cause of the accident was simply "the pilot's inattention to flight instruments during takeoff in conditions of reduced visibility" (p. 8).

This mishap demonstrates the interaction between visual and vestibular spatial disorientation. Without the presence of visual cues there is no way to confirm the perception of a change in horizontal and vertical motion encountered from vestibular sensations; this is a classic example of the black-hole takeoff illusion.

LaGuardia—1959 (night, weather, impoverished visual cues)

During the dark night of 3 February 1959, another accident occurred at LaGuardia Airport, NY, when an American Airlines, Flight 320, Lockheed Electra flew into the East River (CAB report, 1960). The aircraft crashed 1.5 km (4,891 ft) from the landing threshold for runway 22 (Figure 7.5). The aircraft departed Chicago Midway airport late that evening for a near-midnight arrival in New York City. The pilot had an impressive 28,135 flying hours in nine different commercial aircraft. The weather late that night, however, was not very good. The sky was overcast, visibility was only two miles with light rain and fog, and the clouds were just 400 ft above the ground.

This unfortunate mishap demonstrates the difficulty in determining a root cause in aviation disasters. Many factors played a contributing role, and the safety board concluded that "no one factor [was] so outstanding as to be considered as the probable cause of this accident" (p. 21). The following is from the CAB report:

> The Board believes that a premature descent below landing minimums was the result of preoccupation of the crew on particular aspects of the aircraft and its environment to the neglect of essential flight instrument references for attitude and height above the approach surface. Contributing factors were found to be: limited experience of the crew with the aircraft type, faulty approach technique

> in which the autopilot was used in the heading mode to or almost to the surface, erroneous setting of the captain's altimeter, marginal weather in the approach area, possible misinterpretation of altimeter and rate of descent indicator, and sensory illusion with respect to height and attitude resulting from visual reference to the few lights existing in the approach area. (p. 1)

Referencing the "sensory illusion" the CAB report stated: "the illusion of a safe flight altitude with the limited visual reference available over sparsely lighted areas such as the Rikers Island Channel at night, is not an unknown phenomenon" (p. 20). The report then cited three other mishaps; 1952 crash at La Guardia, 1953 crash in San Francisco Bay, CA, and 1955 crash near Springfield, MO.

Kraft—1978 (black-hole environments)

Kraft described the following four commercial airline accidents that occurred in the mid 1960s, all during visually-guided landings at night. These accidents spurred Boeing to determine the reasons for aircraft impacting terrain short of the runway during attempted visual landings at night (discussed in Chapters 4 and 5).

Chicago—1965 On 16 August 1965, a United Airlines Boeing 727 started a visual descent into Chicago from an altitude of 24,000 ft for 6,000 ft. The aircraft, for an unknown reason, never leveled off at 6,000 ft and crashed into Lake Michigan 30.6 km (19 miles) off shore at 9:21 PM local time. Unfortunately all 30 occupants were killed. The reported probable cause was: "the board is not able to determine the reason for the aircraft not being leveled off at its assigned altitude of 6,000 ft" (aviation-safety.net, 27 Jun 08).

Cincinnati—1965 On 8 November 1965, an American Airlines aircraft attempting a night approach and landing into Cincinnati, OH, impacted a ridgeline just prior to the runway. The pilots were carrying out a visual final turn to align themselves with the landing runway when they crashed into the Ohio River Valley. Job (2006) reviewed this mishap to explore its significance 25 years since its occurrence. Job presented three primary areas that contributed to the accident: a possible mis-read of the altimeters, workload during the final portion of the approach, and the pilots' haste in expediting their descent and visual approach. Job also discussed the visual illusions of the lights along the Ohio River bank as the pilots looked towards the more elevated runway lights in the distance, possibly presenting a false impression of being higher above the landing runway. This false perception of height may have induced the pilots of Flight 383 to descend prematurely. The aircraft hit the embankment 225 ft below the runway elevation of Cincinnati's airport, resulting in only 4 survivors of the 62 on board (Job).

Salt Lake City—1966 In 1966, a United Airlines aircraft descended too steeply during a night approach over dark terrain/water and landed short of the runway near Salt Lake City, UT.

Tokyo—1966 In 1966, 133 people were killed when an All Nippon Airways Boeing 727 commercial airliner crashed 10.5 m (6.5 miles) short of the Tokyo airport into the Tokyo Bay. It was night and the aircraft descended from 7,010 m (23,000 ft) in a right turn toward the runway and hit the water at a speed of 444.5 km/h (240 knots).

Florida Everglades—1972 (night, impoverished visual cues, no horizon)

This mishap is a tragic and often referenced accident in terms of crew coordination problems, channelized attentional issues, and a lack of communication between crew members and air-traffic controllers (http://www.pilotfriend.com/disasters/crash/eastern401.htm). In Chapter 6, night was referenced as a meteorological condition that contributes to spatial disorientation and this mishap was specifically mentioned in that it probably would not have occurred had it been during daylight hours. The pilots became distracted with a nose-gear cockpit lightbulb failure and gradually descended into the Everglades. They were flying over featureless terrain on a dark night with no cues to depict a discernible horizon. The autopilot was inadvertently shut off and, with no ambient cues regarding the aircraft's vector below the horizon, the pilots' failed to notice their descent rate and orientation relative to the terrain.

Pago Pago—1974 (night, weather, black hole)

A commercial airline accident occurred 30 January 1974 at Pago Pago International Airport, American Somoa. A Pan American World Airways, Boeing 707 aircraft, Flight 806, had departed Auckland, New Zealand for Pago Pago and crashed 1,025 m (3,365 ft) short of runway 05 at 11:40 PM local time (National Transportation Safety Board [NTSB] report, 1977). Only 5 of the 101 persons on board survived the crash. The 52-year-old pilot of Flight 806 had 17,414 flying hours and his 37-year-old first officer had 5,207 flying hours. Landing runway 05 at Pago Pago brought the aircraft in over a portion of the island and high terrain, landing toward the water. This tragic mishap occurred at night in a challenging final approach terrain environment.

There were Visual Approach Slope Indicator (VASI, similar to PAPIs) lights on the 2,743 × 46 m (9,000 × 150 ft) runway (length/width ratio of 60), but the first officer who survived the crash did not recall seeing them. VASIs are lights configured on the side of the landing runway that relay glide-path information to the aircrew in a qualitative sense of being above glide-path (2 white lights), on glide-path (white and red lights), or below glide-path (two red lights); refer back to Figure 1.4 in Chapter 1.

This mishap occurred during the transition from instruments to visual conditions when, within the last mile to the runway, the determination was made to continue the approach and to land the aircraft. According to the cockpit voice recorder, however, the pilot reported the "runway in sight" at 9.3 km (5 DME, distance measurement equipment, approximately 5 nm) and the crew stated three more times that the runway was in sight during the course of the approach (NTSB). The instrument approach to runway 05 at Pago Pago was over water until the last 3.1 km (1.7 nm), where it was then over uninhabited jungle. The safety investigation report concluded:

> The probable cause of the accident was the flightcrew's late recognition and failure to correct in a timely manner an excessive descent rate which developed as a result of the aircraft's penetration through destabilizing wind changes. The winds consisted of horizontal and vertical components produced by a heavy rainstorm and influenced the uneven terrain close to the aircrafts approach path. The captains' recognition was hampered by restricted visibility, the illusory effects of a "blackhole" approach, inadequate monitoring of flight instruments, and the failure of the crew to call out descent rate during the last 15 seconds of flight. (p. 27)

(It needs to be mentioned that the NTSB Acting Chairman of the accident investigation dissented with the "probable cause" stating that he favored listing "wind shear" as the major factor in the mishap.) Note the multi-causal description by this particular safety investigation board in describing how the mishap occurred. Yes, visual-spatial disorientation was just one of many cited contributors; however, the number of times the pilot reported the field in sight makes one question how loyal to his instruments he was keeping. Thus, the pilot's attempt to keep visual perception of the runway rather than fly using instruments may have prevented him from breaking the chain-of-events.

This accident illustrates a situation in which rain and wind on a dark night over featureless terrain reduce the quality and quantity of visual cues available to a pilot and can lead to a black-hole illusion. This accident may also have been the first formally cited by the NTSB as having the black-hole illusion as a causal factor —the pilot failed to perceive the descent rate toward the terrain. This featureless-terrain illusion is similar to the illusion reported by the pilot in the 1941 water-landing accident at San Juan. The misperception of height and distance is a common occurrence in black-hole approach accidents (Cocquyt, 1953; Mertens and Lewis, 1982; Parmet and Gillingham, 2002; Perrone, 1984; Previc, 2004). Although visual cues are impoverished and perceptual capabilities limited, pilots trust out-the-window cues rather than relying on instruments because they place too much confidence in their visual perception even when cues are lacking (Gray, 2006).

Godson (1978) questioned the NTSB findings in his book, *Clipper 806: The Anatomy of an Air Disaster*, which details the accident at Pago Pago. He concurs with the dissenting vote of the investigation board member that it was a windshear problem due to the thunderstorm that was the probable cause. Godson also questioned how much the rain limited visibility because of the number of times the field was reported in sight and the ease of seeing the runway lights even though they were only illuminated at 10 percent. Also, Godson brought forth the argument that the VASI were probably *not* working. He attributed this conclusion to the crews' failure to voice the VASI indications at any point along the approach once inside Logotala Hill.

Godson (1978) described the challenging landing environment for pilots flying into Pago Pago. A pilot cannot overshoot the runway and land long due to the end of the runway abutting the beach and water. Undershooting the runway is equally dangerous due to the terrain. Logotala Hill sits just 2.7 km (9,000 ft) from the beginning of the runway and has an elevation of 119 m (390 ft) above the runway's elevation. Thus, on this particular non-precision approach, the pilot must maintain a safe altitude above Logotala Hill (232 m, 760 ft minimum) then, once clear, aggressively descend in time to see the runway environment and intercept a "normal" visual glide-path to the runway, of 2.5–3.0 degree glide-path. According to the voice cockpit recorder, much discussion centered on the pilot concentrating on instruments and the first officer keeping an eye on Logotala Hill, and then once clear of that point, providing visual reference to the runway. The pilot was also getting feedback on the distance to the runway. Thus, in terms of monitoring the pilot's approach and working together, the impression is that this was done sufficiently.

According to Godson (1978), Pan Am Airlines had their pilots follow special procedures flying into Pago Pago. Pilots were to maintain 305 m (1,000 ft) over Logotala Hill and, once clear, the VASI will show them high, which is expected, because they will be high, that is, steep for the approach. Granted, the intent is to ensure clearance over the hill; however, pilots then have only one mile to descend aggressively and get themselves in a position to land. This procedure forces pilots into an unstabilized approach and further contributes to the pilots' feeling of being steep, if accomplished on a dark night.

Mt Erebus—1979 (featureless terrain, whiteout, no horizon)

On 28 November 1979, Air New Zealand Flight, 901 while conducting a sightseeing tour flight of the Antarctica, impacted the side of Mt Erebus, killing all 237 passengers and 20 crew members aboard (www.nzhistory.net.nz/culture/erebus-disaster/crash-of-flight-901). This highly publicized tragic event had numerous contributing factors and hotly debated issues. For instance, the expected and briefed flight route was different from what was programmed into the DC-10's inertial navigation system. This deviation should only have been a factor if the weather was poor and the altitude of the sightseeing aircraft was below 3.9 km (13,000 ft). On this day, unfortunately, the pilots were flying below 3.9 km; while the weather was not terribly poor, visibility was limited due to geographic and environmental factors as discussed later.

Another organizational contribution to this mishap was the lack of radar controlled airspace for monitoring the aircraft's specific route of flight as well as expected flight patterns. The McMurdo Station (a non-radar traffic control location) allowed aircraft to descend below the minimum safe altitude of 4.9 km (16,000 ft) because the normal route of flight was approximately 45 km (28 miles) away from the peak of Mt Erebus. Thus for years aircraft flew the wrong route but it became the norm, then unbeknownst to the pilots; the correct route was loaded into the plane's navigation system—setting them up for failure.

What this mishap has in connection with visual perception is the contribution of the visual environment. Although Flight 901 descended in visual meteorological conditions and was flying a sightseeing route using visual landmarks for guidance, they had become geographically disorientated. The pilots misunderstood their position to be over McMurdo Sound with the Ross Ice Shelf visible on the horizon in front of them. Unfortunately they were over a different body of water—Lewis Bay—and heading directly for Mt Erebus at a mere 0.5 km (1,500 ft) above sea level (http://www.nzhistory.net.nz/media/photo/erebus-flight-path-map). The visibility at the time of the crash was 23 miles; however, the horizon was obscured by clouds and blended with the ice/snow resulting in a "sector whiteout." There was no contrast, no discernible horizon; the impoverished environmental cues combined with the geographic disorientation of the flight crew set them up for failure and they had no awareness that the accident that was about to occur.

The Air New Zealand DC-10 did have a ground-proximity warning system installed and it alerted the crew with an alarm. The Captain initiated a climb with a significant addition of engine thrust, but within six seconds they impacted terrain. Basically, the crew didn't have a chance—they were flying toward a white mountain with white clouds above and no horizon. As presented in Chapter 2, visual perception is strongly driven by expectation, and in this case the aircrew perceived their location to be far from the mountain, thus their misinterpretation of the visual scene's cues (http://www.pilotfriend.com/disasters/crash/anz901.htm). To make matters worse, the terrain they impacted was up-sloped, making it difficult to out climb the rising terrain even if they had reacted sooner and with a more aggressive pitch-up.

Cases of Self-Reported Disorientation—1989 (black hole)

In 1989, an experienced pilot, flying as a copilot in Australia, described his experience with visual Spatial Disorientation (Bennett, 1989):

> It was the darkest night you could imagine … .We could see the lights of the runway in the distance. The airport is out in the middle of nowhere and away from any town lighting. It was so dark, and the night so clear, that the runway

> lights appeared as though they were suspended in space, no other detail at all. I remember commenting just how eerie it was … The only visual cues were the runway lights themselves. But the attitude and angle to the runway still appeared normal. Something was not right … As he turned onto final, we were about a mile and half from the threshold … Suddenly the runway lights started to run quickly up the windscreen. We were 2500 m from the threshold and 250 ft above the ground. I immediately called "go around we're too low." Bill started to react as I screamed my first expletive. (p. 21)

On the basis of this quotation it seems reasonable to infer that the black-hole illusion caused the pilot to fly a 1.7 degree glide-path to touchdown. Such a shallow glide-path is unsafe, even during daylight visual conditions and is worthy of any expletive imaginable.

Moosonee Airport, Ontario, Canada—1990 (night, black hole)

A twin-engine commercial aircraft, Beechcraft C99, crashed 11 km (7 miles) short of the runway at Moosonee Airport, in Ontario, Canada (Flight Safety Foundation, 1993). The *Accident Prevention* edition cited the Transportation Safety Board of Canada as stating the cause of the accident was due to pilot as well as crew issues, in that the pilot:

> Inadvertently flew the aircraft into trees, during a condition of visual illusions, as a result of inadequate crew coordination in that neither pilot effectively monitored the altimeter … contributing to the occurrence were the absence of approach lighting, the lack of company crew-paring policy, the captain's unfamiliarity with black-hole illusions and the seating position of the captain. (p. 1)

The terrain surrounding the airport was flat with 8 m (25 ft) tall trees, and on a dark, cloudy night provided few visual cues to a pilot to estimate height and distance. The Canadian safety board called it a "featureless visual environment" (p. 2). Another interesting finding from this accident was that the captain's seat-height position was below the anthropometrically engineered design eye-reference point. The pilot significantly reduced his field of view outside the aircraft because of the lower sitting height.

Muskogee—1991 (size/shape constancy)

In 1991 it was reported that three aircraft landing at the same airfield on the same night all fell victim to featureless-terrain illusion (Ercoline, Weinstein, and Gillingham, 1991). The pilots of three US Air Force Lockheed C-5 Galaxy aircraft flying into Muskogee, OK, all experienced similar illusions and landed shorter than the previous aircraft (Previc, 2004). A C-5 is a huge military strategic transport aircraft, similar in size to a Boeing 747. The first two aircraft, although landing shorter than desired, at least landed on the runway. The third and final aircraft impacted the terrain 457 m (1,500 ft) short of the landing runway and suffered substantial damage. The aircraft had previously landed at a runway measuring 3,658 × 91 m (12,000 × 300 ft, ratio 40) and runway at which the accident occurred was 2,195 × 46 m (7,200 × 150 ft, ratio 48). Three visual illusions were involved that night. The first was a size-constancy problem caused by the different dimensions of the two runways. The second illusion was labeled "erroneous global perception" regarding confusion with the runway approach lights in terms of where the landing runway surface began. A runway has an overrun, an area of pavement not part of the landing runway, and runway markings and lights define the end of the overrun and the beginning of the landing runway. Finally, because of the hazy, dim appearance of the runway lights, the pilots judged that they were farther away from the runway than they actually were—leading the pilots into believing they had plenty of altitude available for a continued descent.

Ordeal in the Arctic—1991 (night, featureless terrain-black hole)

On 30 October 1991, another black-hole mishap occurred when a Canadian Air Force C-130 en route from Thule, Greenland, crashed 19.3 km (12 miles) short of the runway at Alert, on Ellesmere Island, the northernmost island in the Canadian Arctic, shown in the map, Figure 7.6. In 1993, Robert Mason Lee published a book entitled *Death and Deliverance* that described the accident and the struggle for survival of the crew and passengers after the crash. This book led to a movie called *Ordeal in the Arctic* (1997). Four of the eighteen on board died during the crash and one more, the pilot, died of exposure prior to their rescue (http://troywoodintarsia.com/alert/cc130crash.html, 10 June 08). Controlled flight into terrain resulted when the pilot abandoned his instruments and proceeded visually to the runway in a black-hole visual environment. Figure 7.7 is a photograph taken of the wreckage in the summer of 2006 and depicts the barren terrain surrounding the Arctic island.

Nearly every aviation mishap is a result of a series of events that on their own do not necessarily cause the accident but when linked together can produce an accident. In the case of the Ellesmere Island accident, Lee (1993) suggested that the black-hole illusion was only one of several problems faced by the aircrew. The pilot who was flying the aircraft did not have complete faith in his inexperienced crew, his copilot and navigator. The copilot and navigator failed to assist the pilot by confirming altitude and distance, as well as cross-checking available navigational aids and informing him of the rising terrain between their current position and the runway. In addition, the pilot was overconfident, having often carried out the same flight and arrival procedures. He was attempting to expedite his descent to allow him to land, off-load, and become airborne again in minimum time to make room for other inbound aircraft. The entire crew had abandoned their instruments (even as a back-up) and confidently descended by visual reference alone. Flying into such a remote location, at night, with zero cultural lighting, no

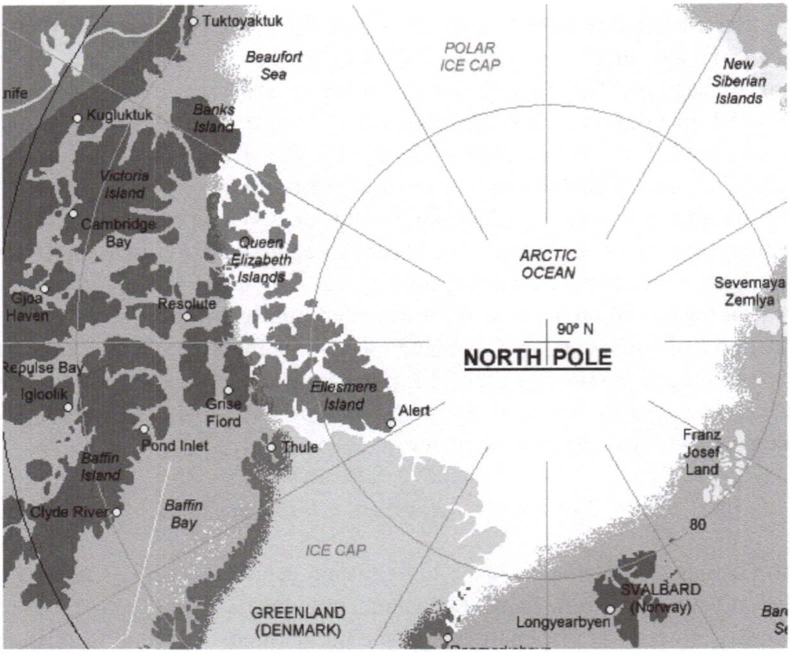

Figure 7.6 Map of Canadian Arctic Islands, Ellesmere Island

Source: From http://www.athropolis.com/map2.htm.

Figure 7.7 Wreckage of C-130 near Alert, Ellesmere Island (photograph taken by Frank Edison)

Source: http://troywoodintarsia.com/alert/cc130crash.html

horizon and aiming only for the bright lights of a distant airfield, perceptually set up the pilots for failure. Lee wrote: "he was flying blind, no differently than if he were attempting to walk across a darkened room towards a point of light without bumping into furniture" (p. 33).

In viewing the movie, *Ordeal in the Arctic*, the depiction of the black-hole approach was very convincing in that no horizon or lights of any kind were visible except for the airfield in the distance. The aircrew in the movie stated that the lights were "shinning like a birthday cake." The lure of the lights of the airport certainly was very strong and enticed the pilot into a visual approach. As was mentioned in Chapter 4, pilots prefer to fly visually, even with only focal-vision environmental-cues available. Many pilots today would make the same error in trusting their vision to accomplish the approach despite an obvious lack of ambient cues; the only impediment to a safe and successful approach and landing was the darkness of night.

Visual-Spatial Disorientation Experience—1993

The first author had his own experience of a black-hole approach and landing in 1993 while flying a Lockheed C-5 Galaxy into a California Air Force Base (AFB). Visual-spatial disorientation almost resulted in failing to land on the runway:

> As the aircraft commander flying a mission for the USAF, our C-5 departed from Dover, AFB, DE, and went to Seattle, WA. The next leg late that night went from Seattle to Travis AFB, CA, outside Sacramento. It was an extremely dark, overcast night with intermittent rain showers. The approach to runway 21 landing toward the southwest brought the aircraft in over the darkest part of the valley with featureless terrain underneath. The illuminated runway in the distance was the only visual cue in the environment. The copilot was flying, and as I monitored his approach, I found myself mesmerized by looking out at the suspended runway image in the blackness of the night. I felt as though we were coming into the runway too steep and had to fight the desire to inform the copilot to shallow his glide-path (the copilot was flying an instrument approach). While I wanted to believe the visual cues in looking outside the aircraft I knew I had to trust my instruments and look inside. The illusion of appearing steep was so strong that it nearly overcame the knowledge of flying on the correct glide-path provided by the instruments.

Recent Mishaps and Incidents Attributed to Visual Spatial Disorientation

Many aviation accidents are considered "classics" in that they happened decades ago. The aviation community may have a false sense of security believing that, due to superior education and technology, those types of older-generation accidents cannot occur in this modern age of aviation. This assumption is not valid. Visual misperceptions have not changed since the earliest days of powered flight. Many of the following accidents have at least one visual misperception factor similar to the earlier mishaps. Pilots of all decades have placed too much trust in their visual capabilities.

Columbia—1995 (night, impoverished visual cues)

This mishap was alluded to in Chapter 6, during the discussion of night as a contributor to spatial disorientation. A commercial aircrew was late for their scheduled landing at Cali, Columbia, December 1995, and during their descent they were asked if they wanted to accept the south landing runway rather than maneuver around to the other side of the airfield to land to the north. This last minute change of landing runways was similar to what happened to the 2002 aircrew that landed at Tallahassee, FL.

In Columbia in 1995 there was no radar control; thus pilots were required to fly specific routings and provide communication back to air traffic controllers of their exact location and altitude. In changing to the south landing runway, the pilots of the commercial airliner had to descend quickly and fly a new routing. However, they were unfamiliar with the arrival procedures and entered the incorrect navigational aid into their computer system (the mode error). Due to night conditions, the pilots failed to notice the mountainous terrain beneath them and, as they navigated their descent in preparation to land, they impacted terrain.

LaGuardia—1996 (impoverished visual cues, lights, haze/fog, contact lenses)

This accident has more detail than others due to the multi-causal areas of discussion: not only visual spatial disorientation but the use of mono-vision contact lenses worn by the pilot. In October, 1996, a McDonnell Douglas MD-88 operated by Delta Airlines impacted an approach light structure and the end of the runway deck during an approach to landing on runway 13, at LaGuardia Airport, NY (NTSB report, 1996); see Figure 7.5. No fatalities occurred, but the aircraft sustained considerable damage. The experience of the crew as very solid; the captain was 48 years old and had a total of 10,024 flying hours and the first officer was 38 years old and had totaled approximately 6,800 flying hours.

The approach to runway 13 brought aircraft in over the water to landing. The pilots flew an instrument approach and, on seeing the runway, attempted to land under entirely visual guidance. Although they broke out of the clouds, rain obscured their visibility to the runway and fog obscured the pilots' perception of the entire length of the runway. Neither of the pilots recalled referring to the Visual Approach Slope Indicator Lights, VASIs (or even seeing them) during the transition to visual conditions though, according to the investigation report, they were working properly (Walters and Sumwalt, 2000). In this accident only seconds elapsed between the end of a normal approach and controlled flight into terrain.

The sequence of events was as follows (NTSB):

> At 4:36:46 PM, the first officer stated, " … a thousand feet above minimums." (Minimums are the lowest an aircraft is allowed to descend unless visual contact is made with the landing/runway environment.)
> At 4:37:24 PM, the first officer advised the captain that he was " … starting to pick up some good ground references."
> At 4:37:57 PM, the first officer stated, "200 ft above [minimums] … speed's good, sink's good." (Use of the phrase *sink* refers to vertical velocity of ft/min descent rate, in this case was within acceptable parameters.)
> At 4:38:10 PM, the first officer called out, "one hundred above" followed by "I got the (REIL [runway end identifier lights]) … approach lights in sight."
> At 4:38:13 PM, the first officer advised the captain, "You're getting a little bit high … a little bit above [the] glide slope … approach lights, we're left of course."
> At 4:38:25.6 PM, the first officer stated, "speed's good" then added, "sink's seven hundred." (Due to the lag time of a vertical velocity indicator [vvi], although the vvi displayed 700 ft/min sink rate, actually the aircraft's sink rate according to the flight data recorder was 1,200 ft/min.)
> At 4:38:31 PM, the first officer cautioned, "a little bit slow. A little slow." As the throttles came up to increase power the first officer was saying, "nose up" while the ground proximity warning system (GPWS) sounded a "sink rate" audible warning.
> At 4:38:36.5 PM, the FDR encoded impact.

According to the NTSB report, the captain stated during a post-crash interview, the approach looked normal until 4 to 5 seconds prior to impact when, "all of a sudden, [the] aim point shifted down into the lights" (p. 6).

Unlike the 1991 Canadian C-130 black-hole illusion that induced the pilots into an unsafe approach that covered several miles, this mishap occurred within a short period of time. Once clear of the clouds and a few hundred feet above the ground the pilot experienced the following four visual problems that contributed to spatial disorientation: absence of ground features (over water); rain on windscreen; atmospheric haze/fog; terrain with few lights to provide height cues (NTSB report, 1996). These visual conditions caused the pilot to erroneously perceive his position to be higher than it was, thus inducing a too-steep descent.

The NTSB report then described that, while the captain of Flight 554 was attempting to transition from instrument to visual conditions, the aircraft's windshield wipers were on the highest setting due to the heavy rain conditions. The limited visibility combined with restricted slant range viewing of 914 m (3,000 ft) created a skewed/shortened image of the landing runway compared to what one would normally expect a 2,134 × 46 m (7,000 × 150 ft) long runway to appear like. The report cited "size constancy" problems in that the shorter runway may appear farther away to the pilot due to the restricted visibility. It is interesting that the NTSB report would comment on this without explanation because, unlike the previous illusions (absence of ground features and atmospheric haze) which induce the pilot into a lower than normal approach, the size constancy illusion in this case would induce the pilot to perceive his position to be too low and make a higher than normal approach—which was not the case.

The final major factor of this particular mishap that has yet to be addressed concerns the fact that the flying pilot, the captain, was wearing mono-vision contact lens. Dismukes, Berman, and Loukipoulos (2007) describe mono-vision contact lenses as providing the correct focus for far objects to one eye and the correct focus for near objects to the other eye, thus eliminating the need for bifocals. The FAA, however, prohibits the use of mono-vision contact lenses for the fear of visual impairment (FAA Pilot Safety Brochure, 2002). The NTSB report describes binocular visual cues accurate for objects out to 183 m (600 ft) while monocular visual cues are good for greater distances. (Recall that various research has remarkably different distances for reliable binocular depth perception cues as discussed in Chapters 2 and 4. For instance, stereopsis may be physiologically capable up to 200 m; however, its effectiveness is very limited at those farther distances.) The report stated:

> The safety Board concludes that the captain's use of MV [mono-vision] contact lenses resulted in his (unrecognized) degraded depth perception, and thus increased his dependence on monocular cues (instead of normal three-dimensional vision) to perceive distance. However, because of the degraded conditions encountered by flight 554, the captain was not presented with adequate monocular cues to enable him to accurately perceive the airplane's altitude and distance from the runway during the visual portion of the approach and landing. This resulted in the captain's failure (during the last 10 seconds of the approach) to either properly adjust the airplane's glidepath or to determine that the approach was unstable and execute a missed approach … A flying pilot with normal depth perception might have perceived the airplane's increasingly excessive sink rate earlier and either slowed the rate of descent to make a normal landing possible or performed a missed approach. However, the captain did not have normal depth perception and did not recognize that anything was wrong with the approach until about 4 seconds before the accident, when the "aim point shifted down into the lights." (p. 59)

The NTSB report summarized the many different distance estimates provided by the three binocular-vision cues of retinal disparity, convergence, and accommodation. In contrast to the NTSB report, Dismukes et al. (2007) believed that the discussion this mishap ought to promote is for stabilized approach criteria, rather than on the pilots' confusion and the contact lens issue. Although the contact lens issue is important, stabilized approaches carry a higher priority for aviation safety. They concluded that:

> The use of monovision contact lenses may well have contributed to the accident by further impairing the captain's processing of visual information that was already substantially impoverished and conducive to illusions; however, existing scientific knowledge is not sufficient to determine with certainty how much the contact lenses contributed to this accident. (p. 244)

One NTSB question was why did other pilots land successfully at LaGuardia during that time but not Flight 554? Dismukes, et al., suggested three reasons for the fate of Flight 554:

1. Workload and risk factors involved with Flight 554s un-stabilized approach and attempts to salvage a landing.
2. Many previous mishaps have occurred given similar perceptual issues confronting aircrew that did not have mono-vision contact lens factors involved.
3. Probabilistic reality of highly experienced aircrew and their variability in performance from approach to approach accomplished in impoverished conditions.

Nova Scotia—1996 (terrain illusion)

A Canadian Airlines Boeing 767 had its tail skid and aft fuselage damaged during a landing on 8 March 1996 (Accident Prevention, 1998). According to the safety investigation:

> The crew responded to a visual illusion with an unwarranted power reduction between the minimum descent altitude and touchdown … The upslope illusion led both crew members to believe [that] the aircraft was higher than it actually was, and the crew did not respond to the visual cues from the precision approach path indicator [PAPI] lights, which showed the aircraft to be too low. (p. 1)

No one was injured in the incident; however, lessons were learned regarding proficiency of the pilot performing back-course localizer approaches, crew resource management training, the appropriate call-outs by the pilot monitoring the approach, and the visual illusion of an up-sloped runway during a night landing. The landing runway at Halifax, Nova Scotia, Canada, runway 06 is 2,708 × 61 m (ratio 44); however, the upslope is 0.77 percent, within 0.03 percent of the maximum upslope allowed (Accident Prevention, 1998). Although the runway ratio is low, that is, a short/wide runway, that much of an up-slope in the beginning of the runway can induce a pilot into overestimating their glide-path, prompting a more aggressive descent than normal. The Boeing aircraft was on a 168 m/minute

(550 ft/minute) rate of descent and then, near the runway, the descent rate increased to 260 m/minute (850 ft/minute). This mishap is one more example that experience does not make one immune from visual misperception; the pilot flying had 17,300 flying hours and the copilot, first officer, had 14,000 total flying hours.

Virgin Islands—1997 (night, featureless terrain—water, black hole)

On 8 February 1997 a Cessna flying for Air Sunshine from ST Croix, US Virgin Islands en route to St Thomas, US Virgin Islands, crashed into the Caribbean Sea 3 miles southwest of the Charlotte Amalie/Cyril E. King Airport (NTSB report, 1998). Two passengers were killed while three others survived. The accident occurred at night while flying a visual approach to the island's runway over dark, featureless water. The NTSB report concluded:

> Evidence suggests that the absence of visual cues caused by the combination of dark sky and darkness over the water produced a "black hole" effect in which the pilot lost visual sense of the airplane's height above the water. As a result, the pilot misjudged the airplane's distance from the island and height above the water. (p. 1)

The 43-year-old pilot survived the accident and stated afterwards that he had failed to push the distance measuring equipment (DME) hold button that would have provided his distance from the tuned navigational frequency at the landing airfield. Just prior to contact with the water, the pilot was attempting to diagnose a landing-gear malfunction. As described above, when flying at night over dark, featureless water, depth cues are eliminated and possibly the horizon as well. It would be difficult to maintain proper aircraft attitude due to the lack of ambient vision cues while distracted by trouble-shooting a gear malfunction without the use of autopilot for altitude hold.

The ensuing safety recommendations from the NTSB to the Federal Aviation Administration consisted of requiring passenger-carrying flights to operate under *Instrument Flight Rules* (IFR). Regulatory guidance for flight operations does not solve the basic problem of the black-hole illusion. Pilots on an IFR flight-plan may inform the controlling agency that the airfield is in sight and request to "proceed visually." Had the Air Sunshine flight been flying an instrument approach rather than a night-visual approach it is possible that the accident might not have happened. However, it seems that, when some pilots see the runway, they often choose to proceed visually, regardless of their instrument flight plan. Kern (2002) stated that an air traffic controller's (ATC) request to "report the runway in sight" is a cognitive trap leading aviators to abandon their instruments and proceed purely visually. By asking that question to a pilot and the pilot acknowledging it affirmatively, air traffic control is relieved of aircraft separation and navigation responsibilities. Although this allows aircraft to proceed with less spacing into an airfield, it increases the pilot's task load during a critical and often already task-saturated phase of flight (as previously mentioned, only 4 percent of total flight time, but 52 percent of all accidents).

Guam—1997 (night, black-hole environment, terrain)

In 1997, a fully-functioning $60 million Boeing 747–300 impacted Nimitz Hill, 6.1 km (3.3 nm) short of Runway 06L, at A.B. Won Guam International Airport, Guam (NTSB report, 1997). Of 254 persons on board, there were 228 fatalities. Flight 801, operated by Korean Air, departed Kimpo International Airport, Seoul, Korea, at 11:53 PM local time for a middle-of-the-night arrival into Guam. Figure 7.8 is the crash site relative to the landing runway in the distance as well as the navigational aid.

The aircrew for Flight 801 was very experienced. The captain was 42 years old and had been flying for Korean Air for ten years; prior to that he had been a pilot in the Korean Air Force and had a total of 8,932 flying hours. The captain had made the trip between Seoul and Guam eight times as a captain in a Boeing 727 aircraft and had flown the same route as a captain of a 747 just one month prior to the accident. The first officer was 40 years old, also a former Korean Air Force pilot, and had 4,066 total flying hours. The flight engineer on board (the aircrew member that monitors the aircraft systems such as fuel, electrics, hydraulics, etc.), was 57 years old and had a total 13,065 flying hours combined as a former Korean Air Force navigator and a commercial flight engineer.

Preparing for the final approach and landing into Guam, the flight crew of 801 discussed the fact that the Instrument Landing System (ILS) glide-slope was

Figure 7.8 Crash site at Nimitz Hill, Guam

Source: http://aviation-safety.net/photos/displayphoto.php?id=19970806-0&vnr=10&kind=C

unusable and that they would fly the localizer approach. An ILS is a precision instrument approach that provides both glide-slope information and lateral guidance, whereas a localizer is referred to as a non-precision approach because it only provides lateral guidance information. To accomplish the localizer approach using the lateral guidance information, vertical altitude restrictions paired with distance markings are provided to incrementally step-down the aircraft. The final Minimum Decision Altitude (MDA) is the restricted descent altitude allowed unless visual contact is made with the landing runway environment. Note the published step-down fixes in Figure 7.9 compared to Flight 801's actual vertical descent path. Figure 7.9 also shows the location of the navigational aid on Nimitz Hill, the VOR, at a location prior to the runway as opposed to being colocated with the runway. Navigation aids placed off the airfield add one more complexity for the pilot to unravel in terms of spatial awareness when interpreting distance from the runway as opposed to distance from the navigational aid. Finally, both Figures 7.8 and 7.9 depict the terrain elevation above that of the runway's elevation, making the approach dangerous if flying a shallow glide-path.

Despite the approach briefing by the captain regarding flying the Localizer approach, the crew became distracted regarding the status of the glide-slope.

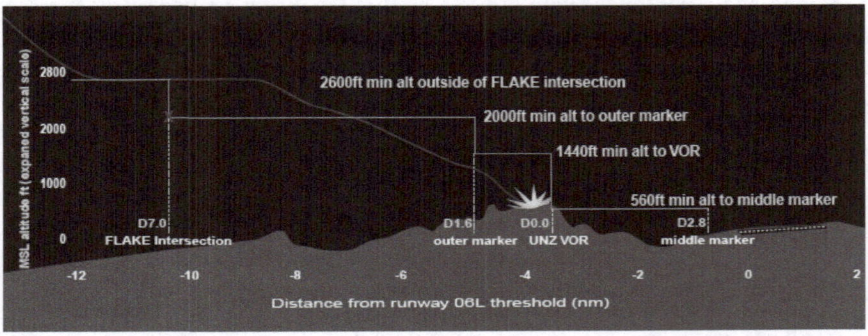

Figure 7.9 Profile view of Flight 801, Guam

Source: Reproduced with permission from Flight Safety Australia.

Another contributing factor in this accident was that, by failing to address missed approach procedures (what to do if they decide not to land or cannot land), they were not mentally prepared to accomplish the required course of action when that course was needed. The NTSB report stated that although the flight engineer reported seeing the airfield from far away, there were rain showers in the vicinity of the airport. Overconfidence in visual abilities and the tendency to want to maintain visual contact may have lulled the aircrew into a complacent instrument approach due to the expectancy of a visual approach.

Although the NTSB found the probable cause of the accident was an improperly briefed and flown non-precision approach, as well as the aircrew's failure to monitor the approach, there was indirect mention of visual-spatial disorientation. The pilot expected to fly a visual approach and consequently failed to adequately brief the instrument-approach procedure, and there was also confusion regarding the status of the navigation systems (NTSB report, 1997). Listed under "findings," the NTSB report stated that the pilot may have lost situational awareness regarding his position relative to the runway, believing he was closer than he actually was (whether this was due solely to a lack of understanding of the correct procedures for the non-precision approach into Guam or due to visual-spatial disorientation was not addressed in the report, demonstrating the difficulty in using the phrase situational awareness).

The black-hole illusion was not specifically mentioned in the NTSB report for this mishap, nor is there any direct reference to visual spatial disorientation. However, the approach late that night brought the aircraft in over the water and then over mountainous terrain and into an area toward Guam called the "black hole" (Ostinga, 2000, p. 23). The report does acknowledge that, in speaking with the captain's first officer about a flight into Guam the month previously, that the captain briefed the "black-hole" area around the Guam navigational aid on Nimitz Hill. Thus, the captain was aware of the visual dangers coming in at night.

It is curious, however, that the accident report did not address environmental conditions for a visual approach in terms of the possibility of a black-hole illusion. The time of the accident was just after midnight; it was a dark night, and visibility was further reduced by rain showers and the approach to landing was over featureless terrain. The landing runway (06L) had a high length/width ratio of 66.7 (3,053 × 46 m, 10,015 × 150 ft), and was up-sloped with a change of 13 m (41 ft) from landing to departure end. The high aspect ratio of a long, narrow runway (Mertens and Lewis, 1983) combined with an up-sloped runway (Kraft, 1978) may have helped induce the pilot to make a shallow approach that, because of the mountainous terrain, is especially dangerous when attempting to land at Guam runway 06L.

Although the NTSB report did not discuss the possibility of spatial disorientation experienced by the crew of Flight 801, there is evidence that the captain did experience it. Neither the Ground Proximity Warning Systems (GPWS) alerts nor the first officer's call-outs to commence a "missed approach" 6 seconds prior to impact seemed to alert the captain of their aircraft's position. One could argue that the captain's lack of response demonstrated a loss of situational awareness as well as spatial disorientation relative to the landing runway. The aircraft's impact site was in the previously acknowledged "black-hole" location, Nimitz Hill, and on that dark, overcast night, all forms of ambient visual cues were removed that could have helped re-orientate the aircrew as to their correct location. In fact, as previously discussed, given the up-sloped runway and featureless terrain, any visual information gained from the runway lights may have actually prompted a continued descent.

Kuwait 1999 (night, black-hole environment, featureless desert terrain)

In 1999, a US Air Force C-130 aircraft flying a night visual approach crashed 881 m (2,890 ft) short of runway 15R at Al Jaber Air Base in Kuwait, resulting in 3 fatalities (Accident Investigation Board [AIB] report, 1999). The cause of the accident was stated to be pilot error. The pilots were cited for failing to follow directives as well as complacency in flight operations that resulted in spatial disorientation during the visual approach. These factors contributed to a loss of situational awareness regarding the aircraft's excessive descent rate. The AIB report noted that the aircraft's glide-path on final approach was initially 3 degrees but soon became too steep (6 to 7 degrees glide-path) and failed to return to the normal glide-path prior to landing. A steep glide-path increases the descent rate and, if uncorrected, results in the aircraft going well below the safe/desired glide-path of 3 degrees. Although not specifically stated in the AIB report, flying at night in a featureless desert environment might have induced a black-hole illusion and the approach described is similar to a black-hole concave approach. The AIB report stated that the pilot erred in judgment by not monitoring aircraft instruments more closely.

In several ways the accident described above resembles previous accidents that were attributed to visual spatial disorientation. The aircrew should have flown an instrument approach, but their preference and perceptual set/expectancy was to fly a visual approach. Consequently, when they visually acquired the runway 8–9.6 km (5–6 miles) out they confidently proceeded visually, as was the case for the Pago Pago and Guam accidents. And then, like the Canadian C-130 BH accident, the bright lights of the distant runway reinforced their decision to proceed visually. What was different with this accident, however, was the presence of a low fog deck that was not detected until the aircraft entered the fog at a point 1 km (0.63 miles) from the runway, 38 m (125 ft) above the ground and with a 518 m/min (1,700 ft/min) descent rate. At that moment the aircrew lost sight of the runway during their visual approach (AIB report, 1999). The pilot initiated a go-around but did not have sufficient altitude to avoid ground impact. The AIB report noted that the pilot's spatial disorientation might have been caused by the lack of ambient visual cues:

> This lack of peripheral visual cues may have contributed to the pilot's spatial disorientation and failure to recognize his transition point to the normal glide-slope. This condition creates considerable difficulty for the pilot, by requiring focal vision alone to accomplish what is normally accomplished with both focal and ambient vision. (p. 21)

A former C-130 pilot shared his experiences flying in and out of the same airfield in Kuwait and commented on how that mishap influenced his flying (C. Hays, personal communication, 2009).

> The instrument approach to Kuwait International was not preferred by the Kuwaiti controllers. Even when requesting the approved instrument approach

> into Kuwait International the controllers would either disregard your request and issue the visual clearance to the preferred runway or not answer subsequent radio calls. Based on the possible threat and the mentality of getting the troops to where they needed to go, I often flew the same visual approach. On one mission, we checked the reported weather and received permission for the night visual approach approximately 6 miles out. While scanning for the runway I saw a black cloud obscuring the runway. I immediately reduced power and asked for the instrument approach. We were given intercept vectors well above glide slope and inside the final approach fix. We eventually went missed approach but it took us almost to the MAP [missed approach point] to positively identify our position. I think that this approach, flown only a few months before the Kuwait crash, could have easily ended for me in a similar outcome if the cloud had not been so prominent or if I had not transitioned to an instrument approach. Following the Kuwait C-130 mishap I started to cross-check my vertical velocity more often throughout the landing and also mentally computing a 3-degree glide slope in order to back up my visual perception.

St. John's—1999 (night, black hole)

An Air Canada Airbus A320 flying a night approach into St. John's Newfoundland airport impacted 76 m (250 ft) short of the landing threshold (Transportation Safety Board of Canada report, 1999). No visual slope-indicator lights were present, and pilots who had previously flown into that runway reported black-hole conditions.

New York—2000 (night, no horizon, weather)

On the evening of 30 March 2000, Delta Airlines Flight 106 departed John F. Kennedy airport bound for Germany carrying 212 passengers and 13 crew members (NTSB report, 2000). The captain of Flight 106 had 19,500 flying hours and the first officer had 9,000 total flying hours. Despite these impressive totals of flying experience, the aircrew became spatially disorientated during the departure leg. The weather that night consisted of some scattered and broken clouds, but most noteworthy was the absence of the moon and horizon. According to the NTSB report, the first officer stated that there was no horizon, stars or moon, and all he saw was darkness. The first officer was accomplishing the takeoff and departure and was hand flying the aircraft, the autopilot was disengaged. While executing a left turn both pilots were also busy with navigational duties and within seconds the aircraft somehow entered a 60-degree right turn. It took both pilots to recover the aircraft back to straight and level flight.

There were no mechanical or structural problems with the aircraft. This incident appears to have resulted from complete disorientation by the flying pilot turning the wrong direction and it going unnoticed by the pilot monitoring him. The lack of visual cues made the visually controlled action of flying the departure nearly impossible; instruments should have been the primary means of controlling the aircraft. This was merely an incident; no one was hurt and no resources were lost; however, it is a great learning case. Two very experienced pilots flying a normal takeoff and departure for a major airline, exposed to the highest quality training and equipment, still fell prey to spatial disorientation.

Greenland—2001 (night, featureless terrain—black hole)

In 2001, a chartered Dassault Falcon 20 cargo flight struck terrain 8.3 km (4.5 nm) from runway 07 during a night-visual approach, killing all three aboard in Narsarsuaq, Greenland (FSF report, 2004). The Danish Aircraft Accident Investigation Board determined that the pilot elected to fly a visual approach rather than the instrument approach and failed to maintain vertical awareness to the terrain. The Flight Safety Foundation reported that the flight crew focused on visual contact with the runway, experienced a black-hole illusion and consequently flew a too-shallow approach resulting in controlled flight into terrain.

Aspen, Colorado—2001 (night visual approach, mountainous terrain)

In March of 2001, a chartered Gulfstream III with 15 passengers and 3 crew members crashed short of runway 15 at Aspen-Pitkin County Airport, Aspen, Colorado, 34 minutes after official sunset (NTSB report, 2001). The aircraft had departed Burbank, California, knowing that their arrival time would put them near the thirty minutes after sunset landing restriction time. The formal NTSB report concluded that the probable cause of the accident was due to the pilots' operation of the aircraft below the non-precision approach minimum descent altitude without sufficient visual cues of the runway environment. This simple description fails to fully describe the events that took place that evening.

The passengers were late arriving to the Burbank airport and the chain of events for the tragic mishap was set in motion. Other contributing factors to the accident were flight discipline breaches, an inappropriate approach briefing, poor decision-making, external pressures from the passengers/customers to land at the contracted destination, some confusion regarding authorization to fly a night non-precision circling maneuver, procedural errors in flying the instrument approach, and deploying speed brakes while the aircraft was configured for landing. However, of interest to the topic of visual perception was that the pilots in some ways hoped for and expected to fly a visual approach. So for them not to comply with each and every altitude restriction of the non-precision approach was not an issue since they seemingly were in the process of "going visual" during the entire approach; nearly 20 miles out before even starting the approach the pilots were busy looking down at the ground attempting to pick up visual references. What is troubling about going visual in this mishap is that the mountainous terrain, deteriorating weather, and night conditions made "going visual" perceptually near impossible.

There are a number of challenging airports around the world and Aspen is one of them due to the surrounding mountains and its field elevation of 7,815 ft above sea level. Airports with high terrain often require non-precision instrument approaches that at the minimum descent altitude still leave the pilot a couple of thousand feet above of the runway. Consequently, the pilot must either accomplish a very steep descent for a straight-in or maneuver the airplane down using a circling, spiraling descent to lose altitude while maintaining a position directly overhead the runway and avoiding the surrounding terrain. Eight days prior to the mishap an inspection of the arrival procedures and approaches into Aspen concluded that a night circling maneuver was unsafe because of the difficult-to-visually-acquire terrain-features due to degraded visual cues (FSF, 2002).

The crew of the Gulfstream III flew the approach under the pressure of time to land prior to or within an acceptable sunset-window as well as pressure from their passengers not to go missed-approach and divert to Rifle (a small airport 54 miles away). Not only was the sun setting on the mountain-enclosed airport but the weather was getting worse. Prior to their arrival the weather forecast was for good visibility, 3,000 ft scattered clouds and overcast at 5,000 ft but with the possibility of lower ceilings and snow showers. Weather similar to this would not be an issue for a normal instrument approach that brings the pilots to the runway within 90–152 m (300–500 ft) above the ground, but the mountainous Aspen approach leaves the pilot still 727 m (2,385 ft) above the ground (hence the need for the circle).

Five minutes prior to controlled flight into the terrain the tower issued a weather advisory that the visibility was down to 2 miles. Despite the deteriorating weather conditions the pilots continued their "visual" descent (3 miles is required to accomplish the particular type of instrument-to-visual approach). The cockpit voice recorder documented the pilots' efforts to look outside and pick up visual references, thus giving credence to how little the pilots attempted to follow instrument procedures to meet their expectation of a visual approach. Technically the pilots should not have descended below the minimum descent altitude unless visual contact was made with the runway environment. Herein lies the problem … they expected to see and continued searching for visual cues while they descended in dark and snowy conditions. In the end the pilots may have confused some lights on a highway for the runway lights. The plane descended at a dangerous rate given mountainous terrain and degraded visual cues. The aircraft struck the ground, instantly killing all onboard 732 m (2,400 ft) from the runway, 92 m (300 ft) right of centerline at a spot 30 m (100 ft) above runway elevation.

Similar to other mishaps, if the pilots had been provided sufficient visual cues this accident might not have happened. Walters and Sumwalt (2003) reviewed this accident in *Professional Pilot* and focused on the flight-discipline problems. Granted, there were aspects of flight discipline in the crew's decision-making and risk assessment; however, the pilots also failed to respect their environment. Consideration must be given to the lack of illuminated ambient cues available for credible visual perception.

Afghanistan—2002 (night, misperception of height and distance, black hole)

A US Air Force C-17 cargo aircraft hit terrain 2,000 ft short of the runway while attempting a night landing at Kandahar, Afghanistan in January, 2002 (AIB report, 2002). After hitting the ground the plane became airborne again and climbed to a safe altitude to allow the crew to assess their situation and the damage to the aircraft. Although there were no injuries, the plane was substantially damaged at a cost of approximately $3.7M. The descent rate of the C-17 was 366 m/min (1,200 ft/min) and, just prior to impact, the aircrew initiated a go-around. Human-factors analysis of the mishap revealed a lack of external visual cues that prevented the perception of a too-steep approach and of imminent impact with ground prior to the runway. According to the AIB report, a black-hole illusion was created by the dark surroundings and dim runway lighting combined with the pilots' failure to use the aircraft's landing lights. Confusion as to the color of the runway threshold lights (lights marking the end of the overrun and the beginning of the runway) also contributed to the mishap.

Utah—2002 (horizon blending with terrain)

In November 2002, an F-16 fighter aircraft crashed onto the Utah Test and Training Range during maneuvers while participating in a 4-ship surface-attack tactics sortie simulating delivery of laser-guided bombs (AIB report, 2002). The experienced fighter pilot died of injuries and the aircraft was destroyed. According to the investigation report, channelized attention combined with a visual illusion caused the pilot to lose spatial awareness.

The visual illusion was a result of the very unusual environmental conditions encountered that particular day. The salt flat, a flood control evaporation site for the Great Salt Lake, was covered with 5–7 cm (2–3 inches) of clear smooth water and acted as a mirror, causing the ground and sky to blend without a discernable horizon, thus removing a crucial ambient vision cue. One of the pilots in the formation that day told the AIB investigators that he had never seen such conditions during his eight years experience of flying over that type of terrain. There was also a scattered layer of clouds, and their shadows on the reflective water created depth-perception problems. These visual conditions contributed to misperception of both altitude and attitude during maneuvering of the high-performance jet aircraft.

An interesting finding during the investigation of the accident was that another pilot in the formation experienced similar visual illusions during the same maneuver. Although the terrain, horizon, and sun location had combined to induce these two pilots to make erroneous visual judgments, the difference between the two outcomes was that one pilot initiated his turn approximately 304 m (1,000 ft) higher than the other, giving him more margin for correcting his visual misjudgment prior to ground impact (AIB report, 2002). The second pilot was unaware of his descent until he reviewed the flight profile during his interview after the sortie. The potential for visual illusions over the Utah range had previously been recognized, but "it took a mishap to rediscover the visual illusions as a hazard" (AIB report, 2002, p. 33).

Another factor contributing to this mishap was channelized attention. Task prioritization during intensive maneuvering of high-performance fighter aircraft is extremely difficult but also critically important. For that reason altitude settings are established prior to takeoff to provide audible warning of impending ground impact. The fact that no attempt was made to recover prior to impact or to eject from the aircraft indicated to the AIB investigators that the pilot did not see or hear any "altitude" warning and was unaware of his position relative the terrain. Another term describing this extreme focus of visual and cognitive attention is *inattentional blindness*, whereattending to one stimulus has the effect of filtering out other stimuli (see Chapters 2 and 8 for further discussion of this effect).

A Close Call—2004 (night, misperception of height and distance)

A Flight Safety Foundation publication described a near-CFIT mishap of a commercial aircraft in July 2004 (Gurney, 2006). The intent of the article was to promote the safety feature of a terrain awareness and warning system (TAWS). Because the incident was shared for safety and prevention reasons, the identifying details of airline and airfield specifics were not provided by the publication. While attempting a visually-guided night approach into a major airport at a remote geographic location, the pilot flew the classic black-hole concave approach glide-path. Because details are not provided, one can only guess why the aircraft got so dangerously close to the ground. The incident may have resulted from either a lack of pilot understanding of the appropriate instrument approach procedures and/or the failure of other crew members to monitor and cross-check the pilot's flying. Regardless, the pilot confidently flew toward the runway and its black-hole environment. The airplane's TAWS alert, occurring at just 76 m (250 ft) above the ground, saved the crew, passengers, and aircraft. This warning alerted the aircrew to take corrective actions 2.8 km (1.5 nm) from the runway.

Arizona—2006 (night, misperception of height and distance)

An incident occurred in May 2006, when a two-seat F-16 aircraft came within 6 m (20 ft) of the ground 0.9 km (2,953 ft) from the intended runway (E. Cassingham, Personal communication, 2006). The pilots, an instructor and a student, were flying a practice night visual approach into a dark airfield that lacked an approach lighting system (sequence of lights illuminating the approach portion of the runway), but did have Precision Approach Path Indicator Lights, PAPIs. Contributing to the incident were inappropriate prioritization of tasks and a failure to monitor and challenge the flying pilot. These factors were exacerbated by a featureless terrain that may have allowed the pilot to perceive a "duck under" maneuver (shortening the desired aimpoint) as safe to accomplish. Alternatively, the landing environment may have induced the pilot to fly an incorrect, dangerously shallow glide-path.

Iraq—2006 (featureless terrain, lack of ambient visual cues)

An US Air Force F-16 fighter aircraft crashed while accomplishing a combat maneuver in the early afternoon of November 2006, resulting in a fatality (AIB, 2006). According to the released report, the conclusions of the mishap were:

> By clear and convincing evidence, the cause of the mishap was the mishap pilot's channelized attention manifested by his desire to maintain a constant visual positive identification of targeted enemy vehicles and subsequent target fixation on these vehicles while they were traveling at a high rate of speed. These two factors, when combined, caused the mishap pilot to begin, and then press his attack below a recoverable altitude.

This tragic mishap is exclusive to military operations in terms of operating a high performance jet aircraft in extreme close proximity of the ground and having a primary task not related to ground clearance (see discussion of low altitude flight in Chapter 4). However, built into the specific procedures are altitude checks to ensure safe accomplishment of the maneuvers, regardless of the terrain and environment. The pilot was performing a high-speed, low-altitude weapon's delivery pass, in which a pilot descends at a 20-degree angle towards the target and pulls up at approximately 457 m (1,500 ft) above the ground. In this mishap, on the pilot's first strafing run he descended at a 26-degree angle and pulled up only 61 m (200 ft) above the ground (Rolfsen, 2007). Then setting up for the next pass, he failed to increase his altitude sufficiently and adjust for the first pass's low altitude. This resulted in his second run at a 25-degree angle and at a speed of 440 mph. Consequently, with insufficient altitude for successful maneuver completion, and despite a high-G pull initiated, the aircraft's tail hit the ground and the plane was destroyed.

This accident was titled *Too Focused on the Fight* in a newspaper article by Rolfsen (2007). In terms of visual perception, one questions if any visual cues were available in the environment that could have or should have informed him of his rate of descent toward the ground? This mishap occurred in early afternoon; however, it was in a sparse desert environment, one that may have been featureless terrain. It is quite possible that there were no real cues other than the target itself, and thus, there would have been no referents to provide size, distance, and depth cues to the pilot. The pilot may have been relying solely on focal vision during the target fixation, with no ambient cues to provide orientation regarding the global earth perspective needed for credible spatial awareness.

Irish Sea—2006 (night, impoverished visual cues, featureless terrain—water)

An experienced helicopter flight crew and all five passengers were killed when their helicopter crashed into the Irish Sea on a dark night approach while attempting a landing on a gas platform (United Kingdom Air Accidents Investigation Report).

The helicopter and crew were accomplishing their second flight between numerous gas platforms in the Morecambe Bay, northwest England. Each flight was of fairly short duration moving supplies and people. On the mishap flight the copilot was flying; it was a very dark and there were poor weather conditions with low overcast clouds. The copilot during the landing lost control and requested assistance from the commander (the more senior pilot); however, there was a 4-second time delay before positive aircraft transfer occurred (Investigative Report). Although the actions of the commander should have been sufficient for successful recovery, the aircraft impacted the sea. The Air Accident Investigation report concluded that the copilot's approach angle during his descent to the helicopter landing deck was incorrect due to the limited visual cues that night.

The crew that evening initially took off from Blackpool Airport en route to an unattended drill platform; it was an 11-minute flight (Figure 7.10). Three minutes later they took off for a manned platform, just a 9-minute flight. The crew on their next leg flew back east, southeast heading toward North Morecambe platform, approximately a 7-minute flight.

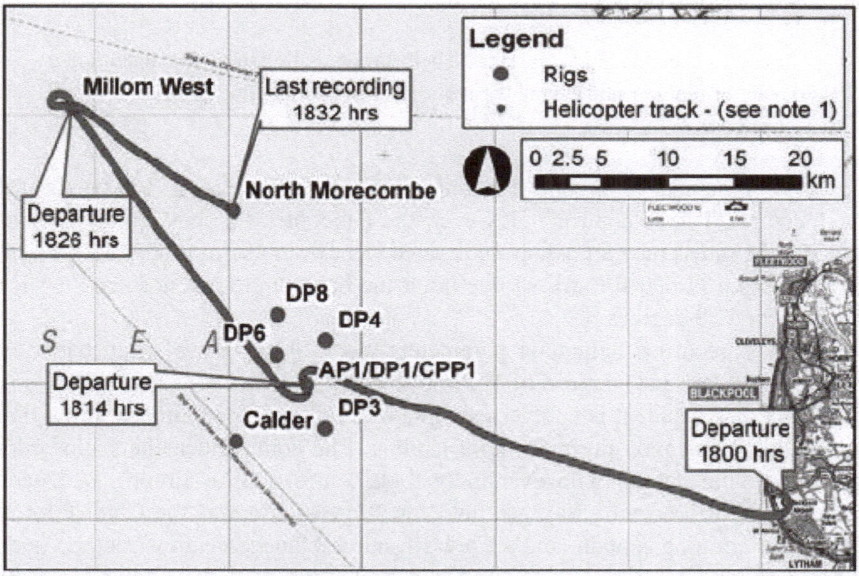

Notes:
1 - The helicopter track was derived from the CVFDR recorded heading and speed data in combination with a wind model. This was adjusted to coincide more accurately with the rigs and intermittent radar returns.
2- All times are in UTC.

Figure 7.10 Route of flight 2006 helicopter mishap

Source: http://www.aaib.gov.uk/sites/aaib/cms_resources/7%2D2008%20G%2 DBLUN%20Section%201%2Epdf

Their altitude was 152 m (500 ft) above the water. As they approached the drill platform the copilot announced that he had a visual with the lights of the landing deck, descended to 82 m (270 ft), then climbed back to 122 m (400 ft) before starting a further descent 30 seconds later. As the helicopter slowed to 55 knots, the following was taken from voice transcripts in the Air Accident Investigation report:

> The commander (pilot not flying) stated, "You get no depth perception, do you?"
>
> The copilot flying replied, "Yea, not on this one, not tonight, no."
>
> Twelve seconds later, the commander asked the copilot, "You all right?"
>
> The copilot answered, "No, I'm not happy, mate."
>
> The commander asked, "We going round?" (Asking if the copilot was initiating a go-around.) At about the same time the engines torques exceeded 100 percent.
>
> The copilot then replied, "Yea, take … help us out … " but the commander failed to react or reply or understand the request … and the copilot again said, "Help us out."

The commander took control stating, "I've got it, I've got it, I have got it, I have control, I have control." However, the helicopter was not in control as it rolled right and as the vertical-speed descent went from 402 m/min (1,320 ft/min) to 515 m/min (1,690 ft/min). At one point the bank angle and nose-down pitch both reached 38 degrees.

The last recorded helicopter parameters were 20 degrees of right bank, 12 degrees nose low and airspeed of 126 knots.

This tragic accident is another example of an experienced aircrew struggling with visual perception during a night landing. The commander, the senior pilot monitoring the copilot who eventually took control of the aircraft, had been flying in the Morecambe Bay gas field for 20 years. He was the Chief Pilot at the base, a training captain, and a Crew Resource Management instructor. Given his experience and duties in training and evaluating other helicopter pilots he would be the pilot everyone else would want to fly with and learn from as he had 8,856 flying hours. He had accomplished 37 night-deck landings in the previous 90 days. The copilot was younger but still experienced with 3,565 total flying hours and had accomplished 7 night-deck landings in the previous 90 days and 107 landings in the previous 365 days. Again, experience does not protect a pilot from visual misperception … it is a physiological limitation of our visual system's perception.

Recall from Chapter 6, Leland (2001) referred to "night visual flying" as an oxymoron. Surface visibility was 3 to 7 km (1.8 to 4.3 miles) with mist and light rain/drizzle. The cloud decks were scattered to broken at 213 m (700 ft) and broken to overcast at 366–457 m (1,200–1,500 ft). These conditions were not conducive to a visual approach as the environment lacked a horizon and any referents for global orientation other than the lights of the landing pad. Estimation of height and distance was nearly impossible by visual reference alone towards a focal target because of the featureless, dark environment. Flying visually at night is perceptually challenging enough and on this night over the Irish Sea, flying in the darkness using visual references was nearly impossible. This was addressed in the formal report:

> The approach was flown essentially by reference to visual cues. In dark, overcast conditions, it is likely that some cues were degraded or absent. For example, without a distinct horizon, the assessment of pitch attitude and approach angle (by reference to the depression of the deck below the horizon) would be compromised. Without textural cues in the ground plane (in this case the sea surface), judgment of pitch attitude and approach angle by inference from textural perspective would also be compromised, as would the appreciation of the range to the deck. (p. 52)

Of concern to the investigators was the inconsistent approach-angle flown toward the landing deck. The report questioned what appeared to be two different phases in the approach and postulated that the copilot had difficulty perceiving the appropriate glide-path angle to the landing deck. The investigation hypothesized that maybe the copilot started the approach too early or too steep, or flew an inappropriate control strategy; or inadvertently changed the pitch attitude. The investigation concluded, "the underlying causes however, most likely stem from the limited visual cues available and the paucity of instrument cross-checks" (Air Accident Investigation report, p. 54).

Noteworthy in this mishap report was how the pilots' visual limitations were articulated in describing the accident scenario. Progress has been made to recognize visual-spatial disorientation as a contributing factor in pilots' controlled actions that lead to accidents.

Mid-air over the Amazon—2006 (small retinal image, closure rate, displays)

This mishap was especially tragic due to the manner by which the events unfolded. William Langewiesche, (son of Wolfgang, who wrote *Stick and Rudder*, the first detailed account of visual flying maneuvers in 1944), detailed this accident in *Vanity Fair* magazine. Langewiesche titled the accident, "The Devil at 37,000 Feet" because of the extremely low odds of aircraft impacting each other at that point in space and given the number of chances available to prevent such an accident. Langewiesche used the phrase, "two arrows touched nose to nose" to create the improbability of the accident (2009, p. 141).

September, 2006, just prior to 5 PM Brasilia local time, a Boeing 737-800 operated by Gol Airlines of Brazil and an American Embraer Legacy 600 business jet collided at 11,278 m (37,000 ft) over the dense jungles in Brazil killing all 154 occupants of the Boeing 737 (NTSB Safety Recommendation, 2007). The aircrew and passengers of the Gol Airlines 737 were completely unaware of the impending impact with the American business jet. The commercial airliner was flying at its assigned and appropriate flight level given their direction of flight. The pilots of Gol Airlines were communicating with their traffic controller. The problem was that the American pilots in the brand new Legacy business jet were flying blind to the air traffic controllers and to other aircraft; the business jet's transponder code was in 'standby' status.

A transponder code provides each aircraft with a specific identifiable radar "name" for air traffic controllers to see and identify the aircraft's location in space. This transponder code allows for tracking and traffic deconfliction. A transponder code also is part of a traffic alert and collision-avoidance system (commonly called TCAS) which allows aircraft to "talk" in a sense between each other regarding where they are and where they are going relative to other aircraft in close proximity. Consequently, by having the business jet's transponder in standby, the aircraft was invisible to air traffic controllers as well as the other aircraft's traffic alert and collision-avoidance systems.

Odds were extremely low that either pilot flying the aircraft involved would have or could have seen the other. First of all, because both aircraft incorporated the TCAS and were flying in radar-controlled airspace, all pilots involved were under the assumption that if there was a traffic conflict they would be informed via the radio frequency and/or their internal alert systems. In such cases pilots cruising at high altitude don't usually spend a lot of time searching for traffic out in front of their aircraft. Even if they had been "clearing for traffic" it would have been very difficult to visually perceive the other airplane due to the small size of the retinal image combined with the rapid closure speed between them.

When interacting with complex systems, operators need some form of notification to shift their attention to a change in status. The transponder in the standby mode is a significant change in aircraft navigational/communication status that should get a pilot's attention. According to Langewiesche (2009) this aircraft used a small, silent visual display on the radio management unit to indicate that the transponder was in the standby mode, as well as a small visual indication on the pilots' Primary Flight Display reading, "TCAS off." There was no central "master caution" warning to draw the pilot's attention (bottom-up processing) to look closely at the cockpit displays. There was no auditory alert to capture their attention for a visual sweep of the cockpit displays, nor were there any color-coded warnings. Because focal vision is consciously controlled, it requires selective attention to draw its purpose; nothing in this case attracted the attention of the pilots. An additional factor was that the Legacy pilots were not familiar with their new jet and its technology. They spent the majority of their flight time "playing" with the switches to learn what does what. Hence, in terms of visual sampling of the cockpit displays, they did not have a well-established cross-check of vital parameters.

The NTSB Safety Recommendation report (2007) stated:

> Using only static text messages to indicate a loss of collision avoidance system functionality is not a reliable means to capture pilots' attention because these visual warnings can be easily overlooked if pilots' attention is directed elsewhere in the flight environment. (p. 4)

The report recommended auditory alerts and more attention-getting visual alerts by using salient colors or perceptible movement cues. Engineering designers of aviation systems often fail to appreciate the visual requirements that exist within the cockpit environment, coupled with the high visual and cognitive demands on the pilots.

Another visual-perception factor operating in this mishap was what Langewiesche (2009) referred to as an altitude-misperception illusion that occurs due to the curvature of the earth when flying at high altitudes and viewing approaching objects. He stated that each pilot would have perceived the other aircraft, " … to be significantly higher until the last few seconds before impact" (p. 137). Recall that a perceptual constant when on the ground is that the horizon cuts our visual field of view in half; it is an invariant in that sense. However, when operating at over 9 km (30,000 ft) above the ground, the horizon does not always visually divide the visual field in half, and thus, it can no longer serve as a reliable cue to estimate co-altitude objects. Therefore, even if the pilots had seen each other, it may not have changed the course of events.

German Air Base (terrain, sloping runway)

At Spangdahlem Air Base, runway 05/23 is a visual illusion trap for aviators (AIB report, 2006). Runway 05 has rising terrain and is up-sloped. Runway 23 has sharply down-sloping terrain, followed by a 65 m (213 ft) culvert with rising terrain up to the down-sloping runway. The approach ends of the runways also have two differently colored surfaces and pilots often misperceive the change in color to equate to different runway areas. Unique obstruction hazards as well as operational practices also contribute to mishaps at the airfield.

The number of mishaps involving multi-million dollar state-of-the-art aircraft is striking, especially given that visual perception problems had been reported, in particular visual misperceptions of altitude and distance. According to Schonauer's 2008 article, at least seven different runway landing mishaps that involved visual misperception have been reported back to the 1970s. The visual spatial disorientation problems at this airfield clearly demonstrate that, although aviation has improved in terms of technological capability, the limitations of pilot visual perception continue to contribute to mishaps.

One of the more recent mishaps at the airfield involved an F-16 pilot flying a visual approach below glide-path and hitting an airfield antenna (AIB report, 2006). The aircraft was damaged to the point that the pilot had to eject and the aircraft was destroyed. The AIB report concluded that, "the following human factors substantially contributed to the mishap: visual illusions, misperception, inattention, and task misprioritization."

To fix the problem with this particular runway, it was recently reported (Schonauer, 2008) that Spangdahlem Air Base has modified the runway layout to help alleviate optical illusions. A dark epoxy coating was added to the ends of the runway, the runways were completely resurfaced, and the navigational antennas were relocated. According to Schonauer, this was the first time such a coating was put on an Air Force runway, but as the article stated, "a unique solution was needed for a unique flight hazard." It needs to be emphasized regarding this particular runway that although actions were taken to improve the actual landing surface and surrounding man-made obstacles were removed, the challenging terrain and runway's slope still remain a source for visual misperception.

Several of the accident reports reviewed in this chapter include the striking comment that, although Precision Approach Path Indicator (PAPI) or Visual Approach Slope Indicator (VASI) lights were available, the pilot attended to his or her visual perception of the scene and didn't notice or properly process the lights, even though those glide-path indicators were functioning correctly at the time of the accident. A question that has received little or no attention is whether pilots have any rational basis for this choice. A rational basis would exist if a pilot believed that the probability that PAPI or VASI lights were signaling erroneous information was so high that the safer option was usually to fly a visual approach. What is the probability of such malfunction? Is it more likely to occur in some airports than in others, and does pilot behavior reflect such a distinction? These are good research questions that deserve a closer look in the near future.

Florida—2008 (night, horizon, night vision goggles)

On January 15, 2008, a United States Air Force F-16 departed Homestead Air Reserve Base, FL, at approximately 6:39 PM local time for a night training mission with the pilot wearing night-vision goggles (Accident Investigation Board, 2008). The pilot successfully ejected 38 minutes into the flight prior to the aircraft impacting the water 126 nm west-southwest of Homestead in the Gulf of Mexico. The pilot had been maneuvering his aircraft in a hard, 90-degree left turn, began to descend, lost sight of a discernible horizon, and became spatially disoriented. The investigation board concluded that the cause of the mishap was:

> The pilot's failure to recognize and recover from spatial disorientation in a timely manner due to inadequate instrument cross check. Additionally, sufficient

> evidence indicates that the nighttime over-water environment, use of NVGs [night vision goggles], and weather conditions limited the visible horizon, substantially contributing to the mishap.

This accident is highlighted due to its recent date of occurrence and the fact that it clearly depicts the interplay between visual and vestibular spatial disorientation resulting in the loss of a nearly $20M aircraft.

The pilot involved had 2,629 flying hours in high performance jet aircraft, and the AIB report emphasized that spatial disorientation can occur to pilots regardless of experience level. The AIB report listed ten human factors that were relevant to this accident. The large number of cited factors demonstrates the difficulty in pinpointing the specific contributing causes of a mishap. The factors and their role in the mishap are presented below (pp. 14–17):

1. Complacency: the visible horizon to the northeast may have created a false sense of security or comfort to the pilot as he maneuvered at night while wearing night-vision goggles.
2. Restricted vision: when flying at night the pilot became spatially disoriented when he maneuvered to the west, the darker section of the night's skyline. Also, a cloud deck was at 6,000 ft [1.8 km] and may have presented a false horizon.
3. Breakdown in visual scan: the pilot failed to execute practiced internal and external cross-checks during maneuvers.
4. Vestibular illusion: during the execution of an 18-second 65-degree nose low turn, any head movement can induce a vestibular illusion.
5. Instrument and sensory feedback systems: the night vision goggles limited the pilot's field of view and resolution; the attitude indicator may have provided inadequate situational awareness due to technical limitations.
6. Habituation: during daylight hours an abeam maneuver is performed at 60 degrees nose-low and may have led to the pilot performing the maneuver at night at nearly the same attitude.
7. Elevator illusion: this illusion occurs when a reduction in descent is perceived as a climb; thus the pilot believes that he has arrested and reversed the aircraft's vertical movement when in fact it is still descending, just at a slower rate. It was during this descent that the pilot's unrecognized spatial disorientation was recognized and a recovery was attempted.
8. Misinterpreted instruments: correct information was displayed to the pilot but it was not interpreted as such. The pilot failed to differentiate the attitude displays of ground and sky due to the extreme nose-low attitude.
9. Gillingham illusion: pilots with restricted visual references try to recover from excessive roll maneuvers by inadvertently inducing more roll while perceiving a constant bank angle. Although the pilot recognized his spatial disorientation condition, due to the lack of visual references available (dark

westerly direction of the night's sky), the board determined it was this time that the disorientation became incapacitating.

10. Temporal distortion: the pilot recalls detailed events within the entire scenario as he attempted to recognize and recover from his disorientation. The pilot initiated an 8.75-G pull to recover; however, he was inverted and the pull only worsened his position. The pilot ejected 3 seconds prior to the aircraft hitting the water.

This report clearly depicts the advances in the safety investigation process regarding human factors issues and the improved specificity of spatial disorientation in formal reports. Aviation safety has come a long way in this regard, as evidenced in the step-by-step analysis of cognitive, visual, and vestibular contributions that set up this particular pilot to succumb to spatial disorientation.

Utah—2009 (night, night vision goggles, limited ambient cues)

Another US Air Force F-16 aircraft crashed during night low-altitude combat maneuvering in June of 2009 at a Utah training range (AIB, 2009). This fatal mishap is a recent example of the vulnerability of the human's visual perception system. This accident is similar to the November 2006 F-16 mishap in Iraq. In both cases the pilots were accomplishing a vertical descent toward the ground at high speed to delivery ordinance, except in this accident the pilot was simulating the ordinance drop (just practicing). In this maneuver the pilots are required to initiate a steep angle descent followed by a high G pull-up maneuver at a certain point to keep safe altitude separation from the terrain. The 2006 mishap occurred during the day in a desert environment, one that lacked ambient visual cues. This accident occurred at night with the pilot wearing night vision goggles.

The AIB determined that the mishap occurred because of the pilot's inability to properly recognize his altitude during the night high angle strafe attack (2009). The investigation board further specified contributing factors to the mishap.

1. Limited total experience in the F-16.
2. Channelized attention on attempting to visually prosecute the attack to the exclusion of visual and auditory cues of more immediate priority.
3. Breakdown in visual scan of flight instruments.
4. Expectancy of the aircraft parameters different from what actually encountered thus altering perception of the target, ground cues, and altitude indications.
5. Inability to distinguish terrain features and proximity to the ground because of low illumination and lack of contrast resulting.

Another topic addressed within the AIB report not listed above was a discussion on "vision restricted by meteorological conditions" (2009). This is basically describing that if daylight conditions been present then possibly the pilot would have been able to determine altitude and distance information from the terrain. The AIB report further explained (p. 20):

> The ground surface was flat with few terrain features, creating a low-contrast background. These factors could have created a "black hole" effect around the area of the target, giving the MP [mishap pilot] few cues to allow him to visually recognize his proximity to the ground. Additionally, NVGs [night vision goggles] restrict the wearer's field of view from 180 degrees to 40 degrees, thus blinding the pilot to peripheral terrain cues that could be used to judge altitude. The MP's vision could have been restricted to the point that, without referencing his instruments, it was difficult or impossible to judge proximity to the ground.

General Aviation

Thus far our discussion has been mostly restricted to commercial and military aircraft; however, all pilots are susceptible to spatial disorientation. A majority of general aviation pilots are Visual Flight Rule (VFR) pilots. VFR pilots are not allowed or trained to fly using only their instruments. When visual cues are occluded by bad weather or darkness (termed Instrument Meteorological Conditions, IMC) a VFR pilot must turn around, land, or fly to a location where Visual Meteorological Conditions (VMC) exist.

A study on spatial disorientation in general aviation from 1970–1975 found that it was the third most frequent cause of fatal accidents in small, fixed-wing aircraft. The second most frequent, and closely connected with spatial disorientation, was "continued VFR flight into adverse weather" (Kirkham, Collins, Grape, Simpson, and Wallace, 1978). These authors reported that spatial disorientation directly contributed to 16 percent of all fatal mishaps. Furthermore, of those accidents in which spatial disorientation was cited as a causative factor, 90 percent involved fatalities. Meteorological conditions played a role in 42 percent of all fatal accidents and 36 percent of those had spatial disorientation as a casual factor. Spatial disorientation was most often experienced by pilots who had 50 flying hours or less (30 percent of such accidents). This is quite different from accidents in commercial and military aviation, where flying experience is not a differentiator in a pilot's likelihood to experience spatial disorientation.

Table 7.5 presents data from a series of NTSB general aviation accident reports for three types of accidents: takeoff, en route, and approach to landing. The data were collected via keyword search for "visual illusion", "spatial disorientation", and "perception" within the accident descriptions. The table clearly shows that visual spatial disorientation and misperception are also important issues for general aviation pilots.

The Nall Report (2008) written by the Aircraft Owners and Pilots Association (AOPA) Air Safety Foundation detailed accident trends and factors in general aviation for the year 2007. The aircraft assessed were fixed-wing general aviation

Table 7.5 General aviation visual misperception data

NTSB General Aviation Visual Misperception Data Sample

	Date	State	Fatalities	Description
Takeoff				
	Date	State	Fatalities	Description
	Aug-86	AZ	2	Night black hole takeoff, SD, Grand Canyon airport
	Sep-88	AZ	4	Night black hole takeoff, SD, Grand Canyon airport
	Apr-90	UT	2	Night departure, SD
	Mar-94	UT	4	Night black hole takeoff, SD
	Nov-95	AZ	2	Night black hole takeoff, SD, Grand Canyon airport
	Jan-96	HI	1	Night departure, lack of visual cues; visual illusion
	Aug-03	CA	1	Night departure, SD
Enroute				
	Date	State	Fatalities	Description
	Jan-84	NM	2	Black hole, night VMC
	Jun-89	MT	4	Black hole, night VMC, SD
	Nov-89	UT	2	Over calm water, lack of visual cues
	Sep-93	UT	1	Over glassy water, Helo
	Oct-93	CA	1	Turning over glassy water, sun & haze, SD
	Jun-98	TX	3	VMC into smoke, haze no visual cues, night
	Jul-99	MA	3	Night VFR, SD over water
	May-00	NH	2	VFR into IMC, night, mountainous terrain
	Oct-00	TX	1	VFR into IMC, night no visual cues, SD
	May-02	AZ	2	Night VMC, mountainous terrain, CFIT
	Oct-03	MN	0	Glassy water, lost depth perception, Helo
	Oct-04	CA	5	Night VMC into IMC, mountainous terrain
	Nov-04	CA	3	Night VMC, mountainous terrain, CFIT
	Jan-05	SC	4	Night VFR, SD
	Feb-05	AK	3	VFR into IMC, snow, white-out SD
	Apr-05	FL	1	Night featureless terrain
	Aug-05	NY	3	Over water, no horizon, night, VMC illusions, SD
	Aug-06	CA	1	Night VMC, mountainous terrain, CFIT
	Feb-07	UT	0	Glassy water, lost depth perception
App/Land				
	Date	State	Fatalities	Description
	Dec-89	CA	0	Black hole approach, CFIT
	Nov-92	WI	2	Dark night, VMC, narrow, upslope rwy, crashed short
	Jan-93	AK	1	Night VMC, snow black hole, 3 miles from rwy
	May-93	NM	4	Night circle approach, no visual cues, CFIT
	May-95	VA	1	Undershot night approach, over water, misjudge alt
	Sep-97	WA	3	Night VMC, misjudged alt & distance, crashed short
	May-00	HI	6	Black hole, mountainous terrain, cnx instrument app
	Mar-01	VA	1	Black hole, night visual illusion
	Mar-02	MD	4	Night black hole, over water
	Aug-02	CA	0	Misperceived alt & distance, upslope rwy, CFIT
	Oct-02	MN	0	Water landing, glassy surface
	Apr-03	IL	0	Narrow runway, high ratio, CFIT shallow GP
	Apr-04	AK	1	VFR into IMC, white-out, SD, Helo
	Jan-05	TX	2	Night black hole approach, hit power lines
	Sep-05	IN	4	Night VMC, hazy, no horizon, SD, lack of experience

aircraft weighing less than 12,500 lbs. The mishap rate for general aviation was 6.47 per 100,000 flying hours or 1 mishap every 15,455 flying hours, third highest in the last ten years. With a total of 1,385 general aviation accidents, 252 were fatal, and the pilot was the major cause in 72 percent of all the accidents and 76 percent in all fatal accidents. Although the total number of mishaps was lower in terms of night fatal accidents than day fatal accidents, the probability of a night fatality was more than twice that of a daytime mishap. Thus, night flying proved the most deadly with 35 percent of all night mishaps resulting in fatalities compared with 16 percent of daytime mishaps resulting in a fatality. Instrument meteorological conditions, IMC, took the most lives, as 78 percent of all fatal accidents occurred in impoverished visual conditions. Further breakdown of the data found that 28 percent of all night VMC accidents were fatal and night IMC resulted in 71 percent fatalities. Next are two mishaps within the area of general aviation.

Catalina Island—1996

Kern (2002) reported on a general aviation accident in March of 1996, in which a pilot unfamiliar with the unique runway environment failed to perceive the proper visual glide-path. The runway at Catalina Island, California, sits atop a plateau, thus making visual approaches perceptually challenging. Landing on a runway with up-sloping terrain immediately prior to it induces the perception of being too high, that is, too steep, on final. To correct this misperception the pilot may over-correct to a position too low for a safe approach. The result in this particular case was that the pilot had to accomplish three approaches because he kept coming in too steep and fast for a safe landing. His final landing still ended up unsafe, with a hard touchdown, bounced in the air with the plane landing a second time, damaging the propeller. Also depicted in Table 7.5 are numerous en route accidents and approach-and-landing mishaps that all involved the latent (pre-existing) condition of limited visual cues, flying in an impoverished visual environment.

JFK Jr.—1999

A well-publicized aviation accident occurred when John F. Kennedy, Jr., his wife and wife's sister all received fatal injuries on 16 July 1999. JFK Jr., as the pilot, succumbed to spatial disorientation during a night flight from Essex County Airport, NJ, to Martha's Vineyard, MA (to drop off his sister-in-law), with the intention of continuing to Hyannis, MA; see Figure 7.11 (Ostinga, et al., 1999).

JFK Jr. was not instrument-rated, and thus had to remain in visual meteorological conditions (VMC), and fly via visual flight rules (FVR). He was working towards his instrument rating but had not yet completed all of the required ground-training and the flight phase (NTSB, 1999). For instance, he had completed 12 of 25 instrument lesson plans and between May 1998 and July 1999 had accumulated only 21 hours of night flight. JFK Jr. had made the same flight numerous times, even at night and, according to the NTSB report, his instructor was confident he

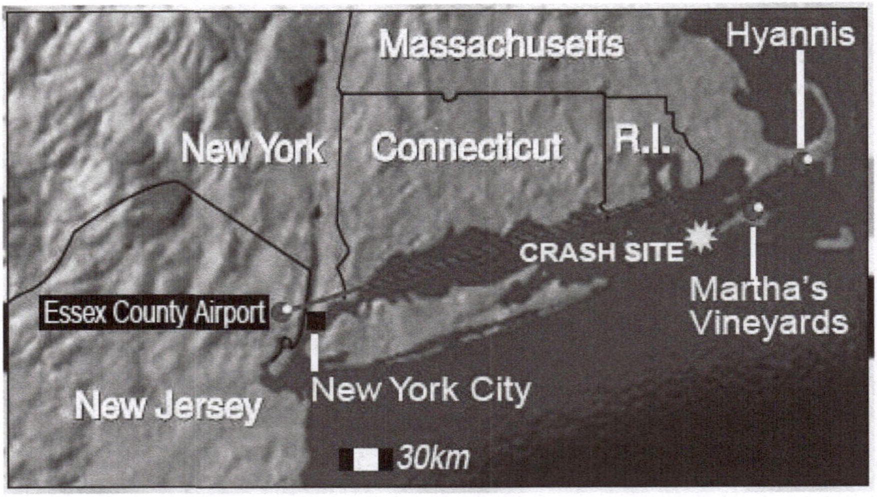

Figure 7.11 Route of flight

Source: Reproduced with permission from Flight Safety Australia.

could complete the flight at night as long as a visible horizon was present. This is a very important distinction made by his flight instructor … the need for a visible, distinct horizon. If a pilot is still struggling to learn and become proficient with instrument flying, the horizon is an essential visual cue for spatial orientation.

His total flight experience was only 310 flying hours, 55 of those at night (NTSB, 1999). Krause (2003) reported that the pilot in the 15 months prior to the accident had flown the route between New Jersey and Massachusetts approximately 35 times, 17 without an instructor accompanying him, and at least 5 flights at night. Thus his familiarity with the route and terrain below may have established overconfidence in his abilities.

Radar data showed that the aircraft was at an altitude of 1676 m (5,500 ft) for the majority of the flight; however, as shown in Figure 7.12, when at 11.3 km (7 miles) from shore, it descended to 670 m (2,200 ft) and then climbed back up to 792 m (2,600 ft). The aircraft then entered a turn and descended rapidly until water impact.

According to the NTSB report the visibility that evening was anywhere from 6.4 to 9.6 km (4 to 6 miles) with haze. Other pilots who flew in that same vicinity that evening reported that it was very dark night, with no visible horizon. Lights could be seen when flying over land and looking straight down, but slant-range visibility was poor when over the water. One pilot reported that he thought that Martha's Vineyard had a power failure, because he could not see any lights at all. Another pilot stated that he had no visual reference but failed to see any clouds or fog. The Hyannis weather forecast had 9.6 km (6 miles) visibility and clear skies and there were no meteorological warnings issued for the route of flight; however,

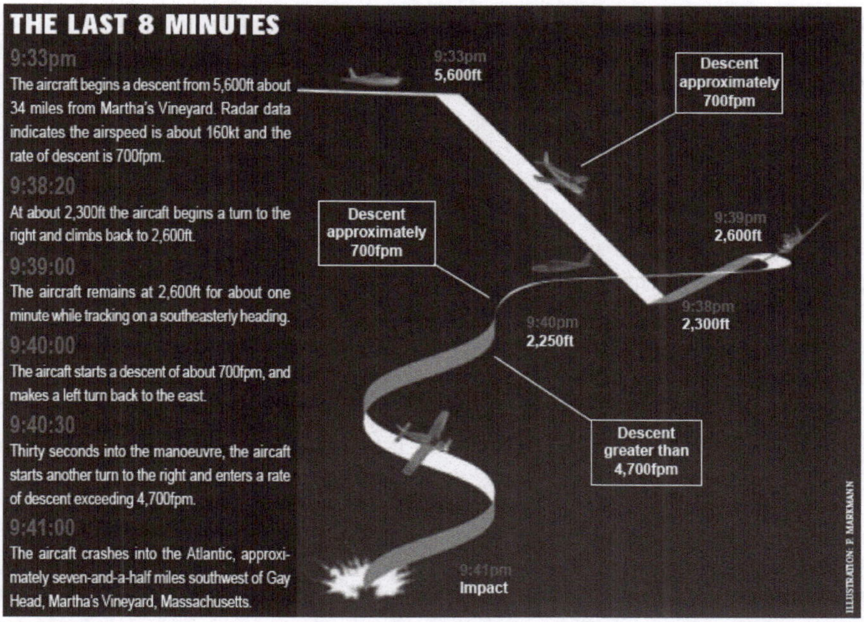

Figure 7.12 Depiction of final moments of spatial disorientation

Source: Reproduced with permission from Flight Safety Australia.

some felt that due to a heat-wave, haze was reducing visibility along the entire east coast (Ostinga et al., 1999).

The NTSB report listed several types of spatial disorientation that JFK Jr. might have experienced, many of them vestibular. However, night-flying hazards were presented in the NTSB report in terms of visual-spatial disorientation. The Federal Aviation Administration airplane Flying Handbook, FAA-H-8083-3, Chapter 10, stated in reference to night flying:

> Under no circumstances should a VFR night-flight be made during poor or marginal weather conditions unless both the pilot and aircraft are certified and equipped for flight under … IFR … Crossing large bodies of water at night in single-engine airplanes could be potentially dangerous … because with little or no lighting the horizon blends with the water, in which case, depth perception and orientation become difficult … During poor visibility conditions over water, the horizon will become obscure, and may result in a loss of orientation.

As pointed out by the Staff Writers for Flight Safety Australia, Ostinga et al (1999), although this accident was surely tragic, a positive outcome may be "greater awareness among visual pilots about the hazards of night VFR [visual flight rules]" (p. 24). The reality is that there are countless examples of aviation accidents and incidents in which some form of spatial disorientation was involved in the chain of events. Possibly night VFR ought to require some form of instrument rating/qualification to ensure pilots have the ability to orientate themselves via their instruments and not have to rely on unreliable visual cues.

Conclusions: Aviation Mishaps

This review of research and accident reports documents the continuing incidence of visual misjudgment as a causative factor in aviation accidents. As the opening quotation states, pilots *do not believe that the accident about to occur is at all possible.* Aviation safety is founded on the notion that a wide dissemination of the conclusions of mishap reports can ensure that investigations of previous accidents can help prevent future accidents by (1) providing a basis for the continuous improvement of pilot education and training, and (2) the sharing of information among the entire pilot community.

Our intent in sharing the above mishap descriptions with the reader is not to portray any actions of previous aviators in a negative light. Aviation safety is founded on the investigation of previous accidents to prevent future ones. Pilot education and training must include all types of aviation lessons learned to improve operations and mitigate risk. Although no accident scenario will be perfectly replicated, the elimination of latent errors found in previous mishaps may help save lives and resources in the future. Thus, our contribution to aviation safety is the presentation of the mishaps in such a manner as to impress upon the reader that no one is immune from spatial disorientation, regardless of experience level. It is tragic that it often takes an additional fatality once again to remind the aviation community of the power of misperception and spatial disorientation.

Unfortunately, even the most recent mishap reports (recent defined as last 12 years of "modern" aviation) reveal a trend of pilots "going visual" and trusting their visual capabilities in spite of impoverished viewing conditions. For example, the 1999 C-130 crash in Kuwait, the 2002 accident in Tallahassee, the F-16 incident in 2002, and the Spangdahlem crashes as well as the F-16 mishaps (Iraq 2006 and Florida 2008) and the helicopter 2006 crash into the Irish Sea all suggest that, just as the pilots in the Canadian Arctic in 1991, at Pago Pago in 1974 and in San Juan 1941, pilots gave too much credence to their visual perception. Although visual-spatial disorientation may not have been the major cause of these accidents, it was certainly a contributing factor. Another common aspect of the mishaps is that the pilots were often trying to expedite their arrival in an effort to save time. Accidents at Cincinnati in 1965, Canadian Arctic 1991, and Tallahassee in 2002 demonstrate how this pressure led to poor decision making.

A final major trend to note is the greatly increased amount of night flying (especially in military aviation), which in turn greatly increases the time pilots spend in an environment conducive to spatial disorientation. Note that Gillingham (1992) made a similar observation 17 years ago, and since then technological advancements have even further increased the capability for pilots to fly at night while performing a variety of missions. Gillingham called for research on visual spatial disorientation:

> Research on visual orientation mechanisms is extremely important because so many mishaps are caused by visual illusions, and because several types of solution[s] to the SD [spatial disorientation] problem depend on visual orientation information. An understanding of the principles of visual spatial orientation and visually induced SD is essential for the development of demonstrations of visually generated forms of SD in ground-based training devices and curricula, and for the design of more effectively orientating symbology in HUDs and *helmet-mounted displays* (HMDs). (p. 304)

As this chapter closes, the quote below fittingly reinforces the need for research. Hasbrook (1975) summarized visual illusion problems in aviation and the number of accidents:

> From these dreary facts, it appears that pilot perception of the approach path is often seriously in error. Whether such error is due to a lack of needed visual cues (particularly at night), visual illusions, lack of knowledge of available cues, potential error-producing visual concepts or a combination of these, is a question that has been studied many times, but still awaits an answer. (p. 39)

A solution to this problem is still lacking thirty years after Hasbrook's assessment.

References

Accident Investigation Board (1999). C-130E, S/N 63–7854, Ahmed Al Jaber AB, Kuwait, 10 December, 1999, United States Air Force.

Accident Investigation Board (2002). C-17A, S/N 00–174, Kandahar, Afghanistan, 8 January 2002, United States Air Force.

Accident Investigation Board (2002). F-16C, S/S 88–0397, Hill AFB, Utah, 13 November 2002, United States Air Force.

Accident Investigation Board (2006). F-16CJ, S/N 91–0337, Spangdahlem Air Base, Germany,14 September, 2006, United States Air Force.

Accident Investigation Board (2006). F-16CG, S/N 90–0776, 524th Expeditionary fighter Squadron, Balad Air Base, Iraq, 27 November 2006, United States Air Force.

Accident Investigation Board (2008). F-16C, T/N 87–0347, 93rd Fighter Squadron, 482nd Fighter Wing, Homestead Air Reserve Base, FL, United States Air Force.

Accident Investigation Board (2009). F-16CM, T/N 89-2108, 421st Fighter Squadron, 388th Fighter Wing, Hill Air Force Base UT, United States Air Force.

Air Force Manual 11–217 (2005, January). Instrument Flight Procedures, Volume 1. Air New Zealand Flight 901. Retrieved December 18, 2008 from http://en.wikipedia.org/wiki/Air_New_Zealand_Flight_901.

Aircraft Owners and Pilots Association (2008). 2008 Nall Report. Aircraft owners and pilots association air safety foundation. Retrieved May 13, 2009 from http://www.aopa.org/asf/publications/08nall.pdf.

Barnum, F. and Bonner, R. H. (1971, August). Epidemiology of USAF spatial disorientation aircraft accidents, 1 Jan 1958–31 Dec 1968. *Aerospace Medicine*, 896–898.

Bennett, M. (1999, July-August). Low, slow and short: To commercial pilots come within seconds of losing their lives. *Flight Safety Australia*, 20–21.

Boeing Corporation (2006, May). Statistical summary of commercial jet airplane accidents. Retrieved November 14, 2006 from Boeing.com/news/techissues/pdf1.statsum.pdf.

Boeing Corporation (2008, July). Statistical Summary of Commercial Jet Airplane Accidents. Seattle, WA: Boeing Commercial Airplanes. Retrieved March 13, 2009 from http://www.boeing.com/news/techissues/pdf/statsum.pdf.

Braithwaite, M. G., Durnford, S. J., Crowley, J. S., Rosado, N. R., and Albano, J. P. (1998). Spatial disorientation in US Army rotary-wing operations. *Aviation, Space, and Environmental Medicine*, *69*, 1031–1037.

Buley, L. E. and Spelina, J. (1970, May). Physiological and psychological factors in "the dark night takeoff accident". *Clinical Aviation and Aerospace Medicine*, 553–556.

Bull, L. C. (1997). A-10 crash was pilot error, investigation reveals. Retrieved June 19, 2009 from http://www.ctie.monash.edu.au/hargrave/svoboda.html.

Cheung, B., Money, K., Wright, H., and Bateman, W. (1995). Spatial disorientation-implicated accidents in Canadian forces, 1982–92. *Aviation, Space, and Environmental Medicine*, *66*, 579–585.

Chimonas, E., Diamantopoulos, E., Markou, I., and Stathogiannis, E. (2002). A spatial disorientation survey of Hellenic air Force pilots. Paper presented at the Research and Technology Organization and Human Factors and Medicine Symposium on Spatial Disorientation in Military Vehicles: Causes, Consequences and Cures. La Coruna, Spain. Retrieved March 17, 2006, from www.spatiald.wpafb.af.mil/hfm/mp-086–07.pdf.

Civil Aeronautics Board (1942). Of the investigation of an accident involving civil aircraft of the United States NC 15376 which occurred in San Juan Harbor, Puerto Rico, On October 3, 1941. Retrieved February 23, 2009 from http://dotlibrary1.specialcollection.net/scripts/ws.dll?websearchandsite=dot_aircraftacc.

Civil Aeronautics Board (1952). Northeast Airlines, Inc., Near La Guardia Field, NY, January 14, 1952. Investigations of aircraft accidents 1934–1965; SA-399, File No. 1–0038. Retrieved March 23, 2009 from http://dotlibrary1.specialcollection.net/scripts/ws.dll?websearchandsite=dot_aircraftacc.

Civil Aeronautics Board (1953). Western Air Lines, Inc., DC -6B, San Francisco Bay CA, April 20, 1953. Investigations of aircraft accidents 1934–1965; SA-399, File No. 1–0038. Retrieved March 23, 2009 from http://dotlibrary1.specialcollection.net/scripts/ws.dll?websearchandsite=dot_aircraftacc.

Civil Aeronautics Board (1955). American Airlines, Inc., Convair 240, N94234, Near Springfield MO, March 20, 1955. Investigations of aircraft accidents 1934–1965; SA-399, File No. 1–0038. Retrieved March 23, 2009 from http://dotlibrary1.specialcollection.net/scripts/ws.dll?websearchandsite=dot_aircraftacc.

Civil Aeronautics Board (1959). Northwest Airlines, Inc., Douglas DC-6B, N575, Minneapolis, MN, August 28, 1958. Investigations of aircraft accidents 1934–1965; SA-399, File No. 1–0038. Retrieved March23, 2009 from http://dotlibrary1.specialcollection.net/scripts/ws.dll?websearchandsite=dot_aircraftacc

Civil Aeronautics Board (1960). American Airlines, Inc., Lockheed Electra, N6101A, in the East River, La Guardia Airport, NY, February 3, 1959. Investigations of aircraft accidents 1934–1965; SA-399, File No. 1–0038. Retrieved November 15, 2006 from http://dotlibrary1.specialcollection.net/scripts/ws.dll?websearch andsite=dot_aircraftacc.

Cocquyt, P. (1953). Sensory illusions. *Shell Aviation News*, *178*, 19–24.

Dekker, S. W. A. (2005). *Ten Questions About Human Error: A New View of Human Factors and System Safety*. London: Lawrence Erlbaum Associates.

Dismukes, R. K., Berman, B. A., and Loukopoulos, L. D. (2007). *The Limits of Expertise: Rethinking pilot error and the causes of airline accidents*. Hampshire, England: Ashgate.

Ercoline, W. R., Weinstein, L. F., and Gillingham, K. K. (1991). An aircraft landing accident caused by visually induced spatial disorientation. *The 6th Biannual International Symposium of Aviation Psychology*, 619–623.

Federal Aviation Administration (n.d.). Spatial disorientation: Seeing is not believing. Retrieved August 3, 2006 from www.faa.gov/pilots/safety/pilotsafetybrochures.

Federal Aviation Administration (2002, August). Pilot safety brochure: Pilot vision. AM-400–98/2.

Flight Safety Foundation (1993). Fatal commuter crash blamed on visual illusion, lack of cockpit coordination. *Flight Safety Foundation: Accident Prevention*, *50 (11)*, 1–4.

Flight Safety Foundation (1998). Boeing 767 descends below glide path, strikes tail on landing. *Accident Prevention*, *55*, 1–8.

Flight Safety Foundation (2003). Reduced visibility, mountainous terrain cited in Gulfstream III CFIT at Aspen. *Accident Prevention*, *59*, 1–12.

Flight Safety Foundation (2004). Nonadherence to approach procedure cited in Falcon 20 CFIT in Greenland. *Flight Safety Foundation: Accident Prevention*, *61 (11)*, 1.5.

Gibb, R. W. and Olson, W. (2008). Classification of air force aviation accidents: Mishaps trends and prevention. *The International Journal of Aviation Psychology*, *18*, 305-325.

Gillingham, K. K. (1992). The spatial disorientation problem in the United States Air Force. *Journal of Vestibluar Research*, *2*, 297–306.

Godson, J. (1978). *Clipper 806: The Anatomy of an Air Disaster*. Chicago: Contemporary Books.

Gray, R. (2006). Vision in flying, driving, and sport. In M. R. M. Jenkin and L. R. Harris (Eds.), *Seeing Spatial Form* (pp. 121–151). Oxford: Oxford University Press.

Gurney, D. (2006, July). Learning from experience: Night vmc. *Flight SafetyFoundation—Aviation Safety World*, 40–42.

Hasbrook, H. (1975, November). The approach and landing: Cues and clues to a safe touchdown. *Business and Commercial Aviation*, 39–43.

Hodgson, D. A. (1971, February). Pilot vision during final approach-and-landing in turbojet transport operations. *Aerospace Medicine*, 205–208.

Holland, D. A. and Freeman, J. E. (1995). A ten-year overview of USAF F-16 mishap attributes from 1980–89. *Proceedings of the Human Factors and Ergonomics Society 39th Annual Meeting* (pp. 30–34).

Holmes, S. R., Bunting, A., Bostock, S., Brown, L., Hiatt, K., Braithwatie, M., and Harrigan, M. (2002). Preliminary survey of spatial disorientation in UK military pilots and navigators. Paper presented at the Research and Technology Organization and Human Factors and Medicine Symposium on Spatial Disorientation in Military Vehicles: Causes, Consequences and Cures. La Coruna, Spain. Retrieved March 17, 2006, from www.spatiald.wpafb.af.mil/hfm/mp-086–07.pdf.

International Civil Aviation Organization (2002). Line operations safety audit (LOSA). Document 9803, AN/761.

Job, M. (2006, January-February). Descent into the valley of death. *Flight Safety Australia*, 40–42.

Jones, D. G. and Endsley, M. R. (1996). Sources of situation awareness errors in aviation. *Aviation, Space, and Environmental Medicine*, *67*, 507–512.

Kern, T. (2002). *Controlling Pilot Error: Approach and Landing*. New York: McGraw-Hill.

Khatwa, R. and Helmreich, R. L. (1998, November–December, January–February). Analysis of critical factors during approach and landing in accidents and normal flight: Data acquisition and analysis working group final report. *Flight Safety Digest, Killers in Aviation: FSF task force presents facts about approach-and-landing and controlled-flight-into-terrain accidents*, 1–77.

Kirkham, W. R., Collins, W. E., Grape, P. M. Simpson, J. M., and Wallace, T. F. (1978, September). Spatial disorientation in general aviation accidents. *Aviation, Space, and Environmental Medicine*, 1081–1086.

Kraft, C. L. (1978). A psychophysical contribution to air safety: Simulator studies of Visual illusions in night visual approaches. In H. L. Pick, H. W. Leibowitz,

J. E. Singer, A. Steinschneider, and H. W. Stevenson (Eds.), *Psychology: From Research to Practice* (pp. 363–385). New York: Plenum Press.

Krause, S. S. (2003). *Aircraft Safety: Accident Investigations, Analyses and Applications*. New York: McGraw-Hill.

Langewiesche, W. (2009, January). The devil at 37,000 feet. *Vanity Fair*. Retrieved February 6, 2010 from www.vanityfair/magazine/2009/01/air_crash200901

Lee, R. M. (1993). *Death and Deliverance*. Toronto: Macfarlane Walter and Ross.

Leibowitz, H. W. (1988). The human senses in flight. In E. L. Wiener and D. C. Nagel (Eds.), *Human Factors in Aviation* (pp. 83–110). New York: Academic Press.

Leland, D. (2001). Night VFR … An Oxymoron. Retrieved 23 March 2009 from http://www.aero-news.net/columns/avsoapbox.cfm.

Lessard, C. S., Mathews, R., and Yauch, D. (2000, Mar–Apr). Effects of rotation on somatogravic illusions. *IEEE Engineering in Medicine and Biology*, 59–65.

Lyons, T. J., Ercoline, W., O'Toole, K., and Grayson, K. (2006). Aircraft and related factors in crashes involving spatial disorientation: 15 years of U. S. Air Force data. *Aviation, Space, and Environmental Medicine*, *77*, 720–723.

Lyons, T. J. and Freeman, J. E. (1988). Spatial disorientation (SD) mishaps in the US Air Force—1988. *Aviation, space, and Environmental Medicine*, *61*, 459.

Matthews, R. S. J., Previc, F., and Bunting, A. (2002, April). USAF spatial disorientation survey. Paper presented at the Research and Technology Organization and Human Factors and Medicine Symposium on Spatial Disorientation in Military Vehicles: Causes, Consequences and Cures. La Coruna, Spain. Retrieved March 17, 2006, from www.spatiald.wpafb.af.mil/hfm/mp-086–07.pdf.

McGrath, B. J., Rupert, A. H., and Guedry, F. E. (2002). Analysis of spatial disorientation mishaps in the US Navy. Paper presented at the Research and Technology Organization and Human Factors and Medicine Symposium on Spatial Disorientation in Military Vehicles: Causes, Consequences and Cures. La Coruna, Spain. Retrieved March 17, 2006, from www.spatiald.wpafb.af.mil/hfm/mp-086–07.pdf.

Mertens, H. W. and Lewis, M. F. (1982). Effect of different runway sizes on pilot performance during simulated night landing approaches. *Aviation, Space, and Environmental Medicine*, *53*, 463–471.

Mertens, H. W. and Lewis, M. F. (1983). Effects of approach lighting and variation in visible runway length on perception of approach angle in simulated night landings. *Aviation, Space, and Environmental Medicine*, *54*, 500–506.

National Transportation Safety Board (1977). Aircraft accident report, Pan American World Airways, Inc., Boeing 707–3215, N454A, Pago Pago, America Samoa, January 30, 1974.

National Transportation Safety Board (1996). Descent below visual glidepath and collision with terrain, Delta Air Lines flight 554, Laguardia Airport, NY.

National Transportation Safety Board (1997). Controlled Flight into Terrain, Korean Air Flight 801, Nimitz Hill, Guam.

National Transportation Safety Board (1998). Safety Recommendation, A-98–87.

National Transportation Safety Board (1999). Accident occurred Friday, July 16, 1999 in Vineyard Haven, MA. NTSB Identification: NYC99MA178.

National Transportation Safety Board (2000). Incident occurred March 30, 2000 in New York City, NY. NTSB Identification: IAD00IA032.

National Transportation Safety Boarad (2001). Aspen, Colorado, March 29, 2001, retrieved August 25, 2009 from http://www.ntsb.gov/ntsb/brief2.asp?ev_id=20010412X00738andntsbno=DCA01MA034andakey=1.

National Transportation Safety Board (2007). Safety recommendation A-07–35 to 37, May 2, 2007.

Neubauer, J. C. (2000, Mar–Apr). Classifying spatial disorientation mishaps using different definitions. *IEEE Engineering in Medicine and Biology*, 28–34.

Ostinga, J. (2000, Jul–Aug). The wrong approach. *Flight Safety Australia*, 23–29.

Ostinga, J., Wolff, M., Newman, D., and White, S. (1999, Nov-Dec). What killed JFK Jr? *Flight Safety Australia*, 22–29.

Parmet, A. J. and Gillingham, K. K. (2002). Spatial orientation. In R. L. Dehart and J. R. Davis (Eds.), *Fundamentals of Aerospace Medicine* (pp. 184–244). Philadelphia: Lippincott, Williams, and Wilkins.

Perrone, J. A. (1984). Visual slant misperception and the 'black-hole' landing situation. *Aviation, Space, and Environmental Medicine*, *55*, 1020–1025.

Previc, F. H. (2004). Visual illusions in flight. In F. H. Previc and W. R. Ercoline (Eds.), *Spatial Disorientation in Aviation, Progress in Astronautics and Aeronautics, Vol. 203* (pp. 283–322). Reston, VA: AIAA.

Rolfsen, B. (2007). Too focused on the fight. *Air Force Times*, April 16, 2007 edition.

Schonauer, S. (2008). 52nd fighter wing modifies runway at Spangdahlem to reduce optical illusion. Stars and Stripes European edition. Retrieved 2 March 2008 from http://www.stripes.com/article.asp?section=104andarticle=59891andarchive=true.

Shappell, S. A. and Wiegmann, D. A. (2001). Unraveling the mystery of general aviation controlled flight into terrain accidents using HFACS. Paper presented at the 11th International Symposium on Aviation Psychology, Columbus, OH.

Sipes, W. E. and Lessard, C. S. (2000). A spatial disorientation survey of experienced instructor pilots. *IEEE Engineering in Medicine and Biology*, *19*, 35–42.

Transportation Safety Board of Canada (1999). Report Number A99A0131, Airbus landing short. Retrieved September 25, 2006 from www.tsb.gc.ca/en/ reports/air/1999/a99a0131.

United Kingdom Air Accident Investigation Report. Report number 07/2008, EW/C2006/12/03. Air Accidents Investigation Branch. Retrieved 23 January, 2009 from http://www.aaib.gov.uk/sites/aaib/home/index.cfm.

United States Air Force Safety Center (2009). Raw mishap data provided from Kirtland Air Force Base, New Mexico.

United States Naval Aviation (2008). Review of Class A spatial disorientation episodes FY2000–2007.

United States Naval Aviation Safety Center (2009). Raw mishap data provided from the Navy's safety center.

Veillette, P. R. (2004). Controlled flight into terrain takes highest toll in business jet operations. *Flight Safety Foundation, Flight Safety Digest.*

Veronneau, S. J. H. and Evans, R. H. (2004). Spatial disorientation mishap classification, data, and investigation. In F. H. Previc and W. R. Ercoline (Eds.) *Spatial Disorientation in Aviation, Progress in Astronautics and Aeronautics, Vol. 203* (pp. 197–241). Reston, VA: AIAA.

Vinacke, W. E. (1947). Illusions experienced by aircraft pilots while flying. *Journal of Aviation Medicine, 18*, 308–325.

Wagenaar, W. A. and Groeneweg, J. (1987). Accidents at sea: Multiple causes and impossible consequences. *International Journal of Man-Machine Studies, 27*, 587–598.

Walters, J. M. and Sumwalt, R. L (2000). *Aircraft Accident Analysis: Final Reports*. New York: McGraw Hill.

Walters, J. and Sumwalt, R (2003, October). Poor cockpit discipline leads to a preventable tragedy. *Professional Pilot*, 168–171.

Webster, N. (2004, May–June). Spatial disorientation. *Approach*, 2–3.

Chapter 8
Aviation's Future: Technological Advancements to Visual Perception

The previous chapters of this book have addressed aviation visual-perception issues that have been plaguing the flying community since its beginning. These have included some of the basic problems of visual perception in dynamic and impoverished environments. This chapter explores new technological developments intended to improve pilots' ability to sense and perceive their environments; these improvements include head-up displays, flight-path displays, night-vision devices (goggles, forward looking infrared systems), synthetic-vision systems, and laser-retinal displays. This chapter will also address unmanned aerial vehicles (UAVs) as an exponentially expanding aerial system that, similar to the just mentioned technological aviation advancements, is changing how humans visually interact with a system. Although these devices have improved some of the visual perception problems described in earlier chapters, unfortunately it will be shown that they have also introduced new forms of pilot error and created different ways for pilots to misperceive their visual environment. This final chapter is simply a brief summary of some of the visual perception issues related to these aviation technological advancements.

Head-up Displays

A head-up display (HUD) projects aircraft status information to pilots to minimize their head-down time and allow more time to view the external scene outside the aircraft. HUDs (illustrated in Figures 8.1 and 8.2) were also designed to remedy the problems associated with pilots shifting lens accommodation as they changed their gaze from close cockpit displays to far outside environmental objects. Such large changes in accommodation are inefficient (because each shift will initially involve mis-accommodation leading to temporary blurry vision) and can lead to eye fatigue. Consequently, many HUDs use a collimated display, (i.e., a display that is focused at optical infinity) to allow the pilot to view the external scene and the flight displays without changing accommodation.

The original intended use of HUDs was for military purposes; however, their use has crossed-over into commercial, business jet and general aviation. In theory, a HUD can greatly reduce the risk of controlled flight into terrain because of its

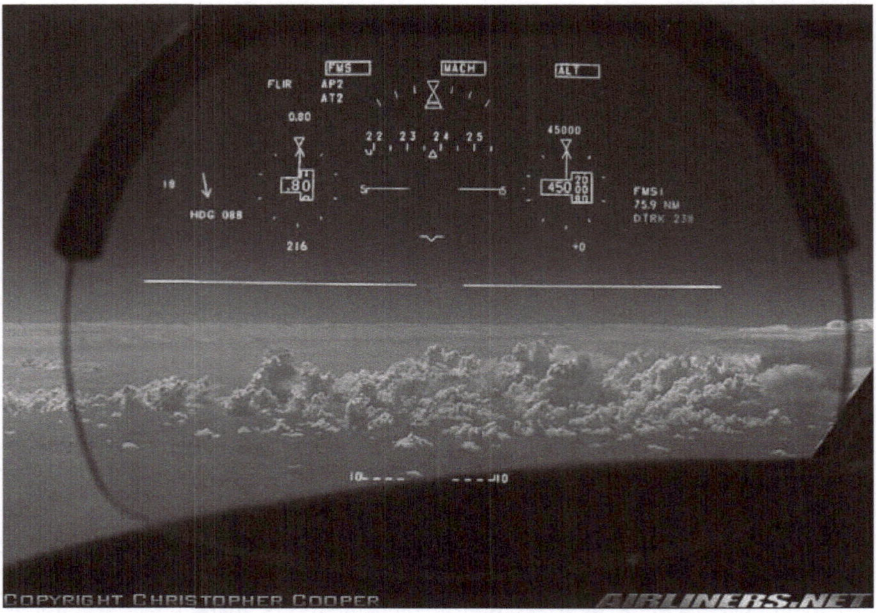

Figure 8.1 Head-up display

Source: With permission from Christopher Cooper. http://www.airliners.net/photo/Gulfstream-Aerospace-G-V-SP/0938438/

presentation of flight information, aircraft status, and flight path vector feedback to the pilot while the pilot maintains attention outside the aircraft. Far too many approach and landing accidents have occurred, and primarily these accidents happen during a non-precision or visual approach to landing. A HUD could potentially mitigate occurrences of mishaps by providing improved spatial orientation and additional vertical and lateral guidance to the pilot during these approaches.

Studies of flight performance have demonstrated improved flight-path maintenance and precision landings for HUDs as compared with traditional head-down displays (HDD) (e.g., Fischer, Haynes and Price, 1980). HUDs have also proven to be particularly valuable in bad weather/poor visibility conditions, permitting takeoffs and landings that were not previously possible (Goteman, Smith, and Dekker, 2007). However, this improved flight performance has come at a cost as HUDs have introduced a host of new perceptual and cognitive problems.

By and large, HUDs have not produced the anticipated benefits associated with accommodation and visual attention. While some studies have supported the accommodation benefits, (e.g., Weintraub and Ensing, 1992), several studies have demonstrated that collimated displays do not in fact shift accommodation to optical infinity (e.g., Hull, Gill and Roscoe, 1982; Norman and Ehrlich, 1986) and may actually increase mis-accommodation. Further Pierce, Geri and Hitt (1998) have shown that collimated displays can lead to large (up to 30 percent) differences in

Figure 8.2 Head-up display during a landing

Source: With permission from Barend Havenga. http://www.airliners.net/photo/Delta-Air-Lines/Boeing-737-800-(simulator)/1367221/M/

the perceived size and perceived distance of objects as compared with the same objects displayed on non-collimated displays.

An even bigger problem is that one of the primary assumptions about pilot visual attention for HUDs has proven to be false. When HUDs were initially developed it was assumed that, because the flight display would be in a pilot's line of sight while viewing the out-of-cockpit scene, he/she would be able to simultaneously see and pay attention to both. Or at the very least, the amount of effort and time required to shift attention between the outside scene and the flight displays would be dramatically reduced as compared to HDDs. However, this has clearly not been the case as several studies have demonstrated a *cognitive tunneling* effect due to attentional capture where pilots completely fail to detect objects in the outside world when viewing the HUD.

The role of attention when using a HUD is crucial to aviation safety in terms of where pilots are consciously putting their focal vision. Attentional capture, as defined by Stuart, McAnally, and Meehan (2005), is "the tendency for the pilot to pay attention to the display at the expense of the events in the outside world" (p. 25). The seduction of the HUD's colors and sharp contrast may prevent a pilot from seeing objects in the outside environment (normally the real world consists of dull colors and low contrasts). Ironically enough, the visual perception aspects of the HUD may actually inhibit improved environmental perception.

Stuart et al. (2005) discussed why pilots may have a bias to focus their attention on the HUD rather than the external scene. One hypothesis regarding limited HUD perception was mentioned above; attentional capture may result because of the saliency or compellingness of the image in terms of color and contrast. Another often listed reason for visual capture is the visual clutter; too many visual cues presented via the HUD imagery occlude the external environment and confuse the pilot.

Larish and Wickens (1991) compared pilots' ability to detect expected and unexpected events on the flight display and in the external scene when using a collimated HUD and an HDD. The results showed that pilots took longer to detect unexpected events (e.g., runway incursions) when a HUD was used. Furthermore, there were no advantages in flight performance associated with the HUD. Larish and Wickens suggested that the initial findings of improved performance with HUDs may have been due to the fact that the image quality and use of symbology was superior to HDDs, rather than the fact that the displays were presented in different locations.

Hofer, Braune, Boucek, and Pfaff (2001) compared pilots' ability to detect unexpected events (e.g., frozen cockpit-displays and runway incursions) with HUDs and HDDs during simulated takeoff and landings. Some of these events were serious enough to be considered as a "potential accident event". Overall, 36.5 percent of the events were missed in the HUD condition and 26 percent in the HDD condition. Results for "potential accident" events alone were even more dramatic: 9/36 of such events were missed when the HUD was used while none were missed for the HDD.

Why are pilots so poor at detecting unexpected out-of-cockpit events like runway incursions with HUDs as compared with HDDs? Intuitively, one might expect the opposite results because the probability that the image of an unexpected object will fall on the pilot's fovea at the instant of incursion is much greater for a HUD than for an HDD. One major factor is that, despite the best efforts of designers, most HUDs do not blend into the outside scene as originally intended. Instead, differences in brightness and the visible frame from the combiner glass serve to make the HUD display clearly distinct from the visual images in outside environment. This becomes particularly evident during maneuvers because the outside scene moves while the frame of HUD remains stationary in the visual field. Research has shown that when observers are required to focus their attention on one aspect of the visual scene (the HUD symbology in this case) they effectively filter out and fail to detect other highly visible elements of the scene (e.g., incursion aircraft on the runway), even when both the filtered and unfiltered parts of the scene are in central vision. This effect is commonly called *inattentional blindness* (IAB) (Mack and Rock, 1998; Simons and Chabris, 1999). This missing of unexpected events was also discussed within the realm of attentional capture by Stuart et al. (2005).

Why is this IAB caused by HUDs any worse than for a HDD? When pilots are focusing their attention on a HDD they cannot see incursion objects because they are out of the useful visual field. Thus, it seems detection rates should be lower. The most likely reason for improved performance with HDDs is that, when using HDDs, pilots are aware that when they are attending to the display they are ignoring the out-of-cockpit scene, therefore they frequently shift their attention between the two. Conversely, with a HUD, pilots are given the false sense that they are attending to the display and out-of-cockpit scene at the same time (because their eyes are pointing at both of them) and, thus, are not compelled to consciously shift their attention between the two as frequently. A similar explanation has been given to explain the problems associated with hands-free cell phone use in driving; just because an operator's eyes are directed towards the outside scene does not necessarily mean he or she will detect critical events.

One possible solution that has been put forth to address these problems associated with HUDs is conformal displays. A *conformal display* is defined as one in which the symbols appear to overlie the objects they represent. For instance, the runway symbol is displayed via the HUD and directly conforms with the runway location in the actual environment. Wickens and Long (1995) examined flight performance and event detection using conformal and nonconformal symbology sets in both HUD and HDD conditions. The flight performance measures included flight-path control and air-speed tracking. Benefits for flight path control were found for the HUD condition (as compared to the HDD) when conformal symbology was used. In addition, conformal displays have been found to be less distracting and require less effort to attend to the environment (Boston and Braun, 1996) and to lead to an increase in flight path tracking accuracy (Fadden, Ververs, and Wickens, 1998). However, when unexpected events in the out-of-cockpit scene are introduced, the probability of non-detection was still greater for a conformal HUD than for a HDD (Wickens and Long, 1995). It has also been suggested that changing the location of HUD symbology, altering intensity and/or reducing clutter may help reduce the attentional capture associated with HUDs (Wickens, 1997; Foyle, Dowell, and Hooey, 2001).

Redundancy issues when using conformal displays are an additional factor that can lead to attentional capture. Stuart et al. (2005) pointed out that pilots focus on the HUD imagery rather than the external scene because it seems that nothing really new is presented externally; everything the pilot needs is within the HUD.

This of course, is far from actuality, but the seduction of the HUD image can overcome a pilot's better sense to remain vigilant to the outside scene.

Despite these research findings, pilots generally have a very positive view of HUDs. In the study by Hofer et al. (2001) described above, pilots reported that the HUD reduced their workload, was easier to use than an HDD, and that it was easier to switch attention from the HUD to the external scene. These positive HUD evaluations occurred despite the fact that the HUD produced an increased number of missed events.

Finally, it is important to note that there is currently no evidence to suggest that the use of HUDs is associated with higher accident or incident rates in flight operations than HDDs. LeBlaye, Roumes, Fornette, and Valot (2002) reviewed accident and incident databases for cases involving HUDs. None of the official databases examined contained incidents involving HUDs. Of the 100,000 ASRS (Aviation Safety Reporting System, a voluntary pilot reporting process for safety incidents) reports collected since 1990, only 16 reports were related to HUDs and only five actually identified problems with HUDs. The identified problems involved a delay in the pilot noticing that the symbology did not match the actual situation, a symbol disappearing during a flight phase, and attentional capture by the HUD symbology. Thus, there appears to be a disconnect between the research and real-world usage of HUDs.

The FAA developed guidelines for certifying HUDs for civilian use (FAA, 2002), and during this certification process, 22 HUD design issues were identified by FAA experts. These issues were broken down into the following categories: location and format design of flight information, display effectiveness to support the intended task, HUD effectiveness in displaying and guiding recovery from unusual attitudes, consistency, discriminability of HUD symbology, and pilot physiological stress associated with HUD optical design. Clearly much more research is needed to investigate these issues.

Flight Path Displays

An important variant of the HUD is the flight path display (FPD). These "tunnel" or "highway in the sky" (HITS) displays provide the pilot with imagery that shows the predicted future flight path of the aircraft. It has been proposed that FPDs will be critical in the next generation of aviation for reducing air traffic congestion because they allow aircraft to follow more precise and complex approach-paths. Conventional glass cockpit displays do not intuitively portray both the lateral and vertical flight profile. Further, current technology requires the pilot to assimilate large amounts of fairly abstract data to properly interpret the current and anticipated (future) aircraft situation (Etherington et al., 2000). Flight-path displays will integrate all that information and display it in an easily understandable manner. The three elements that make up a FPD are a preview tunnel of where the aircraft will be in the future, a predictor symbol, and a 3D perspective (see Figures 8.3

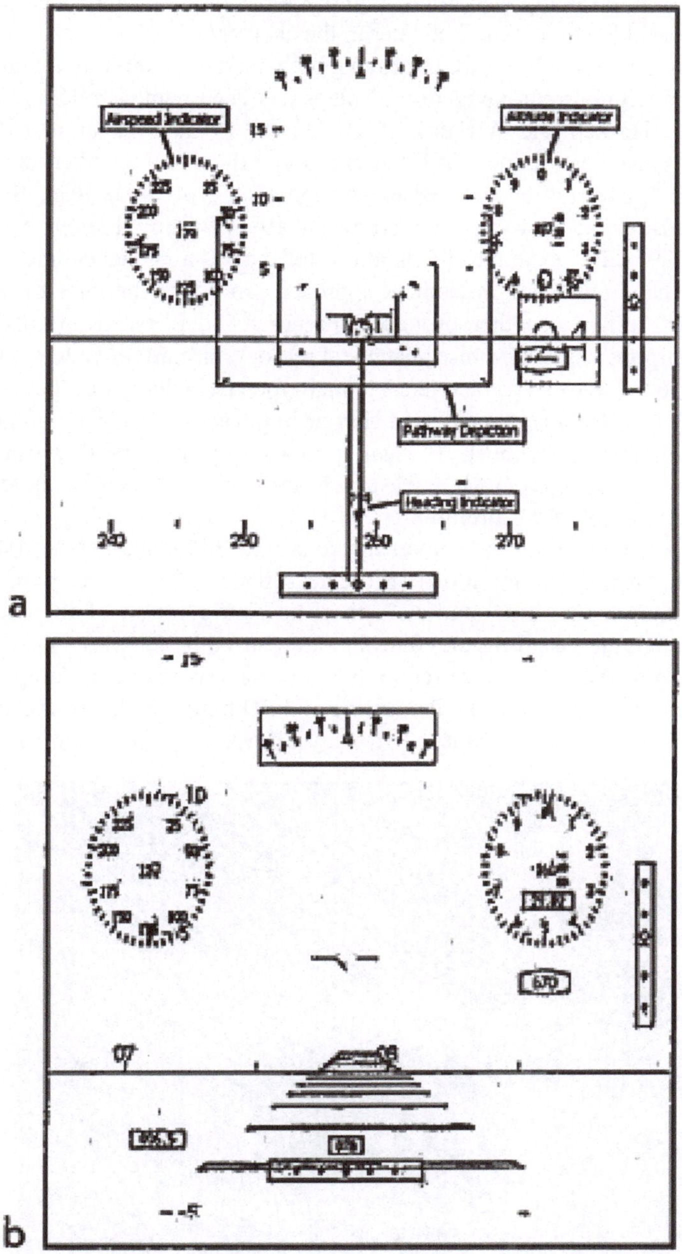

Figure 8.3 Flight path display depiction

Source: Reprinted with permission from Williams, K. W. (2002).

and 8.4 for examples of typical FPDs; Figure 8.4 has the FPD embedded within a Synthetic Visual System, described later in the chapter).

The effectiveness of FPDs for improving both horizontal and vertical guidance as compared with current navigation displays has been demonstrated in several studies (e.g., Haskell and Wickens, 1993; Jensen, 1981; Wickens and Prevett, 1995). It has also been found that FPDs are more effective than other navigation displays for flying nonstandard instrument approach procedures (e.g., Barrows, Enge, Parkinson, and Powell, 1996; Grunwald, 1996; Reising, Liggett, Solz, and Hartsock, 1995). For example, FPDs allow for curved approaches into airports that do not have published instrument approach procedures because of dangers presented by terrain or other obstacles. Figure 8.5 depicts an approach into Portland's airport, where for noise abatement reasons, aircraft are required to fly a particular ground track over the river. A tunnel-type visual display of the approach could help pilots fly this maneuver in rich or impoverished viewing conditions. Currently this type of "published visual approach" can only be flown in visual meteorological conditions (VMC). Clearly, FPDs have many benefits in terms of visual-motor control of the aircraft.

Mulder and colleagues have investigated several FPD parameters to include tunnels and tunnel configurations (Mulder, 2003a; 2003b). One particularly interesting finding of these studies is that the configuration of the FPD can actually change the visual-motor control strategy used by the pilot (see Chapter 5). In these studies, pilots were required to intersect a curved flight approach-path in a simulator using a FPD. Presenting an FPD tunnel with a frame allowed pilots to use a tau (time-to-contact) strategy afforded by the expansion of the

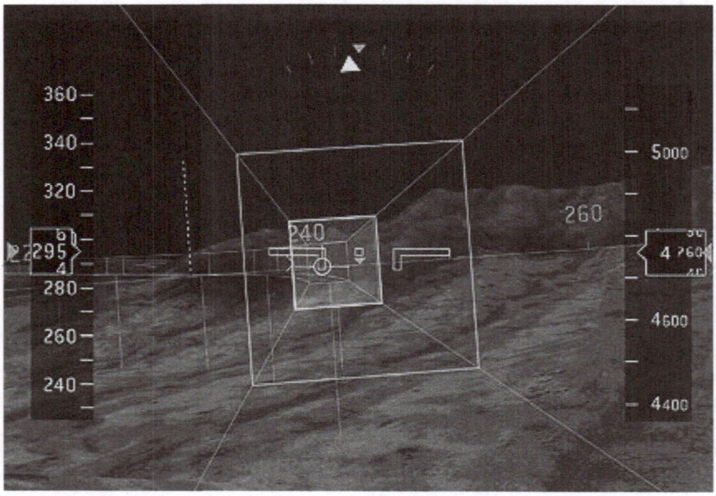

Figure 8.4 Flight path display photograph

Source: Photo courtesy of Rockwell Collins.

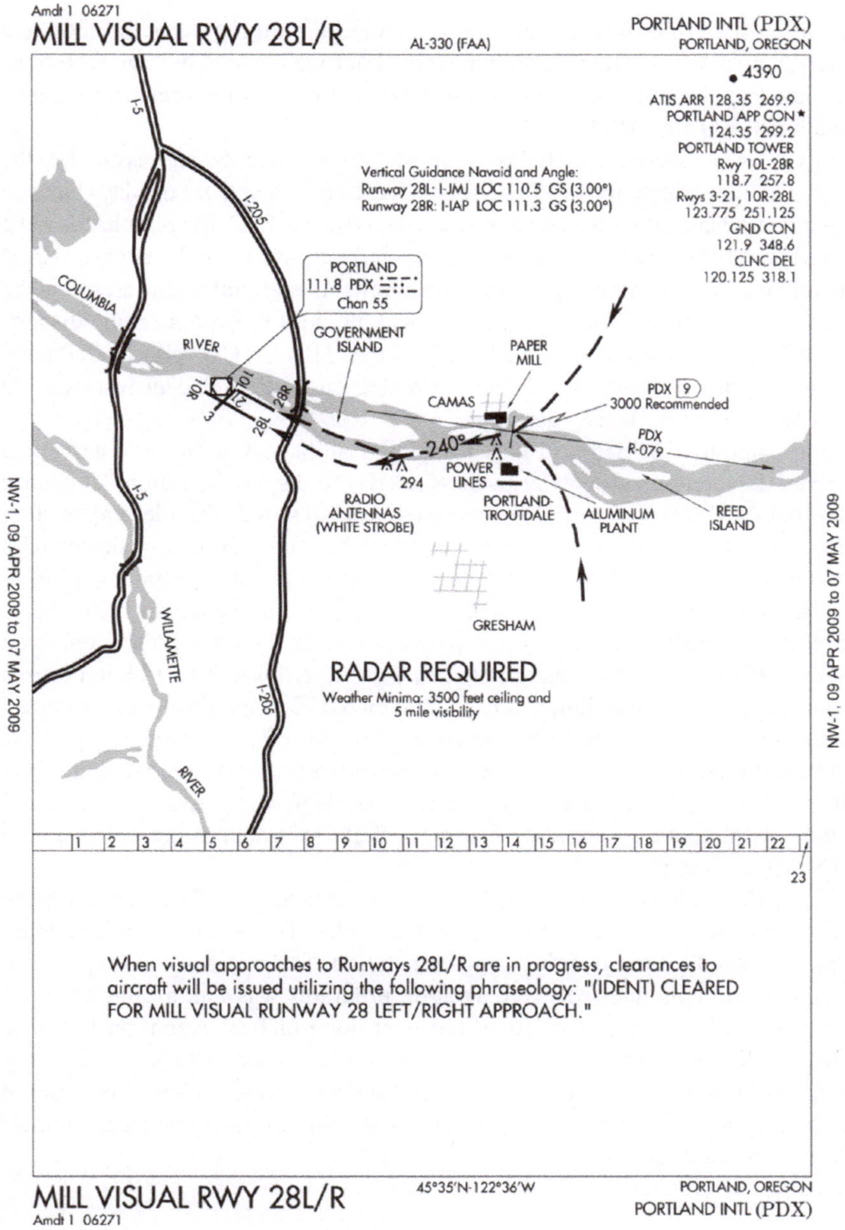

Figure 8.5 Visual approach into Portland International Airport

tunnel frames to determine the time-to-go to the curve. Presenting only the tunnel outline induced the use of the tangent point (i.e., the innermost point of the curve), a strategy used by drivers when taking a curve. Display formats presenting the tunnel frame yielded more consistent timing behavior and were preferred by pilots. Presenting only the tunnel outline resulted in large variations in conducting the maneuver accompanied by a poor path-following performance and a high pilot workload.

Given the research on HUDs discussed above, one critical issue for the successful implementation of FPD displays is the effect that these displays have on pilot attention and situational awareness. When using a FPD, the pilot needs to be aware of (and potentially shift attention between) the position of the intended flight path relative to the current position of the aircraft, other aircraft in the area, and the primary flight information (e.g., airspeed and altitude) presented on the display. Do FPDs produce similar IAB effects as other HUDs? Can pilots effectively utilize the primary flight information presented on a FPD? Recent research has only begun to answer these questions.

For example, Williams (2002) tested the ability of pilots to intercept a pathway depicted on a highway-in-the-sky (HITS) display in a flight simulator. Additionally, pilots were tested on their awareness of speed, altitude, and heading during the flight and their ability to monitor other air traffic. Consistent with previous research on control performance, awareness of the current position of the aircraft relative to the intended path was significantly better for the HITS as compared to traditional navigation displays. However, there were also significant practice effects, suggesting that pilots could develop this ability with traditional displays if given more training. When using the HITS display, pilots detected an average of 3.25/6 other aircraft, indicative of an IAB effect. Similar to the HUD findings, the amount of time the pilot spent looking outside the cockpit while using the HITS display was significantly less than when using conventional aircraft instruments. Awareness of primary flight information presented on the HITS display was poor.

Thus, although there appear to be clear benefits to using FPDs, they introduce some of the same problems associated with other HUDs and therefore more research is need to improve their implementation. The first author's experience with FPDs via computer-software aviation programs was that a pilot becomes consumed with 'making it through' the next hoop or box. Attention becomes focused solely on that task at the expense of other aviation tasks … the very definition of attentional capture. Granted, this assessment is based on limited experience, however, it is indicative of the seduction the airborne target tunnels present to a pilot's focus.

Night Vision Devices

Given the visual perception problems associated with night flying it was only a matter of time before technology provided the pilot with a means to see better at night. However, a major problem with the implementation of this technology is that pilots often think night vision devices can "turn night into day." This fallacy has led many pilots once again into an overconfident state regarding their visual perception capabilities. Falconer and Todd (2007) stated it best:

> It may be tempting for some operators to believe that NVGs enable them to conduct flights they were previously unable to do, but this is not the case. NVGs should be an operational enhancing device, not the sole means to conduct the flight. (p. 37)

Night-Vision Goggles

Night-vision goggle design characteristics Night-vision goggles (NVGs) and other night-vision devices (e.g., head or helmet-mounted devices) greatly enhance the ability to conduct night operations and are used extensively in both rotary-wing and fixed-wing operations within the military as well as by a growing number of general aviation pilots. NVGs provide an intensified image of scenes illuminated by ambient energy in the red and near-infrared portions of the electromagnetic spectrum (approximately 600–900 nm; recall from Chapter 2 that the human visual sensation range is 380–760 nm). The luminance of the NVG image is up to 1000 times higher than the original night scene (Tredici and Ivan, 2008). However, it must be emphasized that some ambient or environmental light needs to be present for the night vision device to amplify … consequently, these devices do not turn absolute darkness into light.

The basis of night-vision devices or night-vision imaging systems is founded on the idea of amplifying existing environmental light sources already available, thus providing an intensified image. As shown in Figure 8.6, light enters the device and the photons of light are converted into electrical energy by a photocathode.

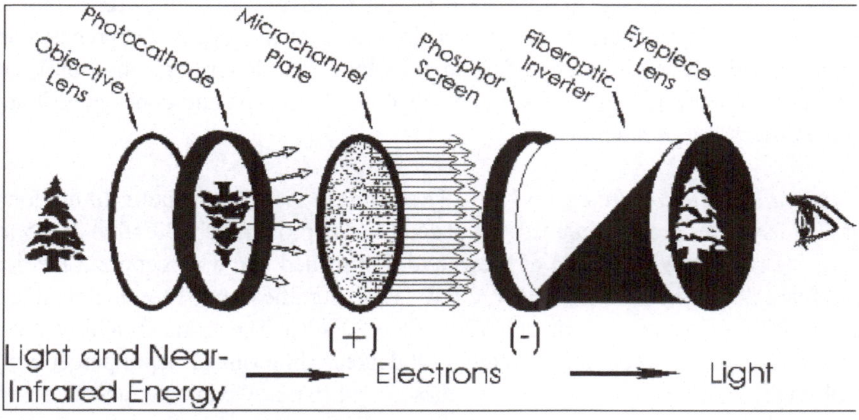

Figure 8.6 Components of night vision goggles

Source: From US AFMAN 11-217 Vol II (1998).

The electrons then strike a microchannel plate, which has bent tubules, causing the electrons to impact these bent tubule walls, separating and accelerating the electrons. The electrons then strike a phosphor screen, causing the screen to illuminate and produce an image. An increase in both the total number of electrons and their accelerations result in the intensified imagery of the scene. Image quality is improved with compact goggle-parts because it reduces dispersion of the photons and electrons.

Night vision devices employ an automatic brilliance-feature which acts to maintain constant image brightness by decreasing the intensifier gain in response to input light-levels exceeding a defined threshold. An intense light-source, emitting energy in that portion of the electromagnetic spectrum to which the goggle is sensitive, can produce a veiling glare and obscure the entire image. To prevent this glare and to protect the image intensifier assembly from permanent phosphor burns, the automatic brilliance-control regulates the amplification level.

A development in the technology of night vision goggles came about due to incompatible cockpit lighting in the mid 1990s. Pilots need to be able to see under their goggles and view their cockpit instruments as well as view the HUD image. However, sometimes a bright cockpit instrument light became illuminated while the pilot was using night-vision goggles, and the resulting glare of the instrument light would wash out and obscure the image of the external scene. Further, wavelengths of light from HUDs also interfered with the NVGs. To help reduce the lighting conflicts, a minus blue filter was added to the goggle to prevent certain wavelengths of light (especially those used in the HUD display) from entering the goggle.

The NVG image is similar to a black-and-white video or television screen in that it is monochrormatic; however, it uses shades of green because of the selected phosphor display (Tredici and Ivan, 2008). Green was selected because of the visual system sensitivity to green as a relatively low-level light source. Because both the rods and cones are sensitive to the wavelengths we perceive as green, and the NVG image is of an intermediate light-level (e.g., mesopic), an intermediate dark-adaptation state is reached and both rods and cones contribute to the processing of the image.

Perceptual limitations due to NVGs The primary visual perception limitations of NVG systems are the reduced visual acuity of approximately 20/40 to 20/25 at best, the reduced visual field of view, and diminished depth perception (Tredici and Ivan, 2008). In terms of acuity, 20/40 is still significantly better than unaided night vision mediated by rods only (20/200 to 20/400). The reduced field of view cuts the normal 180 degrees to roughly 40 degrees. In terms of depth perception, with night vision goggles monocular cues are all that are available, and they are degraded due to contrast and acuity limits. Fortunately, this limitation is less problematic as pilots generally rely on monocular depth cues more than binocular depth cues even in good lighting. However, given all the misperceptions that pilots are prone to in both rich and impoverished visual environments when not using the goggles, it should not be surprising that there are the several additional visual perception issues associated with the use of NVGs.

Recall from Chapter 2 that focal vision is excellent for detail and color, and that ambient (peripheral) vision is required for global orientation. Unfortunately, the extremely limited field of view when using NVGs essentially eliminates ambient visual cues. Additionally, the "soda straw" view through the goggles leads to the use of scanning or head sweeps of the environment, which in turn requires increased cognitive effort to piece together the focal vision scans into a coherent spatial scene. These problems all combine to enhance the susceptibility of the pilot to spatial disorientation.

Aviation challenges when using NVGs There are many flying challenges that occur with the use of NVGs. What follows is a listing of NVG problem areas described by US AFMAN 11–217 Vol II (1998).

1. Flight into weather, that is, clouds, may not be perceived if the clouds are thin enough. Perception of denser clouds is possible with NVGs; however, if dense enough, once in the cloud what little light available may be lost and the cloud may occlude terrain on the other side if at low altitude. Fog leads to similar problems during landing or low-level flight.
2. Light rain or mist may not be directly seen by the pilot when using NVGs; however, these atmospheric conditions decrease contrast and affect distance estimation and depth perception. Heavier rain is perceived with the NVG, but it can inhibit visual perception of the night scene.
3. Snow, like heavy rain, may inhibit NVG performance; however, when there is snow on the ground, the terrain may become easier to discriminate due to the contrast levels as well as helping in reflection of star/moon light.
4. Sand, dust, and smoke inhibit NVGs by blocking light sources and occluding the view of the environment.

Overall, any environmental condition that reduces the ambient light source or illumination level inhibits the potential of the NVG to amplify the visual scene. The most dangerous condition occurs when the change in environmental conditions is subtle; then the contrast slowly degrades and the pilot unknowingly makes visually guided decisions based on unreliable cues.

The NVG problems listed above are not too different than those that occur when a pilot has unaided vision. There are also numerous additional issues a pilot must handle due to the device itself … things that are associated with the technology of the night vision device. For instance, the goggle itself may induce image shading, dark or bright spots, but one of the most common vision degradation is "scintillation" or a sparkling rain of the NVG image due simply to low light-levels in the environment. Remember, the device needs some form of light to amplify. Many of the problems noted above can be discovered during a pre-flight inspection of the goggle prior to use, but if encountered during flight they will provide one more source of potential error in visual perception.

Forward Looking Infrared Systems

Forward-looking infrared systems or FLIRs are also used to operate in impoverished viewing environments. A FLIR was developed on the basis that all objects warmer than absolute zero emit heat and it is possible to differentiate between objects of less than one degree difference or those objects emitting heat at different rates (US AFMAN 11–217, 1998). Consequently, a FLIR image is created by using the thermal properties of environmental objects and displays them to the pilot on either a HUD or HDD. Similar to a night-vision device using amplifying photon technology the image is monochromatic, but in this case can be either gray or green. Table 8.1 compares night vision goggle and FLIR technologies. Note that the combination of both technologies would maximize pilot perception of an impoverished environment.

Table 8.1 Comparison of night vision goggles and forward looking infrared devices

Night Vision Goggles	Forward Looking Infrared
Use reflected energy (visible light near IR)	Use emitted energy (mid or far IR)
Images reflective contrast	Images thermal contrast
Requires at least some illumination	Totally independent of light
Penetrates moisture more effectively	Penetrates smoke and haze
Attenuated by smoke, haze, and dust	Attenuated by moisture (humidity)

Source: Adapted from US AFMAN 11-217 Vol II (1998).

Accidents and Mishaps Associated with the Use of Night Vision Devices

Braithwaite, Douglass, Durnford, and Lucas (1998) assessed military rotary-wing operations and all spatial disorientation mishaps between May 1987 and September 1995. From a total of 223 night-aided vision accidents they found that 131 had been classified as having spatial disorientation as a major or subsidiary component of the accident sequence. (Recall that spatial disorientation is underreported; thus, given the dates of the accident samples, it is safe to assume that an even greater number than 131 actually had some form of spatial disorientation and at least 50 percent of those involved aspects of visual misperception.) They found that 52 percent of the night-vision goggle accidents and 60 percent of the AH-64 FLIR accidents involved spatial disorientation as a significant factor. Thirty-one mishaps occurred when environmental illumination was insufficient or impoverished to allow proper NVG light-amplification. A majority of these mishaps occurred over desert terrain … featureless environments. Table 8.2 lists the summarized findings. Note the prevalence of insufficient visual cues and visual illusions as well as the presence of "overconfidence."

Table 8.2 Night vision device spatial disorientation data

Factor	Number of Cases	% of all NVD SD
Insufficient visual cues	100	85
Visual Illusions	3	2.5
Misjudged clearance obstacle/aircraft	47	40
Misjudged altitude	65	55
Misjudged speed/closure	50	42
Overconfidence	50	42

Source: Braithwaite et al (1998).

Unmanned Aerial Systems

Unmanned aerial vehicles (UAVs), unmanned aerial systems (UASs), or more recently the preferred name has become remotely piloted aircraft (RPAs), regardless of the acronym, they are the future of aviation. Currently they are being used extensively by the military for surveillance purposes as well as for delivery of ordinance with no risk to the operator because of the operator's displaced position from the aircraft as it is operated. Other agencies are using these unmanned systems to patrol borders and survey areas of the geography for movement of personnel. As mentioned in the beginning of this chapter, presented here in Chapter 8 are just brief discussions on visual perception issues of different aviation technological advances. Consequently, what follows is a short depiction of UAS issues. There are several recent books that address UAS issues in a more detailed and comprehensive manner (e.g., Cooke, Pringle, Pedersen, and Connor, 2006).

Visual perception in UASs requires an entirely new perspective of the human-system-environment model due to the fact that the operator is not colocated with the aircraft or the environment. Figure 8.7 depicts this new relationship. Note the difference from Figure 1.1 in Chapter 1. Whereas in traditional aerospace system models the environment houses both the pilot and the aircraft, now the pilot is not in the same environment that the aircraft is in, yet nearly all the same integration requirements exist between the pilot and the aircraft.

Currently, unmanned aircrafts are highly automated with respect to their mission in terms of navigation, lateral and vertical control, and takeoff and recovery fields. However, it is anticipated that in the near future, the UAS mission will be less automated and the humans controlling them will have more autonomy in regulating the UAS mission and actions.

Even with automation, pilots must visually control some unmanned craft inputs, and many of these must be made based upon environmental perception. Thus, the human inputs are vital in the loop for successful mission accomplishment. To accomplish the human-input tasks, the real-time environment in which the aircraft is operating must be presented to the pilot. In order for the pilot to successfully control the UAS, coherent decision-making strategies will be needed in terms of making control inputs based on deterministic information from navigation and system displays as well as

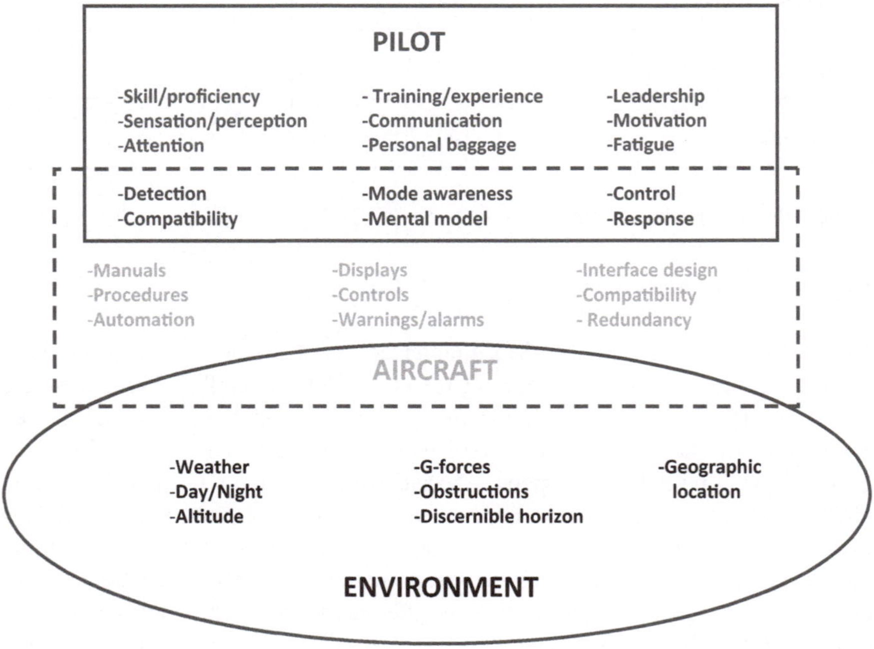

Figure 8.7 Interaction between pilot, unmanned aircraft, and environment

corresponding decision-making strategies via probabilistic information from video (synthetic and real) environmental displays. Similar to the realities of using night-vision goggles, there are operator challenges such as a limited field of view (e.g., the US Air Force's Predator provides the pilot with a 30-degree field-of-view display screen), delay in the information being presented (due to the satellite to computer link), and reduced contrast and resolution of the video image. A unique challenge with UAS control is the absence of other sensory cues (auditory, somatosensory, and vestibular) that can help a pilot understand the state of the vehicle within its environment.

As described in previous chapters, vestibular information can be a double-edged sword; the vestibular information can provide valuable input regarding acceleration and tilt, but, due to rapid adaptation, vestibular signals can become unreliable and lead to spatial disorientation. Thus, it might seem to be beneficial that, when flying a UAS, there are no vestibular signals related to the vehicle. However, there is a strong possibility that the disconnect between vestibular and visual input may cause a pilot to become disorientated because of the lack of pilot movement or because of incongruent signals if the operator of the UAS is on a moving platform (e.g., if the UAS pilot is sitting in the back of a moving truck or on board a ship). These scenarios are very realistic and pose problems to the spatial orientation system.

Because the unmanned aircraft "pilot" must perceive visual cues indirectly, the resultant cognitive resources involved in forming a mental model and maintaining situational awareness are much higher. The demand on the visual sense is extremely high due to the fact that all the information being provided is visually via video feeds, text-messaging between pilots and ground controllers, as well as visual computer interaction. The perceptual challenges combined with the high cognitive load have resulted in many UAS mishaps.

For example, Williams (2004) summarized unmanned aircraft mishaps and incidents within the US Army, Air Force, and Navy. His data consisted of US Army UAS mishaps between 1995 and 2003, US Navy UAS mishaps from 1986 to 2002, and US Air Force UAS mishaps between the years 1999 and 2003. Williams initially separated the mishaps into two groups to include mechanical/system failures and human factors issues. The percentage of mishaps related to human error ranged from 21 percent to 68 percent of the total number, depending upon the type of UAS. The US Army Hunter unmanned aircraft had 15 of the 32 total accidents (47 percent) associated with human factors issues, with 7 of those 15 occurring during the phase of flight when an external pilot was controlling the UAS during a landing. The US Army Shadow UAS does not use an external pilot for landing; the aircraft is designed for an automated recovery and landing system. Consequently, it had a smaller percentage of human factors issues (21 pecent) overall. The US Navy Pioneer unmanned aircraft had 68 of 156 total (28 pecent) accidents categorized as human factors issues, and of those 46 (68 pecent) were landing errors by the pilots. The US Air Force's Predator had 15 accidents/incidents, of which 8 (67 pecent) were attributed to human factors. Details of the types of human factors problems were not addressed beyond mentioning display design, landing errors, and procedural errors.

Gardetto (2009) studied Predator mishaps over a five-year period. He found that the primary human factor involved in the mishaps was the pilots' misperception of their environment, more specifically errors in estimating height above the runway and errors due to integrating two video screens that portrayed different UAS perspectives. The misperception of height is a human limitation that has been present since the beginning of human flight (see discussions in previous chapters). However, the pilot difficulties due to controlling an aircraft while viewing two unique aircraft perspectives is a new problem within aviation. This is a great example of how too much information can become a bad thing for pilot visual perception and cognition.

Self, Ercoline, Olson, and Tvaryanas (2006) also examined UAS flight mishaps, and concluded that visual illusions are a factor in UAS operations because vision is the sense providing the primary source of data. Thus, the visual illusions experienced by manned pilots will also affect unmanned pilots. Specifically mentioned were illusions associated with perceptual constancy (shape/size constancy) and autokinesis, as well as control-reversal problems due to visually manipulating a vehicle from various orientations. The authors articulated the UAV pilot sensation-perception dilemma in that an " … operator is subject to conflicting non-motion ambient and proprioceptive cues from the ground control station environment and motion visual cues from the sensors and displays" (p. 138). Self et al. also noted that current technology limitations in the UAS sensors and displays compound the perceptual challenges. For example, pilots have altitude and distance estimation problems when attempting to interpret environmental cues through a small display screen.

As the UAS becomes a more mature aerial platform the mishap data will begin to tell a more compelling story regarding error trends. One expected future trend is that, as the mechanical systems become more reliable, their contributing factor in mishaps will decrease; unfortunately, the human factor will remain present and increase in its role contributing to mishaps. One attempt to overcome the visual limitations that occur when in flying a UAS with only focal vision displays is to provide a synthetic environment that wraps around the focal vision monitor perspective, that is, synthetic vision systems.

Synthetic Vision Systems

Visual-perception problems have led to many controlled flight into terrain (CFIT) mishaps. Night, impoverished visual environments, visual illusions, masked terrain, and pilot misperception have been the most commonly cited causes of these CFIT accidents during approach to landings and low-level navigation. Kramer, Arthur, Bailey, and Prizel (2005) stated, "limited visibility and reduced situation awareness have been cited as predominant causal factors for controlled flight into terrain accidents" (p. 1). To address these problems, researchers have explored the possibility of allowing pilots to "see" regardless of their environment, that is, the creation of synthetic vision systems (SVSs).

Previously discussed have been the developments of head-up displays, night-vision goggles, and forward-looking infrared systems. SVSs are the combination of all of these technologies plus global-position systems that use known terrain-data bases. SVSs may also incorporate flight-path displays, real-time weather information, as well as traffic-conflict and avoidance data. Any obstacle, object, or specific area that can be defined with geographic coordinates can potentially be displayed to the pilot via the SVS imagery (see Figure 8.8 for some of the inputs for an enhanced vision system).

The result is a synthetic but accurate and credible external view for a pilot to use during any phase of flight (see Figure 8.4, the SVS display houses not only HUD information but also a flight path display, or highway in the sky). An overall safety concept of SVSs is that if pilots can have improved awareness of their environments then the "warn and react" type safety technologies (e.g., enhanced ground-proximity warning systems or terrain avoidance warning systems) become back-up, redundant features rather than the primary means to minimize the risk of CFIT. SVSs may potentially be a solution to overcome airspace limitations. For example, SVSs will enable the design of curved and constant-descent approaches that may be more efficient than navaid-based, straight-line approaches.

Human-factors visual-perception benefits of SVS have included reduced scanning time—that is, more efficient visual search—lower cognitive-load due to integration of disparate spatial cues, and improved visual attention (Calhoun, Draper, Abernathy, Patzek, and Delgado, 2005). The reduced scanning time and improved visual attention aspects support the visual-design principles that advocate the reduction of information-access costs by maintaining essential and

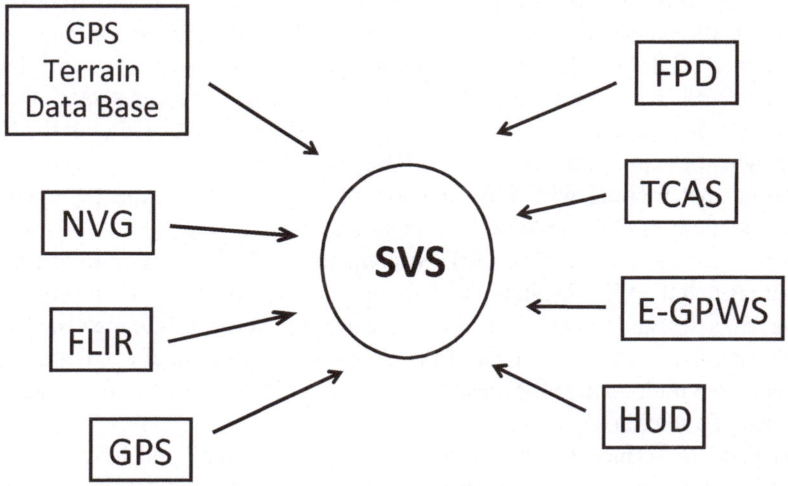

Figure 8.8 Inputs for a synthetic visual system

common cues within a small area to best to minimize the pilot's visual search and mental integration. Thus, SVSs greatly facilitate visual cue sensation, perception, and cognition.

The examination of visual perception in SVS has recently been a very active area of research. For example, one potential problem associated with the SVSs is the use of too much detail or too many types of information for the pilot to attend to for accurate perception. The synthetic display does not need to perfectly recreate every detail of a scene for a pilot to have veridical perception. Researchers are working to identify the sufficient and necessary amount of detail to provide accurate cue interpretation for pilots. Alexander, Stelzer, Kim, and Kaber (2008) examined bottom-up versus top-down contributions toward pilot perception of a SVS. Bottom-up components of an SVS include the display information, including the possible visual clutter that can occur when too much information is displayed. In contrast, top-down components include visual attention, which is influenced by pilot expectation and prior knowledge in perceiving the display, and irrelevant thought processes that have the potential to interfere with efficient processing. The authors concluded that visual-display density influenced bottom-up visual processing while information density affected perception of the display in terms of specific flight phases.

Related to the interaction between top-down and bottom-up processing, Wickens and Alexander (2009) examined attentional tunneling for SVSs. They anticipated that this problem may be even worse for SVS than for HUDs (described above) because 3D displays, particularly with an "immersive" or viewer's-eye perspective, tend to attract attention more strongly than symbolic displays. During an approach-and-landing simulation, pilots were presented with one of five possible unexpected events: (1) radio tower, within the flight path, (2) a blimp in the HITS (Highway in the Sky) pathway, (3) a small track error, (4) an unpredicted loss of visibility, and (5) unexpected severe weather. They found that 71 out of 158 pilots (45 percent) flying with the HITS failed to detect the unexpected event. When the SVS did not contain a HITS display only 2 of 16 (12.5 percent). Inputs for a synthetic visual system) of the unexpected events were undetected. Further research is needed to determine whether these problems associated with the HITS occur when it is presented on a HUD, HDD, and/or within the SVS. The authors proposed that training with SVS should expose pilots to unexpected events to make them aware of the potential for attentional tunneling.

One positive characteristic of SVSs compared to HUDs related to attentional capture is that the SVS displays can better blend image overlays of runways, man-made obstacles, and threats (enemy or noise abatement areas) into the visual scene by altering the color and contrast of the overlays. Recall from above that in HUD displays, the high contrast of these overlays potentially caused visual hindrances in terms of attentional capture.

Beyond the issue of how much information to display and how that might influence attention, some research has focused on the format of the information that is displayed. For example, Schnell, Keller, and Etherington (2009) examined

SVS characteristics such as field of regard, terrain-data density, texture, and shading. Chapter 4 discussed how pilots use their field of view for cue references, such as the landing aimpoint on the runway residing within a certain area of the pilot's windscreen or distance above the glare shield. A synthetic environmental display must present to the pilot the image of the external scene but at a scale that fits within the relatively small SVS display screen. According to Schnell et al., technology allows for the SVS displayed field of regard to be greater than an actual field of view, this results in a minification—more information displayed to the pilot via the display but in smaller image detail.

Schnell et al. (2009) described the benefits and drawbacks of using differently sized fields of regard. For instance, a large field of regard allows for improved lateral orientation for maneuvering; however, the terrain features/details become much smaller and more difficult to discern. In contrast, a smaller field of regard provides more scene detail in the forward direction, but is comparable to focal-only vision with a loss of peripheral environmental cues. Based on their research they recommended that a 60-degree field of regard be used because of its balance between terrain cue minification and global orientation potential. In terms of visual perception and cognitive load, Schnell et al. also make an excellent recommendation to keep a constant field of regard and not allow changes in the minification, because when training with the SVS in visual conditions the pilot perceptually relates the synthetic world with the actual environment. Cue constancy can develop with experience in terms of real-world cue-size, shape, movement related to the synthetic minification of the cues. Schnell et al. concluded, "this learned mapping then becomes very useful for distance and angle estimations when operating in instrument meteorological conditions (IMC) using the SVS" (p. 40).

Other research has compared pilot performance when using SVSs and conventional systems. Using simulators, Jennings, Alter, Barrows, Bernier, and Guell (2003) found that the pilots using the SVS all avoided mountainous terrain during maneuvering whereas the pilots using the conventional system all experienced controlled flight into terrain. The authors concluded that the SVS enhanced situational awareness, reduced workload, and reduced flight technical errors.

Schnell, Kwon, Merchant and Etherington (2004) directly compared flight performance, workload, situational awareness (SA), and visual scan patterns for an SVS display and a conventional glass cockpit-display. The display (shown in Figure 8.9) consisted of the tactical pathway display with synthetic terrain and a 3D map display that presented strategic navigation information. When using the SVS display, the cross-track error, vertical-track error, track-angle error, and flight-path angle error were significantly lower. Furthermore, the workload scores were slightly lower and there were no significant differences in SA across the display conditions. One surprising finding related to SA was that it did not seem to improve the terrain awareness of the pilots. The authors proposed that perhaps pilots trusted the pathway tunnel to such an extent that they did not feel they needed to devote much attention to the aircraft-terrain situation. Finally, scan lengths were significantly shorter when using the SVS, suggesting that pilots can obtain the necessary information with fewer saccades, which over time would reduce eye fatigue. Together these findings indicate there are several advantages to using SVS.

Taking a different focus, Bolton and Bass (2001) investigated the ability of pilots to make spatial judgments (e.g., distance, height, angle) of terrain points in videos of SVS displays. They found that SVSs led to some distortions in the perception of space; in general, pilots underestimated distances and angles. However, these judgment errors tended to be small, suggesting they would be of little consequence in training. Of course these parameters are vital for aviation safety and more research needs to be done in this area of environmental cue perception.

Combining SVSs with other advanced aviation display technology (e.g., NVDs and UAVs) is also a current focus of research and development. Recall that within NVDs and UAV displays the limited field of view is one of the biggest visual perception drawbacks. Researchers are investigating the use of a picture-in-picture (PIP) feature within SVS displays to counter this field-of-view limitation (Calhoun et al., 2005). The SVS with PIP surrounds the actual focal-vision "soda straw" image that the NVG or UAV pilot normally has and provides synthetic ambient cues and a more global perspective of the visual scene. The result is a significantly increased field of view. In sum, the use and further development of SVSs can bring increased SA to the pilot by highlighting objects of interest to the operator as well as providing a rich visual environment even when the actual environment is impoverished or completely degraded.

Laser Retinal Displays

This final topic is introduced so that the reader can better appreciate the possible future of visual displays. A laser retinal display (LRD) removes the "monitor" or "screen" from the pilot's cockpit and presents the image directly onto the retina. It is an extreme form of a head-up display … there is no need to look elsewhere, the image is not only directly in front of the pilot regardless of where his/her head is positioned but directly *on* the retina. "Short of tapping into the optic nerve, there is no more efficient way to get an image into your brain" (Lewis, 2004, p. 24).

Imagine a completely see-through display that contains whatever information is needed by a system operator. The current system requires a head-gear: usually a visor or baseball-type cap to hold the retinal display equipment that projects the image into the eye. The technology uses a very low-powered light-beam that is continuously moving in a scan pattern over the retina. Thus, the dwell time on any one spot of the retina is minimized and there is no danger of any retinal burns from the laser.

Conclusion

An important point to note as we conclude this chapter on aviation advancements and visual perception issues is that technology is not going to remove misperception opportunities. Although the long-standing dangers of unaided flight may be mitigated, new human errors will arise in terms of attentional capture, visual clutter, and minification height/distance perception problems. Education and training are key for the successful employment of synthetic vision systems. It is crucial that pilots only consider the technological advancements as environmental *aids* to enhance their situational awareness rather than allowing themselves to become reliant upon synthetic vision displays as the sole input for their decision-making.

References

Alexander, A. L., Stelzer, E. M., Kim, S., and Kaber, D. B. (2008). Bottom-up and top-down contributors to pilot perceptions of display clutter in advanced flight deck technologies. Proceedings of the Human Factors and Ergonomics society 52nd Annual Meeting, 1180–1184.

Barrows, A. K., Alter, K. W., Enge, P., Parkinson, B. W., and Powell, J. D. (1996). Operational experience with and improvements to a tunnel-in-the-sky display for light aircraft. Proceedings of the ION Conference, Kansas City, MO. September 1996.

Bolton and Bass (2001). Using relative position and temporal judgments to identify biases in spatial awareness for synthetic vision systems. *The International Journal of Aviation Psychology*, *18*, 183.206.

Boston, B. N. and Braun, C. C. (1996). Clutter and display conformality: Changes in cognitive capture. In *Proceedings of the Human Factors and Ergonomics Society 40th annual meeting* (pp. 57–61). Santa Monica, CA: HFES.

Braithwaite, M. G., Douglass, P. K., Durnford, S. J., and Lucas, G. (1998). The hazard of spatial disorientation during helicopter flight using night vision devices. *Aviation, Space, and Environmental Medicine*, *69*, 1038–1044.

Calhoun, G. L., Draper, M. H., Abernathy, M. F., Patzek, M., and Delgado, F. (2005). Synthetic vision system for improving unmanned aerial vehicle operator situation awareness. In J. G. Verly (Ed.), *Proceedings Society for Optical Engineering, Enhanced and Synthetic Vision*, Vol 5802, (pp. 219–230).

Cooke, N. J., Pringle, H. L., Pedersen, H. K., and Connor, O. (2006). *Human Factors of Remotely Operated Vehicles*. Oxford, United Kingdom: Elsevier.

Etherington, T. J., Vogl, T. L., Lapis, M. B., and Razo, J. G. (2000). Synthetic vision information system. In *Proceedings of the 19th Digital Avionics Systems Conference* (pp. 2.A.4–2.A.8). Philadelphia: Institute of Electrical and Electronic Engineers.

Federal Aviation Administration (2002). http://rgl.faa.gov/Regulatory_and_Guidance_Library%5Crgpolicy.nsf/0/8E9FB005423832CC86256C0F007209C1?OpenDocument.

Fadden, S., Ververs, P. M., and Wickens, C. D. (1998). Costs and benefits of head-up display use: A meta-analytic approach. In Proceedings of the Human Factors and Ergonomics Society 42nd Annual Meeting (pp. 16–20). Santa Monica, CA: HFES.

Falconer, B. and Todd, M. (2007). The view ahead: Human factors aspects of night vision goggle (NVG) operations. *Flight Safety Australia*, *56*, 36–37.

Fischer, E., Haines, R. F., and Price, T. A. (1980). *Cognitive issues in head-up displays* (NASA Tech.Paper No. 1711). Moffett Field, CA: NASA Ames Research Center.

Foyle, D. C., Dowell, S. R. and Hooey, B. L. (2001). In R. S. Jensen, L. Chang, and K. Singleton (Eds.), Proceedings of the Eleventh International Symposium on Aviation Psychology, 143:1-143:6. Columbus, Ohio: Ohio State University.

Gardetto, P. (2009). Human factors in predator UAS mishaps and a comparison to manned aircraft: a five year review. Presented at the 2009 Annual Aerospace Medical Association conference, Los Angeles, May.

Goteman, O., Smith, K., and Dekker, S. (2007). HUD with a velocity (flight path) vector reduces lateral error during landing in restricted visibility. *The International Journal of Aviation Psychology*, *17*, 91–108.

Grunwald. A. J. (1996), Improved Tunnel Display for Curved Trajectory Following: Control Considerations. AIAA Journal of Guidance, Control, and Dynamics, 19, 370–377.

Haskell, I.D. and Wickens, C.D. (1993). Two- and three-dimensional displays for aviation: A theoretical and empirical comparison. International Journal of Aviation Psychology, 3, 87–109.

Hofer, E. F., Braune, R. J., Boucek, G. P., and Pfaff, T. A. (2001). *Attention-switching between* near and far display domains: An exploratory study of pilots' attention switching ability with head-up and head-down tactical displays in simulated flight operations (Boeing IR&D Document No. D6–36668). Chicago: Boeing.

Hull, J . C., Gill, R. T., and Roscoe, S. N. (1982). Locus of the stimulus to visual accommodation: Where in the world, or where in the eye? *Human Factors*, *24*, 311–319.

Jennings, C., Alter, K. W., Barrows, A. K., Bernier, K., and Guell, J. (2003). Synthetic vision as an integrated element of an enhanced vision system. Society of Photo-Optical Instrumentation Engineers (SPIE), Bellingham, WA.

Jensen, R. S. (1981). Prediction and quickening in perspective flight displays for curved landing approaches. *Human Factors*, *23*, 355–363.

Kramer, L. J., Arthur, J. J. Bailey, R. E., and Prizel, L J. (2005). Flight testing an integrated synthetic vision system. Society of Photo-Optical Instrumentation Engineers (SPIE), Bellingham, WA.

Larish, I., and Wickens, C. D. (1991). Divided attention with superimposed and separated imagery: Implications for head-up displays (Tech.Rep.ARL–91–4/ NASAHUD–91–1). Savoy: University of Illinois Institute of Aviation.

LeBlaye, P., Roumes, C., Fornette, M. P., and Valot, C. (2002). *Head up displays symbology (HUD): Prenormative study for DGAC/SFACT* (DCSD Tech. Rep. 2/05007). Toulouse, France: DCSD.

Lewis, J. R. (2004, May). In the eye of the beholder. *IEEE Spectrum*, 24–28.

Mack, A. and Rock, I. (1998). *Inattentional blindness*. Cambridge, MA: MIT Press.

Mulder, M. (2003a). An information-centered analysis of the tunnel-in-the-sky display, part one: Straight tunnel trajectories. *International Journal of Aviation Psychology*, 13, 49–72.

Mulder, M. (2003b). An information-centered analysis of the tunnel-in-the-sky display, part two: Curved tunnel trajectories. *International Journal of Aviation Psychology*, 13, 131–151.

Newman, R. L. (2000). HUDs, HMDs, and SDO: A problem or a bad reputation. (Report). Wright-Patterson AFB, OH: Air Force Aerospace Medical Research Laboratory.

Norman, J. and Ehrlich, S., (1986). Visual accommodation and virtual image displays: Target detection and recognition. *Human Factors*, *28*, 131–151.

Pierce, B. J., Geri, G. A., and Hitt III, J. M. (*1998*). Display collimation and the perceived size of flight simulator imagery (AFRL-HE-AZ-TR-*1998*–0058).

Reising, J. M., Liggett, K. K., Solz, T. J., and Hartsock, D. C. (1995). A comparison of two headup display formats used to fly curved instrument approaches. In *Proceedings of the 39th Annual Meeting of the Human Factors and Ergonomics Society* (pp. 1–5). Santa Monica, CA: Human Factors and Ergonomics Society.

Schnell, Keller, and Etherington (2009). Trade-offs in synthetic vision system display resolutions, field of regard, terrain data density, texture, and shading during off-path operations. *The International Journal of Aviation Psychology*, *19*, 33–48.

Schnell, T., Kwon, Y., Merchant, S., and Etherington, T. (2004). Improved flight technical performance in flight decks equipped with synthetic vision information system displays. *International Journal of Aviation Psychology*, *14*, 79–102.

Self, B. P., Ercoline, W. R., Olson, W. A., Tvaryanas, A. P. (2006). Spatial disorientation in uninhabited aerial vehicles. In N. J. Cooke, H. L. Pringle, H. K. Pedersen, O. Connor (Eds.), *Human Factors of Remotely Operated Vehicles* (pp. 133–146). Oxford, United Kingdom: Elsevier:

Simons, D. J., and Chabris, C. F. (1999). Gorillas in our midst: Sustained inattentional blindness for dynamic events. Perception, 28, 1059–1074.

Stuart, G. W., McAnally, K. I., and Meehan, J. W. (2005). Head-up displays and visual attention: Integrating data and theory. In D. Harris and H. C. Muir (Eds.),

Contempaorary Issues in Human Factors and Aviation Safety (pp. 25–46), London: Ashgate.

Tredici, T. J. and Ivan, D. J. (2008). Ophthalmology in aersospace medicine. In F. R. Davis, R. Johsons, J. Stepanek, and J. A. Fogarty (Eds.), *Fundamentals of Aerospace Medicine*, (pp. 349–379). Philadelphia: Limppincott, Williams, and Wilkins.

United States Air Force (1998). United States Air Force Manual 11–217, Volume II.

Weintraub, D. J. and Ensing, M. (1992). Human factors issues in head-up display design: The *book of HUD* (SOAR, CSERIAC 92–2). Wright-Patterson Air Force Base, OH: Crew Station Ergonomics Information Analysis Center.

Wickens, C. D. (1997). Attentional processes in GA head-up display design (Tech. Rep. ARL 97–8/FAA–97–3). Savoy: University of Illinois Institute of Aviation Research Lab.

Wickens, C. D. and Alexander, A. L. (2009). Attentional tunneling and task management in synthetic vision displays. *International Journal of Aviation Psychology*, *19*, 182–199.

Wickens, C. D. and Long, J. (1995). Object versus space-based models of visual attention: Implication for the design of head-up displays. *Journal of Experimental Psychology: Applied*, *1*, 179–194.

Wickens, C. D., and Prevett, T. (1995). Exploring the dimensions of egocentricity in aircraft navigation displays. *Journal of Experimental Psychology: Applied*, 1, 110–135.

Williams, K. W. (2002). Impact of aviation highway-in-the-sky displays on pilot situation awareness. *Human Factors*, *44*, 18–27.

Williams, K W. (2004). A summary of unmanned aircraft accident/incident data: human factors implications. DOT/FAA/AM-04/24, Office of Aerospace Medicine, Washington DC.

Index